The Neuron

The Neuron

New York   Oxford

Cell and Molecular Biology

IRWIN B. LEVITAN

Brandeis University

LEONARD K. KACZMAREK

Yale University

OXFORD UNIVERSITY PRESS

Oxford University Press

Oxford   New York   Toronto
Delhi   Bombay   Calcutta   Madras   Karachi
Petaling Jaya   Singapore   Hong Kong   Tokyo
Nairobi   Dar es Salaam   Cape Town
Melbourne   Auckland

and associated companies in
Berlin   Ibadan

Published by Oxford University Press, Inc.,
200 Madison Avenue, New York, New York 10016

Oxford is a registered trademark of Oxford University Press

Library of Congress Cataloging-in-Publication Data
Levitan, Irwin B.
The neuron : cell and molecular biology
Irwin B. Levitan, Leonard K. Kaczmarek.
p. cm.   Includes bibliographical references.   Includes index.
ISBN 0-19-505832-1
1. Molecular neurobiology.   2. Neurons.
I. Kaczmarek, Leonard K.   II. Title.
[DNLM: 1. Neurons—cytology. 2. Neurons—physiology.
WL 102.5 L666n]   QP356.2.L48   1991
591.1'88—dc20   DNLM/DLC
for Library of Congress   90-7221

9 8 7 6 5 4 3 2

Printed in the United States of America
on acid-free paper

To the Sheilas

Preface

A fundamental goal of neuroscience is to understand the way neurons generate animal behaviors. This requires, among other things, a knowledge of how many neurons are involved in the control of a specific behavior, where these neurons are located, and how they are connected. All this information, however, is of no avail if we do not understand the intrinsic properties of the individual neurons themselves. Just as each individual human provides his or her unique contribution to society, the properties, abilities, and "personalities" of different cells in a neuronal circuit all play roles in the output of that circuit. They also determine how that output alters with time to change the behavior of an animal.

Neuroscience has come of age. Its practitioners are no longer simply a hodge-podge of physiologists, biochemists, and anatomists who stumbled across the brain and decided to study it. Rather it is a mature discipline in its own right. This is evidenced in part by the appearance of several textbooks of neurobiology, some aimed at medical students, others designed to give undergraduates a broad overview of the field. By necessity such treatments must emphasize breadth at the expense of depth, and we have found ourselves frustrated in looking for a textbook that can be used to communicate in detail to students the principles of cellular and molecular neurobiology. Hence this book.

The text reflects our own training, research interests, and biases. It is unabashedly reductionistic, in line with our conviction that understanding the elements of the nervous system—individual neurons and the molecules that regulate neuronal activity—is essential for even the most rudimentary understanding of how the brain works. We recognize that, although understanding the elements is essential, it is not in itself sufficient for a complete description of brain function, and where appropriate we place cellular and molecular principles in the context of the nervous system in which they function. However, for the reasons given above we have preferred to emphasize depth, and this is not intended to be a comprehensive text covering all of neuroscience. Rather it is designed for a first course in cellular and molecular neurobiology, for undergraduate or beginning graduate students who already have a basic grounding in biochemistry and cell biology. Students who have mastered the material in this book can move on readily to more specialized courses in neural systems, development, behavior, and computational neurobiology.

We had fun writing this book. It gave us the opportunity to interact with and tap the expertise of knowledgeable and stimulating colleagues from whom we learned a great deal. Many of them read portions of the manuscript and pointed out (occasionally with great glee) our misconceptions and murky writing style. Eve Marder, Chris Miller, and Jimmy Schwartz undertook the heroic task of reading the entire book, and their suggestions were invaluable. Others who read several (sometimes many) chapters were Spyros Artavanis-Tsakonas, Bill Catterall, Arlene Chiu, Martha Constantine-Paton, Pietro DeCamilli, Dorothy Gallager, Scott Kasner, John Lisman, Joanne Mattessich, John Perkins, Jan Rosenbaum, Larry Squire, and Kate Turtle. We are grateful to all of them. Much of the treatment of resting and action potentials in Chapter 2 was inspired by a wonderful little book titled "Neurophysiology: A Primer," written by Chuck Stevens many years ago (J. Wiley and Sons, New York, 1966). We are also grateful to John Dowling, Paul Forscher, Steven Hunt, Andrew Matus, Tom Reese, Bruce Schnapp, Toni Steinacker, Masatoshi Takeichi, Asa Thureson-Klein, Monte Westerfield, and especially Dennis Landis for providing us with original photographs for some of the figures. Mike Lerner gave us the idea for, and the chemical structures of, the different odorants in Figure 12-14a. Maureen Ferrari and Joyce Chase did yeoman duty at the keyboard and in dealing with endless administrative details, and Paula Shelly helped with the preparation of the index. As always it was a pleasure to interact with our editor Jeffrey House and his colleagues Edith Barry, Donna Grosso, and Susan Hannan at Oxford University Press. Finally we thank our friends and colleagues for their encouragement, for telling us over and over again that a text of cellular and molecular neurobiology is sorely needed.

October 1990   I. B. L.
                                                                                      L. K. K.

Figure acknowledgments

We acknowledge permission from the publishers for the reproduction of published diagrams or photographs in the following figures: Figure 1-2, Springer-Verlag, N.Y., Heidelberg, Berlin, S. L. Palay and V. Chan-Palay, Cerebellar Cortex: Cytology and Organization, 1974; Figure 1-6, Alan R. Liss, Inc., New York, M. D. Landis and T. S. Reese, Differences in Membrane Structure Between Excitatory and Inhibitory Synapses in the Cerebellar Cortex, *The Journal of Comparative Neurology*, 155: 93–126, 1987; Figure 1-9, Elsevier Publications, Cambridge, England, M. P. Sheetz, E. R. Steuer, T. A. Schroer, The mechanism and regulation of fast axonal transport, *Trends in Neuroscience*, 12: 474–478, 1989; Figure 1-12, Alan R. Liss, Inc., New York, M. D. Landis, *The Journal of Comparative Neurology*, 260: 513–525, 1974; Figure 1-14, L. K. Kaczmarek, M. Finbow, J. P. Revel, and F. Strumwasser, The Morphology and Coupling of *Aplysia* Bag Cells Within the Abdominal Ganglion and in Cell Culture, *Journal of Neurobiology*, 10: 535–550, 1979; Figure 5-1, Macmillan Magazines, M. Noda, T. Ikeda, T. Kayano, H. Suzuki, H. Takesima, J. Kurasaki, H. Takahashi, S. Numa, *Nature*, 320: 188–192, 1986; Figure 5-3, The American Physiological Society, C. M. Armstrong, Sodium channels and gating currents, *Physiological Reviews*, 61: 662, 1981; Figure 5-4, Elsevier Science, Cambridge, England, W. A. Caterall, Voltage-dependent gating of sodium channels: correlating structure and function, *Trends in Neuroscience*, 9: 7–10, 1986; Figure 6-1, The Rockfeller University Press, L. Makowski, D. Caspar, W. Phillips and D. Goodenough, GAP Junction Structures II. Analysis of the X-Ray Diffraction Data, *The Journal of Cell Biology*, 74: 629–645, 1977; Figure 7-3, The Physiological Society, I. A. Boyd and A. R. Martin, The end-plate potential in mammalian muscle, *Journal of Physiology,* 132: 74–91, 1956; Figure 7-6, Elsevier Publications, Cambridge, England, S. J. Smith, G. J. Augustine, Calcium ions, active zones and synaptic transmitter release, *Trends in Neuroscience,* 11: 458–464, 1988; Figure 7-7, The Physiological Society, G. J. Augustine, M. P. Charlton, and S. J. Smith, Calcium entry and transmitter release at voltage-clamped nerve terminals of squid, *Journal of Physiology,* 367: 163–181, 1985; Figure 12-7, The Physiological Society, J. J. Art and R. Fettiplace, Variation of membrane properties in hair cells isolated from the turtle chochlea, *Journal of Physiology,* 385: 207–242, 1987; Figure 12-12, Elsevier Publications, Cambridge, England, T. D. Lamb, Transduction in vertebrate photoreceptors: the roles of cyclic GMP and calcium, *Trends in Neuroscience,* 9: 224–228, 1986; Figure 12-15, S. C. Kinnamon, Taste transduction: a diversity of mechanisms, *Trends in Neuroscience,* 11: 491–496, 1988; Figure 13-2, Alan R. Liss Inc., New York, P. Rakic, Neuron-glia relationship during ganglion cell migration in developing cerebellar cortex. A Golgi and electronmicroscopic study in *Macacus rhesus, The Journal of Comparative Neurology,* 283–312, 1971; Figure 13-2, Plenum Press, New York, M. Jacobson, *Developmental Neurobiology,* 1978; Figure 13-4, Annual Reviews, Inc., J. R. Sanes, Roles of extracellular matrix in neural development, *Annual Review of Physiology,* 45: 581–600, 1983; Figure 14-2, The Rockefeller University Press, P. Forscher and S. J. Smith, Actions of cytochalasins on the organization of actin filaments and microtubules in a neuronal growth cone, *The Journal of Cell Biology,* 107: 1505–1516, October, 1988; Figure 14-7, Cell Press, M. Matsunga, K. Hatta and M. Takeichi, Role of N-cadherin cell adhesion molecules in the histogenesis of neural retina, *Neuron,* 1: 289–295, 1988; Figure 14-10, American Association for the Advancement of Science, P. G. Haydon, D. P. McCobb and S. B. Kater, Serotonin selectively inhibits growth cone motility and synaptogenesis of specific identified neurons, *Science,* 226: 561–564, 1984; Figure 15-6, American Association for the Advancement of Science, M. Constantine-Paton and M. I. Law, Eye-specific termination bands in tecta of three-eyed frogs, *Science,* 202: 639–641, 1978; and Figure 15-9, Elsevier Publications, Cambridge, England, M. C. Brown, Sprouting of motor nerves in adult muscles: a recapitulation of ontogeny, *Trends in Neuroscience,* 7: 10–14, January, 1984.

Contents

III Behavior and Plasticity, 299

The Neuron

An introduction to the cellular structure of neurons and glia

O tell me where is fancy bred
Or in the heart, or in the head

Although Shakespeare asked this question near the end of the sixteenth century, the answer had been known, at least to some, for more than two millennia. Many ancient Greek scholars, Hippocrates and Plato among them, appreciated that there is something special about the brain, and argued that the brain is responsible for behavior in humans and other animals. We now take it for granted that the brain is the organ that obtains information about the environment, processes and stores this information, and generates behavior. In addition, it is the brain that is responsible for such vaguely defined aspects of behavior as feelings, aspirations, and abstract thoughts, those qualities we consider—with more than a touch of hubris—quintessentially human. In essence then, it is the brain that makes us what we are, and indeed it can be argued that all other organs are there simply to support the brain. For this reason understanding the biology of brain function is a major goal of modern science. The brain is one of the last great frontiers in the biological sciences, and the unraveling of its mysteries is comparable in complexity and intellectual challenge to the search for the elementary particles of matter or the effort to explore space.

Levels of Organization

One can study the functioning of the nervous system at a number of levels of organization (Fig. 1-1). Biochemists and molecular biologists investigate the properties of molecules that perform tasks important for brain function. Physiologists may study the characteristics of individual nerve cells or col-

Molecules

Individual
Cells

Pairs of Cells
Connected by
Synapses

Networks of
Interacting
Cells

Systems in the
brain that
regulate behavior

Behaving animal

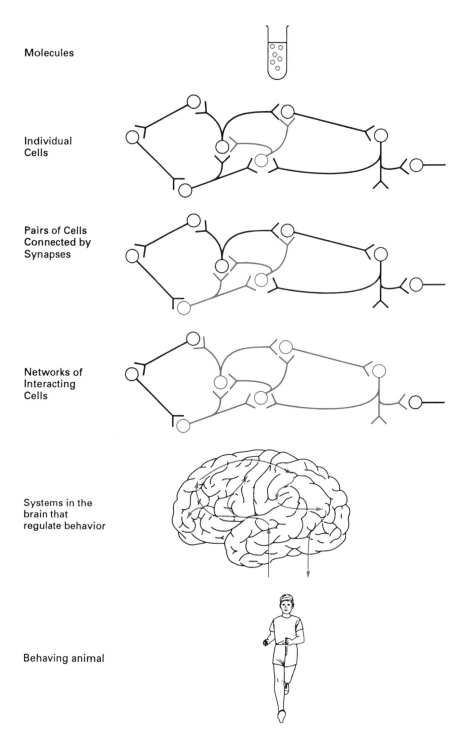

FIGURE 1-1. Levels of organization for studying structure and function in nervous systems. Depending on their background and training, different scientists may take different approaches to the study of the nervous system.

lections of cells that are functionally related. Behavioral psychologists explore patterns of behavior and its modification—learning—in experimental animals ranging from lower invertebrates to humans. Computational neuroscientists attempt to put it all together, to model higher brain functions in terms of the known properties of molecules, cells, or collections of cells.

In this book we will focus on the top part of Figure 1-1, on the cells of the brain and the molecules that control their function. Although where appropriate we will discuss these cells and molecules in the context of the nervous system to which they belong, our major emphasis will be on *signaling,* or *information transfer,* within and between nerve cells. We shall see that such signaling is essential for an organism to (1) sense information about its environment, (2) import this information into its brain where it can be processed, and (3) generate a behavioral response. Chapters 2 through 5 will explore the ways in which nerve cells are specialized for *intracellular signaling,* the movement of information from one part of the cell to another. Chapters 6 through 12 will consider *intercellular signaling,* the mechanisms by which nerve cells communicate with each other and the outside world. Finally, in Chapters 13 through 17 we will address *neuronal plasticity,* changes in the properties of a neuron. This section will include cellular and molecular aspects of *development,* the processes by which nerve cell shape and patterns of connectivity (and hence signaling) are set down during embryonic and early postnatal life. In addition, we will discuss *behavior* and its modification at the level of signaling in individual nerve cells, collections of nerve cells, and behaving animals.

Although our focus will be on cellular and molecular mechanisms of brain function, we emphasize that cellular and molecular neurobiology does not exist in a vacuum. First, many of the mechanisms we will discuss in the context of brain cells have their counterparts in other cell types—that is, there is an emerging awareness of a satisfying unity in cell biology. Second, it is becoming evident that understanding the brain requires study at *all* the levels of organization depicted in Figure 1-1, from behaving human beings to single brain cells and the molecules that regulate cellular activity. No single level is inherently more important than any other, and information from all of them will be necessary for even the most rudimentary understanding of normal and abnormal brain functions. Thus, by its very nature, the problem demands a multidisciplinary attack, one that bridges traditional scientific disciplines and facilitates collaboration between scientists with very different training and experimental approaches. History has amply shown that it is at the boundaries between disciplines where the most significant advances are likely to be made.

The Cellular Hypothesis

We have grown so accustomed to thinking of the brain as a cellular organ that it is easy to forget how recently this view was the subject of intense debate. Around the end of the nineteenth century the great neuroanatomists Santiago Ramon y Cajal and Camillo Golgi argued passionately about whether the brain consists of enormous numbers of discrete cells, or is a continuous syncytium of tissue. The answer to this question is of course of enormous significance for the understanding of how signals spread from one part of the nervous system to another. Cajal made elegant use of a technique, which had been discovered fortuitously by Golgi, for staining tissue. For reasons that are not understood to the present day, this technique, known as silver impregnation, results in the staining of only a small subset of the neurons present in a brain section. As a result, individual entities showed up clearly in tissue sections. Using other staining methods, these same sections would have appeared only as "tangled thickets" (Fig. 1-2). Cajal correctly identified these entities as individual nerve cells,

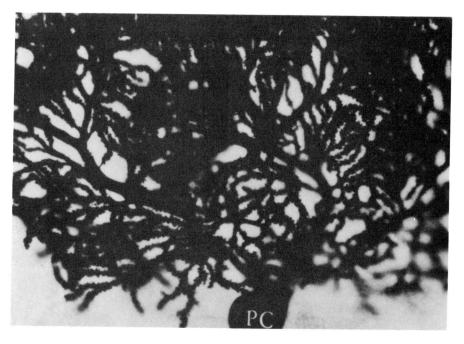

FIGURE 1-2. The dendritic tree of a single neuron. Golgi stains, such as this picture of a Purkinje cell (PC) neuron in the cerebellum, allow the complex shape of a single neuron to be resolved from the "tangled thickets" seen when all surrounding neurons are also stained (from Palay and Chan-Palay, 1974).

although Golgi never accepted this interpretation and continued to put forward his "reticular theory" of a continuous meshwork. The debate was finally won by the advocates of the cellular hypothesis, and it is now universally accepted that the brain, like other organs, is cellular. In retrospect it is clear that one reason for the long confusion over this issue is the complexity of brain tissue. As we shall see in this book, there are many different cell types in the brain, and many of these cells have a complex asymmetric three-dimensional structure that makes it extremely difficult to ascertain where one cell ends and the next begins. Thus our discussion of the cell biology of the nervous system necessarily begins with a description of the different classes of cells found in the brain, and of those aspects of their structure that specialize them for particular functional roles.

The Brain Consists of Neurons and Glia

In the middle of the nineteenth century the German anatomist Rudolf Virchow recognized that cells in the brain could be divided into two distinct groups: (1) neurons and (2) a more numerous group of cells that appears to surround the neurons and fill the spaces between them. Virchow called this second category of cell the *neuroglia*, or nerve glue, the implication being that one of its functions is to hold the neurons in place. Glial cells can themselves be divided into several subclasses based on their appearance in tissue sections. In the central nervous system the two main types of glial cells are the *astrocytes* and the *oligodendrocytes*. As their name implies the astrocytes have a star-like appearance, with numerous long arms radiating out from a central cell body. The oligodendrocytes also have a central cell body, with radial arms that tend to be shorter and more branched than those of the astrocytes. As will be discussed, the oligodendrocytes play an essential role in the functioning of neurons by forming the *myelin sheath* around axons in the central nervous system (see Fig. 1-5). In the peripheral nervous system the *Schwann cell*, another class of glial cell, is responsible for forming the myelin sheath. There has been much speculation about other possible functional roles for glia, but even though they comprise some 90% of the cells in the brain there is surprisingly little definitive information available. For example it is thought that glia might

1. provide structural support for neurons, fulfilling a role played by connective tissue cells in other organs;
2. segregate groups of neurons one from another, and act as electrical insulators between neurons;
3. play a nurturing role, supplying metabolic components and even proteins necessary for neuronal function;

4. participate in the uptake and metabolism of the neurotransmitters that neurons use for intercellular communication;

5. take up and buffer ions from the extracellular environment; and

6. play a role in information handling and memory storage.

There is evidence in the literature for each of these possibilities, and it seems likely that glia have evolved multiple functional roles. Only in the case of myelination, however, is their role well understood.

The Neuronal Cell Body: Neurons Are the Same as Other Cells

As Cajal recognized, the nerve cells or *neurons* are the individual signaling elements of the brain. Although we will argue that the intercellular communication that is the hallmark of brain function makes the brain a unique organ, in many respects neurons (and glia) closely resemble other types of cells. Figure 1-3 is a diagram of the archetypal neuron. It consists of a cell body with processes of different size and shape emanating from the cell body. It is this neuronal cell body or *soma*, an electron micrograph of which is shown in Figure 1-4a, that most resembles cells in other organs. The most prominent organelle in the cell body is the nucleus, which contains the genetic material, DNA. The genomic DNA in neurons is identical to that in other cells in the organism (although in the so-called giant neurons in some invertebrates the genome divides many times without corresponding cell divisions, resulting in as many as 50,000 copies of the genome in the nucleus of some of these cells; the functional consequences of this are not understood). Even though the genome is no different than in other cells, genes are regulated in specific ways that result in the synthesis of a pattern of proteins specific to neurons. Of course, specific gene expression also occurs in all other tissues and accounts for the existence of specific cell types in liver, heart, muscle, and other organs, in addition to the many different types of neurons and glia found in the brain.

The entire neuron, like all other cells, is enclosed by a *plasma membrane*. This is a double layer (or *bilayer*) of phospholipid molecules, which acts as a barrier preventing the contents of the cell from mixing with that of the extracellular space. The plasma membrane is also an effective electrical insulator, hindering the diffusion of charged ions in and out of the cell. This is important, because signaling in nerve cells requires the *controlled* movement of ions across the plasma membrane, a process mediated by specialized proteins located in the membrane.

What other common organelles are to be found in neuronal cell bodies (Figs. 1-3 and 1-4)? The cell body also contains *mitochondria* to supply the cell's energy needs. In fact, because a great deal of energy is required to maintain the transmembrane ionic gradients that are essential for neuronal signaling, neurons tend to be particularly rich in mitochondria. There are *ribosomes,* which are responsible for the synthesis of proteins destined for insertion into membranes, or for secretion; they are located on the membranous sacs of the *rough endoplasmic reticulum.* The rough endoplasmic reticulum is often unusually dense adjacent to the nucleus of neurons, giving rise to the structural feature called the Nissl substance (named after its discoverer). Other membranous components include the *smooth endoplasmic reticulum* and the *Golgi complex,* involved in the processing of proteins

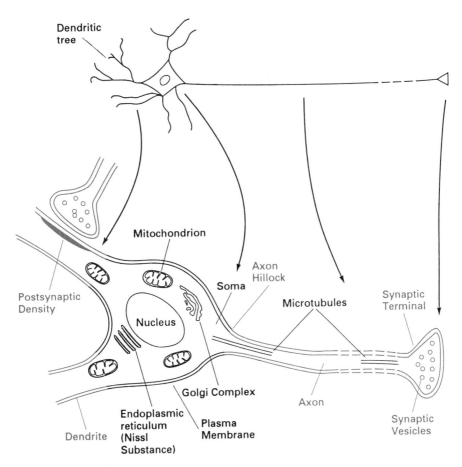

FIGURE 1-3. Ultrastructure of the neuron. Drawing of a typical nerve cell to show its overall shape and characteristic organelles. Structures and organelles that are found exclusively in nerve cells are shown in red.

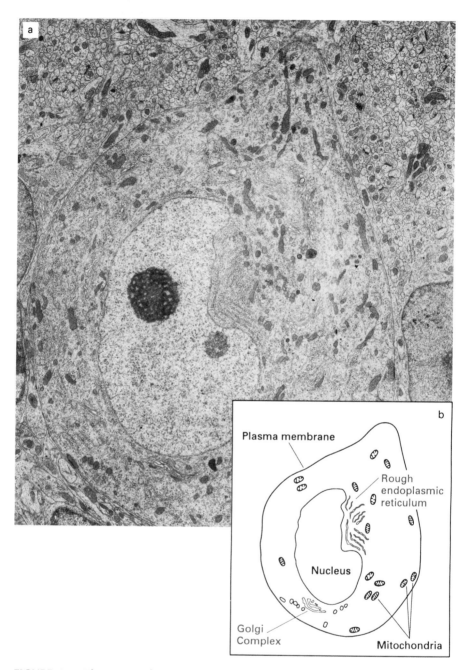

FIGURE 1-4. The soma of a neuron. *a*: A transmission electron micrograph of a section through the cell body of a typical neuron, a Purkinje cell of the cerebellum (courtesy of Dennis Landis). *b*: The diagram identifies some of the intracellular organelles.

for membrane insertion or secretion (see Chapter 6), and *lysosomes* and other granules involved in the breakdown and disposal of cellular components.

We have been listing these organelles of the cell body at a rapid pace simply because the reader with a background in cell biology will already be familiar with their structures and functions. Although there may be some features particularly characteristic of neurons, for example, the presence of the Nissl substance or the density of mitochondria, these are relatively minor *quantitative* differences between neurons and other cells rather than *qualitative* ones. Therefore only the trained observer would readily identify an electron micrograph such as that in Figure 1-4a as a picture of a neuron rather than of some other kind of cell.

Neuronal Processes: Neurons Are Different

Having emphasized those structures that neurons have in common with other cells, in the remainder of this chapter we will discuss what makes the neuron unique. It must be remembered that the essence of nervous system function is *signaling,* or *information transfer,* both *intra*cellularly from one part of a cell to another, and *inter*cellularly between cells. It is a fundamental premise of cellular neurobiology that a great deal will be learned about how the nervous system works by investigating

1. those aspects of neuronal structure that specialize them for information transfer;
2. the mechanisms of intracellular neuronal signaling;
3. the patterns of neuronal connectivity and mechanisms of intercellular signaling;
4. the relationship of various patterns of neuronal connectivity to different behaviors; and
5. the ways in which neurons and their connections can be modified by experience.

All of these will be discussed in this book. We begin now with a description of the *axon,* which is specialized for intracellular information transfer, the *dendrite,* which is often the site at which information is received from other neurons, and that most highly specialized structure of all, the *synapse,* which is the point of information transfer between neurons (Fig. 1-3).

Axon structure. The axon is a thin tube-like process that arises from the neuronal cell body and travels for distances ranging from micrometers to meters before terminating (at a synapse—see below). It originates at a conical-shaped thickening on the cell body called the *axon hillock* (Fig. 1-3). The axon is often (but not always) unbranched until just before it terminates, but it may branch many times in the terminal region. Its diameter remains more or less unchanged throughout its length. Its structure, like that of the dendrite, is formed and maintained by the *cytoskeleton,* an organelle system that is present in all cells but that exhibits certain unique properties in unusually shaped cells such as neurons. The role of the cytoskeleton in the formation and maintenance of neuronal form will be discussed briefly below. As we shall see in Chapters 2 through 5, specialized proteins in the axonal plasma membrane allow the axon to transmit electrical signals rapidly along its length, from soma to terminal.

The myelin sheath. The myelin sheath surrounds many, but not all, axons in the vertebrate nervous system. Although it is formed by glial cells and is not strictly a part of the axon, it is so important for axonal function that we will discuss its structure here in the context of axon structure. The sheath is formed by oligodendrocytes on central nervous system axons and by Schwann cells in the peripheral nervous system. When a myelinated axon is examined in cross section with the electron microscope, the axon is found to be surrounded by concentric circles of alternating dark and light bands (Fig. 1-5a). This structure arises by the tight wrapping of the oligodendrocyte or Schwann cell membrane around the axon during development. The cytoplasm of the glial cell is squeezed out of this region, so that the concentric circles represent layers of closely apposed glial plasma membrane. One can get a feel for the structure of myelinated axons by examining a roll of paper towel end-on. The central cardboard tube represents the axon, and the layers of paper the wrapping of glial plasma membrane.

A single Schwann cell may occupy up to about 1 mm of the length of a peripheral nervous system axon. Since some axons may be up to 1 m or more in length, the myelin sheath consists of a large number of Schwann cells each occupying its small portion of the axon. There are gaps between adjacent Schwann cells of several micrometers, known as *nodes of Ranvier* (Fig. 1-5b). The Schwann cell-covered region between the nodes of Ranvier

FIGURE 1-5. (facing) The myelin sheath. a: An electron micrograph of a cross section through a myelinated axon (modified from Peters et al., 1976). Ax_1, center of axon; nf, neurofilaments; mes_1, internal mesaxon; DL, dense line; IL, interperiod line; T_1, external tongue process. The last four are typical structural features, which are not described further in this text. b: Electron micrograph and diagram of a node of Ranvier. At this region, the myelin sheath is interrupted for a short distance between adjacent glial cells, exposing the axonal plasma membrane to the extracellular space (micrograph courtesy of Dennis Landis).

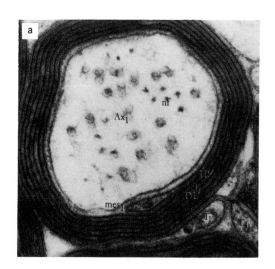

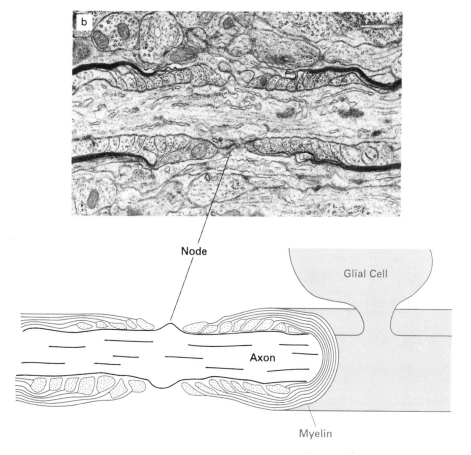

Node

Glial Cell

Axon

Myelin

is known as the *internode.* To carry our analogy further, the myelinated axon can be compared with a large number of rolls of paper towel lined up end-to-end, with small gaps in the paper layer (but of course not in the cardboard tubes) representing the nodes of Ranvier.

This multiple membrane layer, which also happens to be unusually rich in lipid, acts to effectively insulate the axonal cytoplasm (the *axoplasm*) from the extracellular fluid. This means that electrical current can flow across the axonal plasma membrane only at the nodes, and, as we shall see in Chapter 4, this has profound implications for the speed of transmission of electrical signals along the axon. The functional importance of myelin is underscored by the severe impairments in motor function observed in the so-called *demyelinating diseases,* including multiple sclerosis and amyotrophic lateral sclerosis (Lou Gehrig's disease), which are associated with extensive degeneration of the myelin sheath. In addition, in those species in which birth occurs prior to myelination, the newborns exhibit severe deficits in motor performance, and in fact are quite helpless until myelination is complete.

Dendrite structure. Dendrites are neuronal processes that tend to be thicker and much shorter than axons and often are highly branched, giving rise to a dense network of processes known as the *dendritic tree* (Fig. 1-3). In addition, their cytoskeleton differs from that of axons. Dendrites often originate from the cell body, but in some neurons in invertebrates they arise from the proximal regions of the axon. The electron microscope reveals the presence of numerous finger-like projections or thickenings, called *dendritic spines,* that arise from the main shaft of many dendrites (Fig. 1-6). These spines are the synaptic input sites at which the neuron receives information from another cell. Like the axonal membrane, the plasma membrane of the dendrite contains a particular set of proteins that allows the dendrite to carry out its assigned function (see Chapters 9 and 10). In this case, the proteins do not specialize the membrane for rapid electrical signaling, but instead allow it to receive and integrate information from other nerve cells or from elsewhere in the body. However, the role of the dendrite is not *exclusively* the receipt of information. In many nerve cells both input and output of information occur on the same set of dendrite-like fine processes.

Formation and Maintenance of Neuronal Form

The cytoskeleton. How do structures such as axons and dendrites arise? The neuron, like all cells, contains a heterogeneous network of filamentous structures known collectively as the *cytoskeleton.* The major components of this network are the *microfilaments,* the *neurofilaments,* and the *micro-*

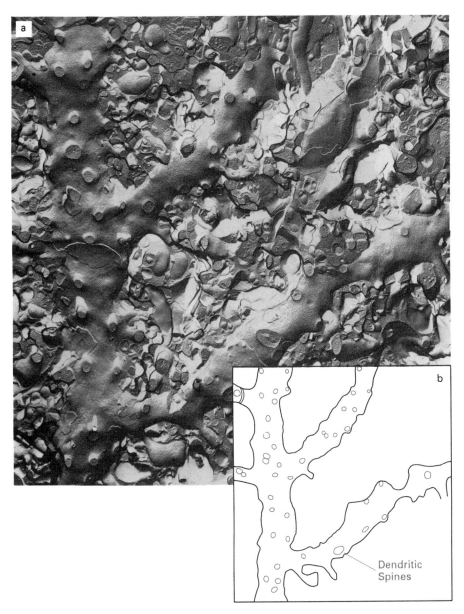

Dendritic
Spines

FIGURE 1-6. Dendritic spines. Diagram and electron micrograph of part of the dendrites of a Purkinje cell from the cerebellum. The micrograph was made by Dennis Landis and Tom Reese (1974) using the freeze-fracture technique, which is described in Chapter 6.

tubules. The function of microfilaments is best understood in skeletal muscle, where they are made up of the proteins actin and myosin. These filaments are present in highly ordered structures, and interact to produce muscle contraction. Actin is also present in axons and, as we shall see in Chapter 14, is particularly prominent in the growing tips of axons, the so-called growth cones, where it may contribute to the regulation of membrane movement.

The neurofilaments are probably the least well understood of the cytoskeletal components with respect to function. They consist of long filaments approximately 10 nm in diameter, intermediate in size between actin filaments (about 5 nm) and the microtubules (about 20 nm). For this reason they fall into the general class of cytoskeletal components known as *intermediate* filaments in nonneural cells. Certain pathological conditions, including Alzheimer's disease, are associated with a profound disorganization of neurofilaments, but it is not known whether the chaotic tangles of neurofilaments seen in the brains of Alzheimer's patients actually cause the progressive senility characteristic of this disease.

Microtubules carry out a variety of functions in different cells. They play an important role in cell movement and are the major component of the mitotic spindle, an organelle that participates in cell division. Microtubules are prominent inhabitants of axons and dendrites. Like the other filaments, they are polymeric structures, made up of large numbers of repeating units of a 50-kDa protein known as *tubulin.* The polymerization of tubulin into microtubules depends on the nucleotide GTP and is promoted by *microtubule-associated proteins* (MAPs), which may also help anchor microtubules to membranes or to other cytoskeletal components. A number of different MAPs exist and these are often differentially associated with axons and dendrites. In addition, the amounts of the various MAPs change in characteristically different ways during neuronal development. Figure 1-7 shows a micrograph of a section through the rat cerebellum, stained with an antibody that recognizes one particular MAP, MAP1. The staining is particularly prominent in the dendrites of the large cerebellar Purkinje neurons. Earlier in development, however, MAP1 is restricted to the axons of these cells. Findings such as these suggest that the MAPs play a crucial role in the generation of structural differences between axons and dendrites.

Axonal transport. When one examines a picture of a neuron with its markedly asymmetric structure (Fig. 1-3), several questions come to mind. One series of questions referred to above concerns the mechanisms by which this structure is formed during development. Another has to do with the maintenance of the structure during the normal everyday life of the neuron. Since portions of the cell may be as far as 1 m or more from the cell body, the neuron must have mechanisms for providing proteins and other neces-

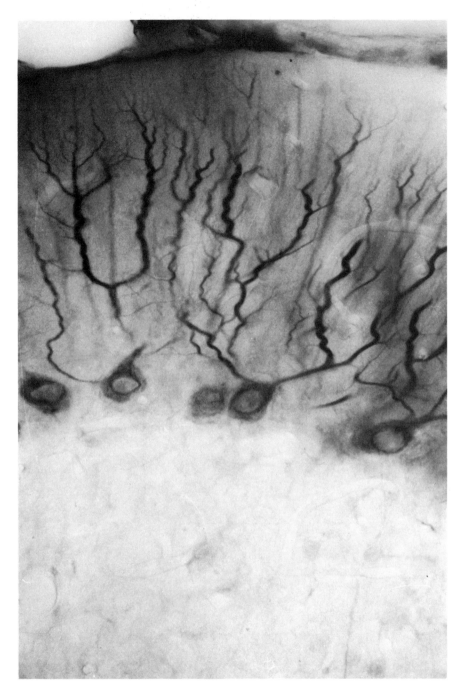

FIGURE 1-7. Localization of MAPs in axons and dendrites. In this experiment carried out by Andrew Matus and his colleagues, a section through a rat cerebellum was stained with an antibody that specifically recognizes one particular microtubule-associated protein, MAP1. The dendrites of the large Purkinje cell neurons stain prominently with this antibody.

sary metabolic materials to its distal regions. It is known that protein synthesis is restricted to the cell body and perhaps the proximal regions of dendrites, and does not occur in axons or their terminals. How then do proteins required for normal membrane turnover, and enzymes for metabolic functions such as neurotransmitter synthesis and degradation, reach their appropriate sites? What is the mechanism for providing energy to sites remote from the soma?

To carry out these functions the neuron has evolved a series of elegant transport systems known collectively as axonal transport. Of course, all cells are faced with the problem of moving cellular components from one part of the cell to another, but, as pointed out above, this is particularly acute for neurons. Hence, their transport systems are highly specialized. Such systems are necessary because in the absence of an active process, a typical protein would require approximately *10 days* to diffuse passively down a 1-cm axon from soma to terminal.

Axonal transport was first studied in detail some 50 years ago. When axons are ligated and examined under the microscope, a variety of vesicular structures can be seen to accumulate in the proximal portion of the axon immediately behind the ligation (Fig. 1-8). It was inferred from such experiments that vesicles are transported down the axon in a proximal-to-distal (soma-to-terminal) direction. Other early studies on axonal transport used parts of the brain in which the cell bodies are remote from the axons and their terminals. The visual system is particularly useful in this regard, since

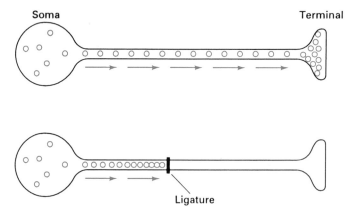

FIGURE 1-8. Vesicles are transported from soma to axon terminal. Vesicles and organelles are synthesized in the cell body and transported (arrows) by an active process down the axon toward its terminal. When the axon is ligated, vesicles are seen to accumulate in the axon on the side of the ligature proximal to the cell body. Experiments of this type were first done by Paul Weiss and his colleagues in the 1930s.

the cell bodies of the retinal ganglion cells are confined to the retina, and their axons travel to the brain in the optic nerve. Accordingly, a number of investigators applied radioactively or fluorescently labeled materials to the eye and asked whether and how these materials are transported along the optic nerve into the brain. The most extensive of these studies used radioactive amino acids that are incorporated into proteins and provide a marker for following protein transport. It has become evident from such experiments that proteins are transported down the axon at different rates, with each rate probably corresponding to different forms of transport. There is a rapid form of axonal transport that is characterized by transport rates of the order of several hundred millimeters per day, and one or more slow processes that transport proteins by 1–10 mm per day.

It has long been inferred that components of the cytoskeleton, such as microtubules, which run the length of the axon, might play a fundamental role in axonal transport (Fig. 1-9a). This has been confirmed by experiments with the microtubule-disrupting drug *colchicine,* which blocks axonal transport. Enhanced contrast video microscopy allows the movement of vesicles and organelles along microtubules to be seen in real time. Among the transported organelles that can be seen are mitochondria, which answers the question posed above concerning the energy requirements of axons and nerve terminals. The various vesicles carry proteins and other materials to be inserted into membranes or to be released at the nerve terminal.

Other experiments have shed light on the molecular machinery, the so-called molecular motors, that drive axonal transport. A 120-kDa vesicle-associated protein called *kinesin* has been shown to interact with both microtubules and vesicles. Using energy provided by the hydrolysis of ATP, kinesin can drive the movement of one relative to the other (Fig. 1-9b). Since the microtubules are effectively fixed in place by other cytoskeletal components and cell membranes, the net effect of kinesin's action is to move vesicles in a proximal-to-distal *(orthograde)* direction along the axon. Organelles can also be transported in a distal-to-proximal direction, that is toward the cell body (*retrograde* transport). This is mediated, at least in part, by another molecular motor, a microtubule-associated protein (MAP1c) also called cytoplasmic *dynein,* that differs from kinesin. A single microtubule can serve as a track for transport in both the orthograde and retrograde directions, the direction of transport being determined by the nature of the motor that binds to the organelle.

Intercellular Communication: The Synapse

We have emphasized that *inter*cellular communication, which results in the passage of information from one part of the nervous system to another, is the essence of nervous system function. It is this information transfer that most clearly distinguishes the brain from other organs, and thus it is not surprising that the neuron has evolved a unique and highly specialized structure, the *synapse*, to carry out this task. The term synapse, derived from a Greek word meaning "connect," was introduced by the British

a

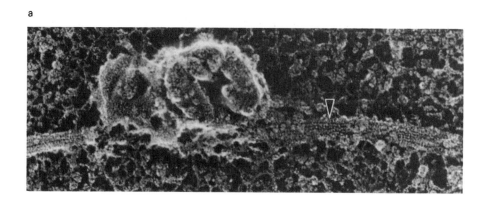

b

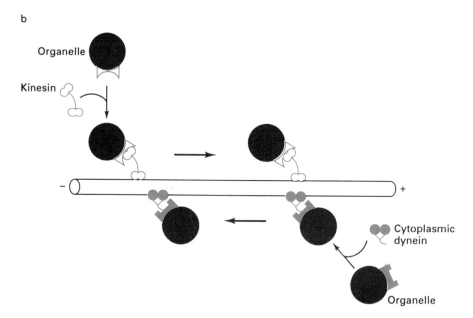

FIGURE 1-9. Vesicles and organelles are transported by an active process along microtubules. *a*: This electron micrograph from an experiment by Bruce Schnapp shows a vesicle sitting on a microtubule (arrowhead) in cytoplasm from an axon of a squid. *b*: The binding and movement of organelles such as vesicles along microtubules are generated by molecular motors called kinesin and cytoplasmic dynein (Sheetz et al., 1989).

physiologist Charles Sherrington near the end of the nineteenth century. Sherrington was studying *spinal cord reflexes* (Fig. 1-10). Such reflexes are invariant behavioral responses to a particular kind of stimulus to the animal; they do not require the brain itself but are mediated via the spinal cord. One often-cited example is the rapid withdrawal of the arm when the fingers encounter a hot stove.

Sherrington's studies on reflexes coincided with the time when the neuron doctrine was becoming well established as a result of the work of Cajal and his followers. He defined a set of neuronal connections, a *pathway,* responsible for the reflex, and noted that information always travels through the reflex pathway in one direction only. More specifically, input is via the *sensory* component of the pathway (black in Fig. 1-10), which provides information about the outside world. Output is through the *motor* component (red in Fig. 1-10), which drives muscles to provide an appropriate behavioral response to the sensory input. Sherrington also became convinced from his detailed anatomical and physiological studies that the pathway is not unicellular, but that there is a discontinuity between the sensory and motor components. In many reflex arcs there may also be an *interneuron* in the spinal cord, which provides the link between the sensory and motor neurons.

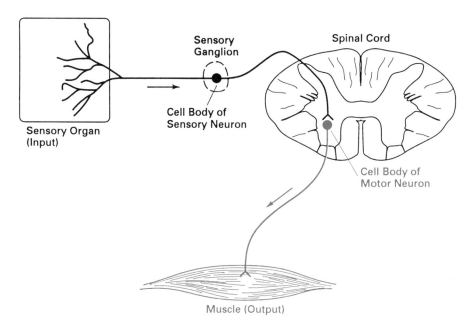

FIGURE 1-10. A reflex arc. The classic reflex arc, with its sensory neuron input and motor neuron output. The arrows indicate the direction of information flow in the reflex pathway.

With remarkable insight Sherrington proposed that at the point of contact between the sensory and motor neurons there might be a structure—the synapse—that allows unidirectional information transfer between the neurons. It is important to emphasize that Sherrington's definition of the synapse was a functional one, and it was to be many years before the anatomical correlate of this functional connection was identified.

Sherrington's pioneering work established the idea that understanding synaptic structure and function is essential for understanding how the brain works. This emphasis on the synapse in cellular neurobiology has been maintained throughout the twentieth century. We know now that the strength of synaptic transmission—information transfer at synapses—is not fixed. This is in contrast to information transfer along the axon, which to all intents and purposes is *all-or-none*. Accordingly, one widely held belief in modern neurobiology is that changes in the properties of synapses underlie the plasticity of nervous system function, including learning and memory (see Part III). In addition, it is becoming evident that malfunctioning synapses are associated with a whole range of debilitating diseases, among which are Parkinson's disease, manic-depressive illness, and schizophrenia. As our understanding of synaptic structure and function has increased, it has become possible to think about designing rational treatments for these diseases, but the relatively limited success to date emphasizes how far we still have to go. It is these considerations, together with the sheer intellectual excitement of trying to understand how intercellular communication works in the nervous system, that continue to motivate research on synaptic transmission.

Two kinds of synapses. Through the first half of the twentieth century there was a bitter controversy about the nature of information transfer at synapses. The school of neuropharmacologists, led by Sir Henry Dale, insisted that synaptic transmission is mediated by a chemical substance liberated from the terminal of one neuron (the *presynaptic* cell), which interacts with and influences the properties of the follower neuron or muscle cell (the *postsynaptic* cell). The first convincing piece of evidence for this idea came from the study of a neuron-to-muscle synapse in the frog. It had been known that electrical stimulation of the vagus nerve leads to slowing of the heart rate. In a classical experiment carried out in 1921, the Austrian pharmacologist Otto Loewi placed a frog heart, innervated by the vagus nerve, in a chamber containing physiological saline, and connected the chamber with another that contained a second, noninnervated heart (Fig. 1-11). The experimental arrangement allowed the saline solutions in which the two hearts were sitting to exchange freely. He then stimulated the vagus nerve, which of course resulted in slowing of the first heart. Loewi noted that, after some delay, the second heart slowed as well (Fig. 1-11).

He concluded correctly that some chemical released by the firing of the vagus nerve results in slowing of the heart rate, and that this chemical diffused through the saline solution to the second chamber where it acted on the second heart. The chemical responsible for this phenomenon was subsequently isolated and identified as acetylcholine, the first chemical *neurotransmitter* to be characterized. Although this is a neuron-to-muscle synapse, it is now clear that neuron-to-neuron chemical synapses operate in the same general way, and that acetylcholine is an important neurotransmitter in the brain as well as in the peripheral nervous system.

The conflicting point of view, put forward most forcefully by the elec-

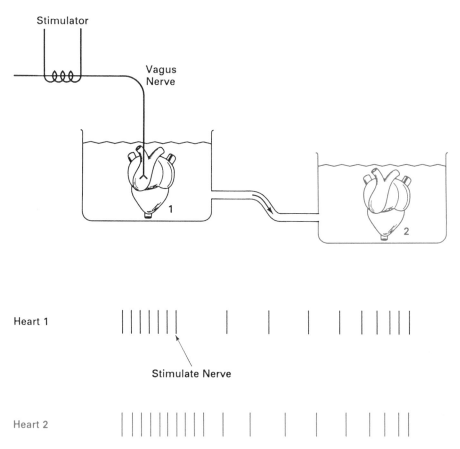

FIGURE 1-11. Chemical transmission at a nerve–muscle synapse. In this famous experiment performed in 1921, Otto Loewi placed an innervated (1) and a noninnervated (2) heart in two separate chambers connected by a bridge of physiological saline. At the bottom is shown the rate of beating of both hearts before and after stimulation of the vagus nerve connected to the first heart.

trophysiologist Sir John Eccles, held that synaptic transmission is electrical and results from the movement of ions from one cell to another via intercellular pathways. Again, a variety of evidence supported this contention. The arguments between the two factions were often acrimonious because each believed that there is a single global mechanism of synaptic transmission. We know now that they were wrong in this belief, and that both points of view were correct. Chemical and electrical synaptic transmission exist side by side in most nervous systems.

Properties of chemical and electrical synapses. Although chemical and electrical synapses both mediate intercellular information transfer, they do so by very different mechanisms. It is instructive to consider those features that distinguish the two types of synapses in the context of the different functional roles that they may play. We know a good deal more about chemical than electrical synapses, but this may reflect historical accident rather than the relative importance of the two kinds of synapses. Convenient experimental preparations for the investigation of chemical synaptic transmission have been available for many years, but only more recently have conceptual and technical advances allowed a more thorough examination of the properties of electrical synapses. It is worth mentioning in this context that much of what we know of chemical synaptic transmission has been gleaned from studies of the vertebrate *neuromuscular junction,* which is a neuron-to-muscle rather than a neuron-to-neuron synapse (see Fig. 7-1). This has arisen in part because the frog sciatic nerve–gastrocnemius muscle preparation is so accessible and convenient, far more so than neuron-to-neuron synapses in the central nervous system (Fig. 1-12). In addition a number of gifted investigators, most notably Sir Bernard Katz and his collaborators in London, have exploited this preparation brilliantly to elucidate the details of synaptic transmission between nerve and muscle cells.

Accordingly the picture many neurobiologists hold of the "typical" synapse is the neuromuscular junction. It is now evident that this is inappropriate in many ways. For example many central nervous system synapses operate on a time scale orders of magnitude slower than the neuromuscular junction, and the molecular mechanisms involved in the transduction of the chemical signal into an electrical response in the postsynaptic cell are very different (see Chapters 9 and 10). Although we will be referring often to studies on the neuromuscular junction in discussing various aspects of synaptic transmission in subsequent chapters, these caveats must be kept in mind.

FIGURE 1-12. A synapse. An electron micrograph and a drawing of a synapse onto the spine of a Purkinje cell dendrite (micrograph from Landis, 1987).

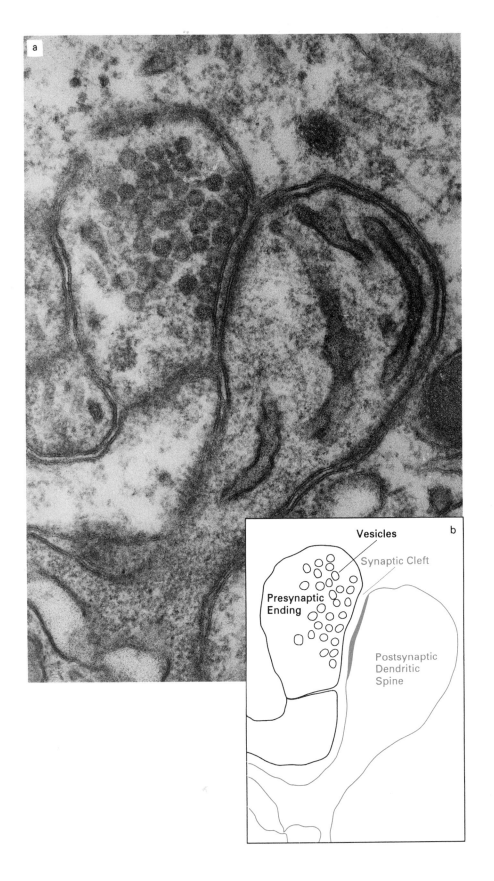

Vesicles

Synaptic Cleft

Presynaptic Ending

Postsynaptic Dendritic Spine

Chemical synapses. The most obvious difference between chemical and electrical synapses is in their structure as seen in the electron microscope. Chemical synapses have an *asymmetric* morphology, with distinct features found in the presynaptic and postsynaptic parts or "elements" of a synapse (Fig. 1-12). The presynaptic ending is a swelling of the axon terminal, containing mitochondria, and, most important, a variety of vesicular structures. As will be discussed in Chapter 6, the vesicles contain the neurotransmitter, for example, acetylcholine, that is released from the presynaptic terminal and produces some change in the postsynaptic cell. As shown in Figure 1-12, the vesicles are often clustered adjacent to the membrane of the presynaptic terminal at its point of closest contact with the postsynaptic cell. This clustering may be related to the association of vesicles with specialized cytoskeletal filaments.

The electron microscope also reveals that the pre- and postsynaptic elements of a chemical synapse are separated by a 200–300 Å gap, the *synaptic cleft* (Fig. 1-12). This gap is somewhat larger than the normal extracellular space between cells, and its presence emphasizes that there are not direct membrane connections between the pre- and postsynaptic cells at chemical synapses. Stains that bind to sugars reveal that the synaptic cleft is loaded with carbohydrate, presumably associated with glycoproteins in the pre- and/or postsynaptic membranes. The function of this extracellular carbohydrate is not well understood, although one idea favored by many investigators is that it plays a role in cell–cell recognition during synapse formation (see Chapter 15).

Just as most presynaptic endings are axon terminals, most postsynaptic elements of central nervous system synapses are dendrites, giving rise to the term "axodendritic" synapse. However, like most rules this one has exceptions. Axosomatic synapses (where the postsynaptic target is a neuronal cell body) and (more rarely) axoaxonic synapses exist, and it is now becoming evident that dendrites can also act as presynaptic elements (dendrodendritic synapses). In fact in some nerve cells there are fine dendrite-like processes at which both input and output of information take place.

The microscope reveals that virtually the entire somatic and dendritic surfaces of most central nervous system neurons are covered with presynaptic terminals. In other words, there can be enormous *convergence,* onto a single neuron, of input information from hundreds or even thousands of presynaptic cells (Fig. 1-13). As we will see in subsequent chapters, neurons are constantly integrating these multiple synaptic inputs, and the location of any given postsynaptic site can be very important in determining its contribution to the neuron's overall activity. When we consider, in addition, that a single presynaptic axon may branch many times and provide input to dozens or hundreds of postsynaptic targets (*divergence*—Fig. 1-13), we can begin to appreciate the complexity of the computations that even the simplest of nervous systems carry out.

In contrast to the presynaptic terminal, the postsynaptic element is usually characterized by the absence of vesicles adjacent to the plasma membrane. There is often a highly electron-dense structure, the *postsynaptic density*, associated with the postsynaptic membrane immediately opposite the accumulation of vesicles on the presynaptic side (Fig. 1-12). The functional significance of this distinctive morphological specialization is not well understood, but it may help to anchor receptors for the neurotransmitter in the postsynaptic membrane, and may contain molecules that are involved in *transduction*, the conversion of the chemical signal into an electrical response in the postsynaptic cell (see Chapters 9 and 10).

Associated with the morphological asymmetry of chemical synapses is an important functional asymmetry: chemical synapses are *unidirectional*. That is, rapid transfer of information only occurs from the pre- to the postsynaptic cell (see Fig. 1-10). As we shall see, this is very different from the *bidirectional* information transfer characteristic of most electrical synapses. In addition, at chemical synapses, there is a delay that may be a millisecond or longer between the arrival of information at the presynaptic terminal and its transfer to the postsynaptic cell. This delay reflects the several steps required for the release and action of the chemical neurotransmitter (Chapter 7). Furthermore, the response of the postsynaptic neuron may outlast the presynaptic signal that evokes it, sometimes by a very long time. The transduction mechanisms that may be responsible for such long-lasting changes in the target cell will be considered in detail in Chapters 9 and 10.

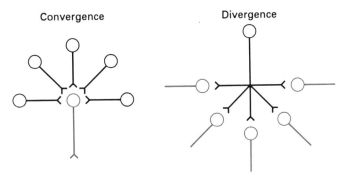

FIGURE 1-13. Convergence and divergence in nervous system function. Any given nerve cell can receive convergent synaptic inputs on its dendrites from a large number—perhaps as many as thousands—of other neurons. The axon of a neuron may branch many times, and thus divergent information may be provided to a large number of postsynaptic targets of this single neuron.

Electrical synapses. The striking asymmetry in structure and function of chemical synapses may be contrasted with the symmetrical morphology and bidirectional information transfer characteristic of electrical synapses. First there are no morphological specializations that allow pre- and post-synaptic elements to be distinguished. Indeed since signals can move in both directions through electrical synapses, each cell may be pre- *or* postsynaptic at different times. Instead of the synaptic cleft that separates the two elements of the chemical synapse, electrical synapses are characterized by an area of very close apposition between the membranes of the pre- and post-synaptic cells. Within these membranes are found *gap junctions,* cell-to-cell pores that allow ions and small molecules to pass freely from the cytoplasm of one cell to the next. Gap junctions exist in many types of cells. One test that is often used to establish whether a pair of cells is connected by gap junctions is *dye coupling;* that is, a low-molecular-weight dye such as *Lucifer Yellow* injected into one of the cells can spread rapidly into the cytoplasm of its synaptic partner (Fig. 1-14). As is the case for gap junctions in other cell types, molecules with a molecular weight up to about 1000–1500 can be transferred through most neuronal gap junctions. The structure and function of gap junctions will be discussed in more detail in Chapter 6 (see Figs. 6-1 to 6-3).

It is the movement of small ions from one cell to another through cell-to-cell gap junctional channels that mediates intercellular signaling at electrical synapses. These channels provide a low resistance pathway for ion flow between the cells without leakage to the extracellular space, and thus signals can be transmitted with little attenuation. Two important functional properties follow immediately from this mode of transmission. First, as referred to above, information transfer can be bidirectional, that is, a *functional* symmetry accompanies the *structural* symmetry. There are examples where the efficacy of electrical synaptic transmission is higher in one direction than in the other (so-called *rectifying* synapses), and, in fact, the first electrical synapse whose properties were investigated in detail, between two large axons in the crayfish, rectifies markedly. In general, however, the rule of bidirectionality holds, and, in fact, it is an important criterion in identifying electrical synapses. The second important functional consequence of this mechanism is that electrical synapses are fast. There is no delay analogous to that seen with chemical synaptic transmission.

The extent of electrical connectivity in the central nervous system is not clear. One functional role for electrical synapses that is widely accepted is in the synchronization of the electrical activity of large populations of neurons. For example, in both vertebrates and invertebrates it has been demonstrated that populations of neurosecretory neurons that synthesize and

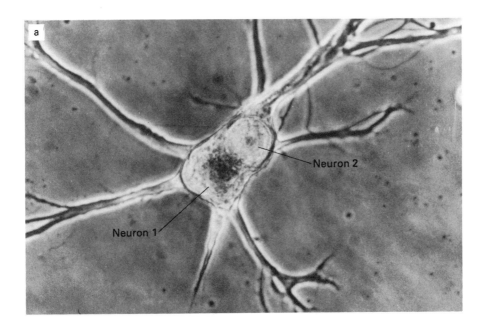

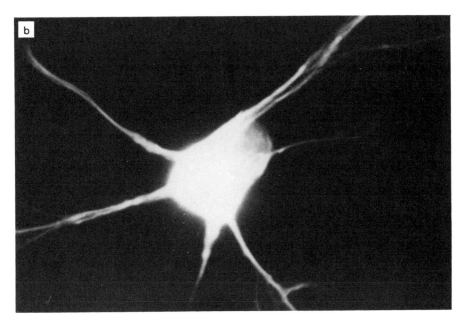

FIGURE 1-14. Dye coupling via an electrical synapse. *a*: A photograph of two adjoining neurons isolated in a culture dish. The lower cell (neuron 1) was injected with the fluorescent dye Lucifer Yellow by one of the authors (Kaczmarek et al., 1979). *b*: The fluorescent image shows that dye has passed into neuron 2.

release biologically active peptide neurotransmitters and hormones are connected extensively by electrical synapses. Simultaneous recording from several neurons within such populations reveals that they all are electrically active at the same time, probably resulting in concerted release of their neurotransmitter. Large numbers of electrical synapses are also found in the retina, where they may influence the processing of visual information.

It has been suggested that electrical synapses are less subject to alteration of their properties than chemical synapses, and thus might provide an invariant mode of intercellular communication. However, it is becoming evident that the efficacy of electrical synaptic transmission can be regulated, perhaps as extensively as that of chemical synapses. Furthermore, as we shall see in Chapter 16, an examination of the properties of small networks of neurons reveals that many neuronal pairs in such networks are connected by *both* chemical and electrical synapses, and that the efficacy of one of the synaptic types may be modulated by the other. Thus there is no question that more detailed information about electrical synaptic transmission will be necessary as we extend our understanding of intercellular communication in the nervous system.

Summary

The brain, like all other organs, is made up of vast numbers of cells. Unlike some other organs, however, there is a wide variety of cell types to be found in the brain. These may be conveniently divided into two main categories, the *neurons,* which are involved in intracellular and intercellular information transfer, and the *glia,* whose role is less well understood. There is extensive structural and functional heterogeneity within each category. In many respects neurons and glia closely resemble cells in other organs, but neurons in particular possess a number of structural features that make them unique. They are asymmetric cells, with morphologically and functionally distinct regions. In this chapter we have focused on the unique structural elements characteristic of neurons throughout the animal kingdom. These include dendrites, which are specialized for the receipt of information from other neurons, and axons, which are specialized for the intracellular transfer of information over long distances (aided by the glia-derived myelin sheath). Finally we have discussed the synapse, the specialized structure that mediates information transfer from one neuron to another. It is the intracellular and intercellular communication that is the essence of nervous system function and that makes the brain so complex and difficult to study and yet at the same time so fascinating for the student of cell biology.

Electrical properties of neurons

We have emphasized in the introductory chapter that information transfer, within and between nerve cells, is an essential element of nervous system function. We have also seen that neurons are highly asymmetric cells, with processes that often extend a considerable distance from the cell body. The basic question underlying the next four chapters is how information is transferred intracellularly, from one part of the neuron to another. Chapter 2 deals with electrical signaling. Neurons, like other cells, exhibit a voltage difference across their plasma membranes. Rapid changes in this transmembrane voltage, called *action potentials,* can be propagated from one part of the cell to another and are used by neurons (but not by most other cells) to encode information. Chapter 3 describes *membrane ion currents.* These are movements of ions across the plasma membrane that are responsible for the transmembrane voltage and the generation and propagation of action potentials. These ion currents flow through *ion channels,* a ubiquitous class of membrane proteins that have evolved to provide exquisite control over the movement of ions across the plasma membrane. We discuss the properties of ion channels in some detail because an appreciation of how they work is essential for understanding electrical signaling. Chapter 4 examines the ways in which the combined activities of different channels give rise to complex patterns of neuronal electrical activity. Neurons differ from most other cells in their particular complement of membrane ion channels, which allow the generation and propagation of action potentials in complex temporal patterns. Finally, Chapter 5 discusses the emerging information about the *molecular structure of ion channels.* Molecular cloning approaches now allow us to identify the structural features of an ion channel that are responsible for a particular aspect of its function. The goal of these approaches is ultimately to understand electrical signaling in neurons in terms of its underlying molecular mechanisms.

Electrical signaling

Although neurons have many features in common with other cells, they are unique because their primary function is to modify and transmit messages from one cell to the next and between different parts of the same cell. In this and the following three chapters we will focus on the nature and mechanisms of the *electrical signals* that neurons use both for intracellular communication and as stimuli for the generation of intercellular messages.

Intracellular Transfer of Information: The Axon

In the first chapter we showed that neurons are highly asymmetric cells and that different parts of the neuron exhibit structural features that enable them to carry out specialized tasks. We will concentrate first on the *axon,* the part of the neuron responsible for transmitting information from one part of the cell to another.

Axons are tube-like structures that arise from the neuronal cell body. They vary widely in size, shape, and other characteristics (Fig. 1-3). Some axons within the central nervous system are only a few microns in length, not much greater than the diameter of the neuronal cell bodies from which they arise. In contrast, axons that run from the central nervous system to other parts of the body are as long as 1 m in humans and even longer in larger animals. It is immediately apparent that whatever the mechanism of axonal information transfer, it must be able to operate over long distances without garbling or losing messages. Axon diameter also varies from less than 1 μm to almost 1 mm, and the diameter of any single axon may be different at different distances from the cell body. Axon diameter is an important factor in determining the speed at which information moves along the axon; whether or not an axon is myelinated also influences speed of transmission.

We will begin with a descriptive treatment of how information is passed along an axon, and then discuss the diverse patterns of electrical activity exhibited by different kinds of neurons. In subsequent chapters we will describe the mechanisms of these phenomena in detail. To summarize the descriptive message in advance, there is a voltage difference across the axonal membrane, and information is carried in the axon in the form of rapid changes in this voltage difference. These voltage changes, which are generally referred to as *nerve impulses, spikes,* or (most commonly) *action potentials,* travel rapidly along the axon from the cell body toward the distal portion of the axon.

Resting Potential and the Passive Membrane Response

Neurons, like all other cells, exhibit a voltage difference known as the *membrane potential* across their plasma membranes. It will become evident in Chapter 3 that the membrane potential arises because the plasma membrane is differentially permeable to different ions. When the axon is at rest, that is, when it is not conducting nerve impulses, the value of the membrane potential is called the *resting potential.* In neurons the resting potential is usually in the range -40 to -90 mV. By convention membrane potentials are expressed relative to the extracellular fluid, that is, negative membrane potentials indicate that the inside of the cell membrane is more negative than the outside. When the membrane potential is less negative than the resting potential the cell is said to be *depolarized;* when it is more negative, the cell is *hyperpolarized.*

It is possible to measure the transmembrane voltage (V_m) by inserting a measuring electrode, connected to an electrometer, into the cell. The electrode can be either a silver wire or a fine-tipped glass pipette filled with a conducting salt solution. When the tip of the measuring electrode (M) is in the extracellular fluid, there is no voltage difference between it and the reference electrode (R), which is also in the extracellular fluid (left side of Fig. 2-1). When the measuring electrode tip is passed through the plasma membrane, there is a sudden negative voltage deflection of some 40–90 mV relative to the extracellular reference electrode (right side of Fig. 2-1), reflecting the negative resting potential (V_r).

With appropriate (and very simple) electronics, one can also inject negative (hyperpolarizing) or positive (depolarizing) current into the cell via the same electrode. We shall see that in the real world neurons are constantly bombarded with physiologically relevant current "injections" as a result of synaptic activity or sensory input. With negative currents, the membrane potential changes in a hyperpolarizing direction, and the size of the change simply mirrors the amount of the applied current stimulus (Fig. 2-2). Such voltage shifts in response to hyperpolarizing current injection reflect *passive* membrane properties. Similar passive responses are seen in response to small depolarizing stimuli (Fig. 2-2).

The Plasma Membrane Is a Capacitor and a Resistor Connected in Parallel

The plasma membrane of a nerve cell, or indeed of any cell, provides a resistance to the flow of ions between the intracellular and extracellular compartments. Accordingly, it can be thought of as an electrical *resistor*

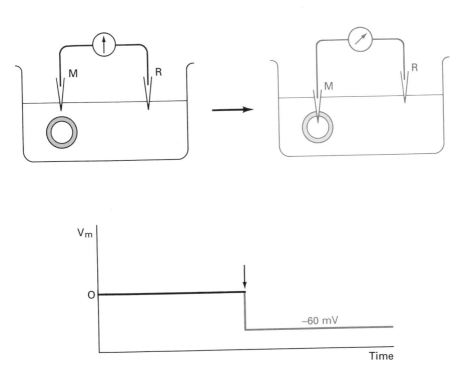

FIGURE 2-1. Measurement of the resting potential. The voltage difference (V_m) across the cell membrane can be determined as the voltage difference between a measuring electrode (M) inside the cell and a reference electrode (R) in the extra-cellular fluid.

with some membrane resistance R_m measured in ohms (Ω). In addition, the lipid bilayer provides an extremely thin insulating layer between two conducting solutions. This allows the membrane to separate and store charge, and thus it is also an electrical *capacitor* with some membrane capacitance C_m measured in farads (F). These considerations allow us to describe the lipid bilayer membrane in terms of an *equivalent electrical circuit* as shown in Figure 2-3a.

What does all this mean for voltage changes in biological membranes? Figure 2-2 shows that the shape of change in membrane voltage does not precisely follow the shape of the injected pulse of current. This is because the membrane capacitance must be charged (or discharged) for the voltage to change, and this does not occur instantaneously; rather it takes some time determined by the membrane *time constant* τ, which is equal to the product of the membrane resistance and capacitance:

$$\tau = R_m C_m$$

As can be seen in Figure 2-3b, the voltage changes exponentially with time (t), according to the equation:

$$V_t = V_0 e^{-t/\tau}$$

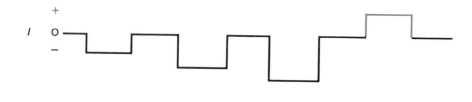

FIGURE 2-2. Passive response of the membrane. Current (I) can be injected into the cell via an electrode, and the resulting change in membrane potential (V_m) from that at rest (V_r) can be measured with the same or another electrode.

The voltage falls to $1/e$ of its initial value (V_0) in a time equal to one time constant. Time constants in biological membranes vary over a wide range, even though C_m per unit of membrane surface area is remarkably constant at about 1 $\mu F/cm^2$ in all membranes examined. Accordingly, the dissimilar time constants must reflect large differences in R_m from one cell to another, and even between separate membrane regions of the same cell. We shall see that different neurons exhibit distinct and often highly complex patterns of endogenous electrical activity in the absence of external stimulation. The membrane time constant plays an important role in determining just what a neuron's endogenous activity is, and how the cell reacts to stimuli.

The Action Potential

All the characteristics of the passive membrane response described above apply to hyperpolarizing stimuli of any size and to small depolarizing stim-

a

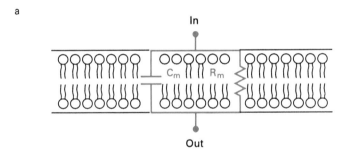

b

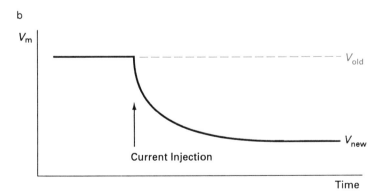

FIGURE 2-3. Passive electrical properties of the plasma membrane. a: The plasma membrane can be depicted as a resistor (R_m) and a capacitor (C_m) connected in parallel. b: Voltage changes do not occur instantaneously.

uli (Fig. 2-2). However, as the depolarizing stimulus gets larger, a critical stimulus strength or *threshold* is reached, below which only a passive response is seen, and above which the response looks very different (Fig. 2-4). The actual level of the threshold will vary from neuron to neuron, but it tends to be in the range 10–20 mV depolarized from V_r. Beyond the threshold, one observes a large change in membrane potential several milliseconds in duration, superimposed on the passive response (Fig. 2-4). The membrane potential depolarizes very rapidly, and then there is a slightly less rapid return to the resting level. Note that V_m does not go to 0 but actually becomes some 50 mV positive, that is, the inside of the cell is briefly positive relative to the extracellular medium.

This *active* response of the membrane when the depolarization exceeds threshold is the nerve impulse or action potential. It is this signal that is responsible for transfer of information from one part of a neuron to another. The threshold is essential to ensure that small random depolarizations of the membrane do not generate action potentials. Only stimuli of sufficient importance (reflected by their larger amplitude) result in information transfer via action potentials along the axon. Another important property of action potentials is that they are all-or-none events; this *all-or-none law*, as it is called, is an essential feature of axonal signal transmission. The all-or-none law is illustrated in the right side of Figure 2-4, which demonstrates that any stimulus large enough to produce an action potential produces the same size action potential, regardless of stimulus strength. In

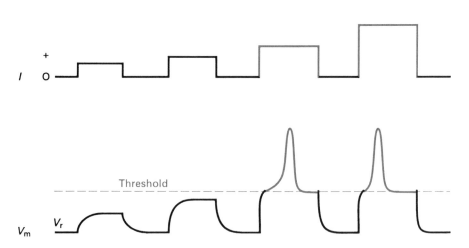

FIGURE 2-4. Active responses to large depolarizing stimuli. Although small depolarizing responses produce a passive membrane response, as in Figure 2-2, the voltage response to larger depolarizing currents is very different.

other words, once the stimulus is above threshold, the amplitude of the response no longer reflects the amplitude of the stimulus. This is *very important*; it means that information about stimulus strength must be represented—*encoded*—in the axon in some other way.

Although the *amplitude* of the action potential is generally independent of stimulus intensity, many of its other properties are not. In particular the *latency*, the time delay from the onset of the stimulus to the peak of the action potential, is a function of stimulus strength. As shown in Figure 2-4, the stronger the stimulus the shorter the delay between stimulus and action potential. We shall see that this *strength–latency relationship*, together with another phenomenon known as the *refractory period*, allows the encoding of stimulus strength in terms of the *frequency* of action potentials in the axon.

For several milliseconds after the firing of an action potential, it is impossible to evoke another action potential no matter how large the depolarizing stimulus; in other words, the axon is *refractory* to stimuli during this time. This *absolute* refractory period is followed by a *relative* refractory period, during which the stimulus must be larger than normal to evoke an action potential (Fig. 2-5). One useful way of thinking about the refractory period is in terms of the threshold. During the absolute refractory period the threshold is essentially infinite, and no stimulus, no matter how large,

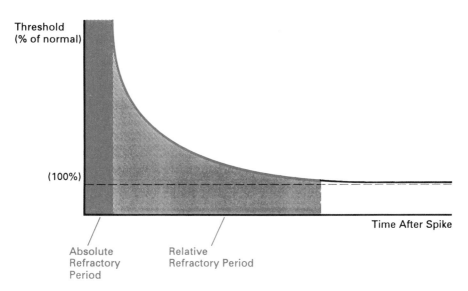

FIGURE 2-5. The threshold is not fixed. For a short period of time after the firing of an action potential, the threshold is much greater than normal.

can exceed it. During the relative refractory period the threshold is larger than normal, that is it requires a larger than normal stimulus to exceed it. The threshold returns to the normal level with a time course shown in Figure 2-5; because only an above-threshold stimulus will evoke an action potential, this curve describes the stimulus strength required to generate a second action potential, as a function of time after the first action potential.

Frequency Coding

Let us consider the response of the axon to a *sustained* stimulus in the light of these concepts. If the stimulus depolarizes the axon above the normal resting threshold, an action potential results. However, even if the depolarizing stimulus is maintained, a second action potential will be evoked only after the threshold has dropped back below the level of the sustained stimulus (Fig. 2-6a). This will take some time, as described in Figure 2-5. The same will be true for all subsequent action potentials during the stimulus. Thus the axon will fire action potentials as long as the stimulus is main-

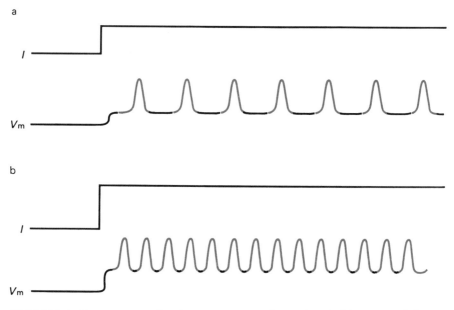

FIGURE 2-6. Frequency coding in axons. a: An above threshold sustained depolarizing stimulus (*I*) produces action potentials at a certain frequency. b: When the depolarizing stimulus is larger, the frequency of action potential firing is greater.

tained, but they will be spaced apart in time. Now consider the response of the same axon to a *larger* depolarizing stimulus (Fig. 2-6b). Again the second action potential will fire only after the threshold drops back below the stimulus level, but this happens more quickly because the stimulus is larger (and consequently the threshold does not have to drop quite as far). Accordingly the second and subsequent action potentials occur with less delay. It can be seen from a comparison of Figure 2-6a and b that the larger stimulus is reflected in a higher frequency of action potentials in the axon. Thus, even though action potential amplitude obeys the all-or-none law and does not reflect stimulus intensity, the phenomena of threshold, latency, and refractory period do indeed allow the encoding of stimulus intensity as a *frequency code* in the axon.

Passive Spread and Action Potential Propagation

Everything we have discussed thus far refers to *local* changes in the membrane potential at a single point in the axon. However we have also emphasized that the axon is specialized to move information from one part of the neuron to another. Thus it is time to ask how nerve impulses spread along the axon from the point of a stimulus.

Although it may seem to be a contradiction in terms, a phenomenon known as *passive spread* plays an essential role in the propagation of the active response. Let us look first at passive spread in terms of hyperpolarization of the membrane potential. Suppose an axon is penetrated by several microelectrodes some distance apart, and a hyperpolarizing current is injected through one. As can be seen in Figure 2-7a, a voltage change is observed at all the electrodes, but it is largest at the stimulating electrode and decreases in amplitude with distance away from this electrode. When the amplitude of the voltage response is plotted as a function of distance from the stimulating electrode, it can be seen to fall exponentially with distance (Fig. 2-7b). In other words the voltage change does spread from one point to another, but it is attenuated with distance, until eventually it becomes so small that it is essentially undetectable. This phenomenon is known as passive spread because it can be seen in a dead axon or even an electric cable with similar properties.

The extent of attenuation of the voltage change is determined by the membrane *space constant* λ, defined as the distance at which a voltage change has fallen to $1/e$ of its initial value. The voltage at some distance d can be described in terms of V_0, the voltage at distance 0, and the space constant:

$$V_d = V_0 e^{-d/\lambda}$$

The space constant can vary markedly from axon to axon, depending in particular on axon diameter and molecular characteristics of the axon membrane.

These considerations for hyperpolarizations also serve to describe perfectly well the passive spread of a *small* (subthreshold) depolarizing voltage change. When the local depolarization exceeds threshold, however, the picture changes dramatically. The above-threshold depolarization of course evokes a very large local voltage change, the action potential. There is decrement of this large voltage change as it spreads along the axon, just as there is for smaller depolarizations or hyperpolarizations. But because the local depolarization is so large, the passive spread is sufficient to depolarize

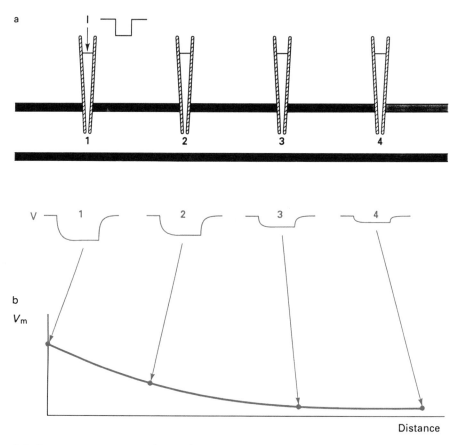

FIGURE 2-7. Passive spread. *a*: When current (*I*) is injected at one point (1) in the axon, a voltage change (*V*) can be measured at that point. Electrodes in other parts of the axon some distance away (2, 3, 4) measure smaller voltage changes. *b*: A plot of the decrement in voltage over distance.

neighboring regions of the axon above threshold (in spite of the attenuation with distance), and a full size action potential is generated at a point adjacent to the original one (Fig. 2-8). This is repeated for each small region of axon until the action potential has swept over its entire length. An often-used and highly appropriate analogy is a firecracker fuse, where ignition of one point on the fuse brings the neighboring segment above its ignition temperature, and this continues until the fuse has burned down to its end.

To summarize this descriptive treatment of axonal information transfer, several important characteristics of axonal membranes allow action potentials to carry information faithfully from one part of the neuron to another:

1. There is a *threshold* for generation of action potentials that guarantees that small random variations in the membrane potential are not misinterpreted as meaningful information.

2. The *all-or-none law* guarantees that once an action potential is generated it is always full size, minimizing the possibility that information will be lost along the way.

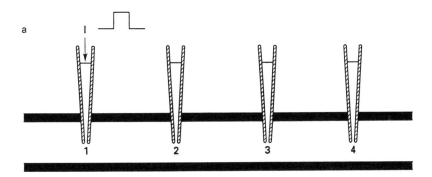

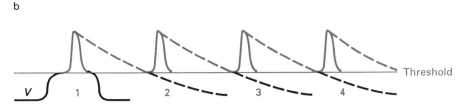

FIGURE 2-8. Propagation of the action potential. Although the large voltage change due to an action potential at one point decreases with distance (dashed lines), the depolarization that spreads to the adjacent region of the axon is still above threshold. Thus a full size action potential is generated at each point in the axon.

3. The *strength–latency relationship* and the *refractory period,* together with the threshold, allow the encoding of information in the form of a *frequency code.*

4. The phenomenon of *passive spread,* which arises simply from the cable-like properties of the axonal membrane, allows the propagation of action potentials along the axon and the transfer of information over long distances within the neuron.

Having discussed the axonal characteristics that are essential for the generation and propagation of action potentials, we will see now that neurons generally do not fire only single action potentials, or trains of action potentials at constant frequency. Rather they may exhibit complex temporal patterns of firing that are appropriate for the particular tasks the neurons must carry out.

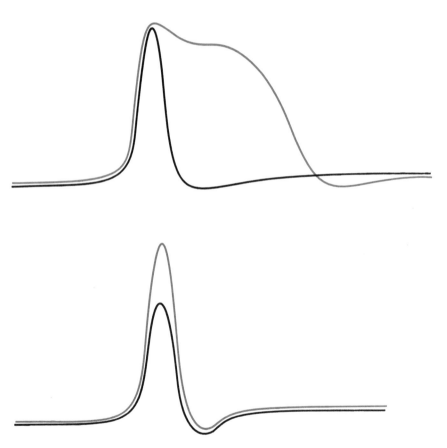

FIGURE 2-9. Different shapes for action potentials in different neurons. When the electrical activity of different neurons is recorded, action potentials of different amplitudes and durations are seen.

Different Patterns of Neuronal Electrical Activity

Cells in different regions of the nervous system are remarkably diverse in their morphology and in their electrical and biochemical properties. In the first part of this chapter we gave a description of action potential generation and propagation in a typical axon. However, even so fundamental a phenomenon as the action potential can vary in shape and size in different neurons (Fig. 2-9). In addition, the *pattern* of action potential firing exhibits great diversity in different neurons (Figs. 2-10 and 2-11). Diversity in electrical and other properties should come as no surprise when one considers the wide variety of different behaviors and physiological functions that neu-

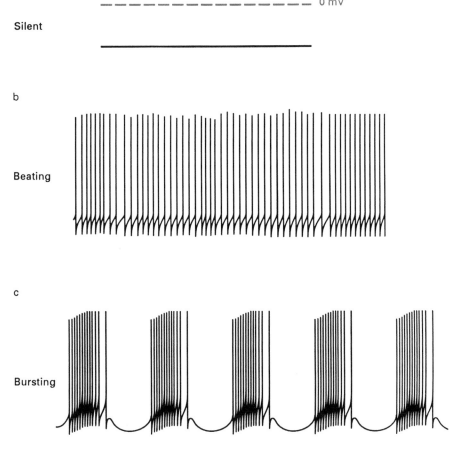

FIGURE 2-10. Different patterns of endogenous electrical activity. Some neurons do not fire spontaneously at all (*a*). Others may beat (*b*) or burst (*c*) in a regular manner.

rons have to control. For example, the neuronal pathways and the individual neurons that control fast visual reflexes or rapid escape behaviors are designed in a very different way from the neurons that control slow behaviors including breathing, feeding, and reproduction. In this section we will describe some of the different types of firing patterns that are encountered in nerve cells. In addition, we will provide some examples of how a *change* in its electrical properties allows a neuron to regulate different types of behavior.

It is important to note that the relatively simple picture we have painted of axonal information transfer becomes more complicated when one moves to the neuronal cell body. For example, some cell bodies do not fire action potentials at all—they are said to be *electrically inexcitable*—and they carry out electrical signaling in more subtle ways. Even in those cell bodies that do fire action potentials, the more subtle mechanisms may still be present,

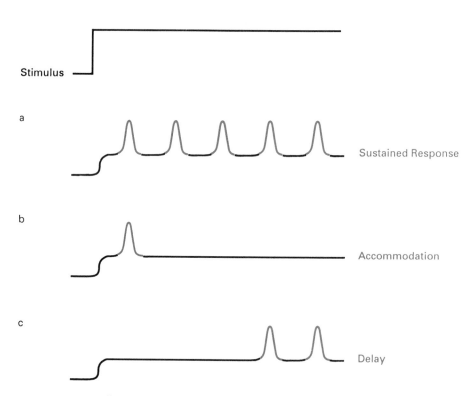

FIGURE 2-11. Different responses to a sustained stimulus. Some neurons respond to a sustained depolarizing stimulus with a sustained response (a), as shown also in Figure 2-6. Other neurons may fire one or a few action potentials and then stop responding (b), or may fire action potentials only after a delay (c).

leading to far more complex patterns of electrical activity than are usually observed in axons.

Silent, Beating, and Bursting Neurons

Although the generation of an action potential is fundamental to a neuron's ability to transmit information, there are many other aspects of its electrical properties that play important roles in shaping neuronal input and output. Some neurons have a steady unchanging resting potential in the absence of external stimulation, that is, they are *silent* (Fig. 2-10a). Other neurons, however, generate a variety of endogenous electrical patterns. For example, some cells fire repetitively at constant frequency, that is, they *beat* (Fig. 2-10b). Such neurons are also sometimes called *pacing* neurons. Although external stimulation can change the firing rate of the cell, or inhibit it altogether, the mechanisms that drive repetitive firing are often intrinsic to the neuron itself and do not require continual synaptic activation or other external stimuli.

Some neurons that fire spontaneously in the absence of external stimulation do not fire at fixed regular intervals but instead generate regular bursts of action potentials that are separated by hyperpolarizations of the membrane, as shown in Figure 2-10c. Such cells are termed *bursting* neurons. This ability of a neuron to burst repetitively is used by the nervous system in at least two different ways:

1. Bursting neurons generate rhythmic behaviors. Many fundamental behaviors, such as breathing, walking, swimming, and the chewing of food, require the continual rhythmic stimulation of a group of muscles. Numerous examples of electrical bursting can be found in neuronal circuits that generate such rhythmic motor outputs. Although in many cases the exact form and timing of the bursts, and even their generation, may be regulated by interactions among several different neurons (see Chapter 16), the ability to generate bursts can also be intrinsic to specific neurons that continue to burst in the absence of external inputs.

2. Bursting neurons are used to secrete neurohormones. Neurons, in addition to acting directly on other neurons or on muscle cells, may secrete hormones into the circulation. Figure 2-10c shows the bursting activity of a nerve cell that secretes peptide hormones in the marine mollusc *Aplysia* (a sea hare). This neuron, like many other molluscan neurons, is large and readily identifiable on the basis of morphological, biochemical, and electrical criteria. Accordingly, these cells can be given names; this one, called neuron R15 (see Fig. 11-5), has been studied extensively as a model bursting neuron. Another example of such bursting activity is found in neurons that are located in the

hypothalamic region of the mammalian brain and that have been termed *magnocellular neurons* (see Fig. 8-10). Individual magnocellular neurons contain either vasopressin or oxytocin, peptide hormones that are used in the control of water retention and lactation, respectively. For reasons that are not yet fully understood it appears that a bursting pattern of electrical activity, such as that in the magnocellular neurons and in R15, is more effective than a steady pacing pattern of firing as a stimulus to the intracellular machinery that causes peptide release (see Chapter 6).

The Response to Sustained Stimulation of a Neuron

Thus far we have been discussing patterns of neuronal activity that are *intrinsic* to the neuron under study. However neurons are often subjected to external stimuli, for example, a continual barrage of synaptic stimuli from other neurons. Experimentally such continual stimulation may be mimicked by a sustained depolarization or hyperpolarization from an intracellular microelectrode. Three different ways that a neuron may respond to a depolarizing stimulus are illustrated in Figure 2-11. The cell may generate action potentials repeatedly throughout the period of stimulation as described in the first part of this chapter, with a constant frequency that reflects the strength of the stimulus (Fig. 2-11a; see also Fig. 2-6). Alternatively a neuron may fire only a single action potential at the onset of the stimulus and remain silent thereafter (Fig. 2-11b). This response is sometimes termed *accommodation* to the stimulus. Finally, a neuron may not fire at the onset of stimulation but may generate action potentials only after a delay (Fig. 2-11c). In this case short periods of stimulation fail to trigger any action potentials in the cell.

The first two modes of response may be found in a variety of cells, for example, those that relay sensory information from the environment to the central nervous system. Accommodation in such sensory neurons would result in *behavioral habituation,* the commonly observed decrement in response during a sustained sensory stimulus (see Chapter 17). The third mode, in contrast, would be expected in cells that respond only to excess stimulation. A clear example of this is provided in motor neurons of the ink gland in *Aplysia*. Inking in *Aplysia* is a defensive response to a strong noxious stimulus, such as a mechanical stimulus that punctures the skin. It is believed that the ink that is extruded makes the surrounding water murky and provides camouflage for the animal, allowing it to hide from the predator generating the stimulus. The neurons that control the ink gland are normally not active, and they do not respond to small or transient stimuli; they begin to fire only when they receive the sustained synaptic input that is generated by a large and prolonged noxious stimulus.

Stimulation May Change Neuronal Electrical Properties

There may also be long-term modulation of neuronal properties in response to more subtle external stimuli than the sustained excitation described above. Few behaviors that are controlled by the nervous system remain fixed throughout the life of an animal. For example, feeding and reproductive behaviors have to be turned on and off at appropriate times. A defensive or escape response to a tactile stimulus may be appropriate at one time and not at another. Even the characteristics of essential physiological functions such as breathing may be altered in response to external stimuli. To a large extent, such changes in the behavior of an animal occur because of changes in the electrical properties of neurons that control those behaviors. Synaptic or hormonal stimulation may produce either short-term or long-term changes in the shape of action potentials, in the endogenous pattern of firing of a neuron, or in the way the cell responds to other external stimuli. In Chapter 11 we shall discuss in considerable detail the mechanisms by which such modulations of neuronal electrical properties are brought about.

Summary

The language of intracellular signaling in nerve cells is electrical. There is a voltage difference known as the membrane potential across the plasma membrane, and information is carried from one part of the cell to another in the form of action potentials, large and rapidly reversible fluctuations in the membrane potential, that propagate along the axon. Since action potentials are all-or-none events, their amplitude carries little information about the stimulus that triggered them; instead, several fundamental membrane properties associated with the generation and propagation of action potentials allow stimulus strength to be encoded in the *frequency of action potential firing*.

Different neurons exhibit different patterns of action potential firing. Some neurons are normally silent. That is, their membrane potential remains at the resting potential unless the firing of action potentials is triggered by some external stimulus, and they return to their nonfiring state when the stimulus is no longer present. However, many neurons exhibit more complex endogenous electrical activity, often firing action potentials in a regular pattern without external stimulus. In some cases it is possible to interpret the pattern of endogenous activity in terms of the particular function that the neuron is assigned in the nervous system.

Finally the electrical properties of a neuron are not fixed but are subject to modulation by input from its environment. This includes sensory information from the outside world, hormones released from other parts of the organism, and chemical and electrical signals from other neurons to which the neuron is functionally connected. Such modulation of neuronal properties is of fundamental significance, because it allows the animal to respond and adapt its behavior in a continually changing environment.

Ion channels
and membrane ion
currents

The electrical activity of nerve cells—indeed of all cells—depends on the movement of charge, carried by small inorganic ions, across the plasma membrane. The phenomena described in Chapter 2, the membrane potential, the firing of action potentials, and the grouping of action potentials in complex temporal patterns, all arise from such transmembrane ion flow. In addition, modulation of the endogenous electrical activity by external stimuli involves *changes* in transmembrane ion flow. But how is it that ions can move across the plasma membrane at all? The lipid bilayer of the plasma membrane is an excellent electrical insulator and is largely impermeant to charged species. It requires an enormous amount of energy to move an ion through the hydrophobic interior of the bilayer, and accordingly the cell must make special provision to allow transmembrane *ion current* to flow.

One way for ions to cross the plasma membrane is via energy-driven pumps or carriers, which use the energy from ATP to overcome the energy barrier imposed by the plasma membrane. Such pumps or carriers are proteins that pick up an ion on one side of the membrane, physically transport it across the bilayer, and release it on the other side. Because energy is expended in this process in the form of ATP hydrolysis it is possible for such *active* transport processes to move ions *against* a concentration gradient.

Although pumps and carriers are essential for many cell functions, they play only a supporting role in electrical signaling in most nerve cells. The stars of this show are *ion channels*, a ubiquitous class of specialized membrane proteins that span the plasma membrane. These form hydrophilic pores through which ions simply flow from one side of the membrane to the other down their concentration gradients. In the second part of this chapter we will describe ways of measuring the activity of ion channels and discuss how their activity regulates the flow of membrane ion current. But let us first review the fundamental physicochemical concept of the *equilibrium potential*, which is essential for understanding electrical phenomena in biological membranes.

Ionic Equilibria and Nernst Potentials

The rate of flow of an ion across the plasma membrane is determined by

1. the *concentration gradient,* the difference in the concentrations of the ion on the two sides of the plasma membrane;
2. the *voltage difference* across the plasma membrane; and
3. the *permeability* of the ion, the ease with which it moves through its ion channel across the plasma membrane.

The simplest example. Consider the case (Fig. 3-1) of a plasma membrane that separates two aqueous solutions, representing the inside and outside of a cell, each containing only a single generic ion X^+ that can permeate the membrane readily (the counterion Y^- cannot cross the membrane). Let us suppose further that the concentration of X^+ on one side of the membrane (*inside* the cell) is 10 times as high as it is on the other side (*outside* the cell). In other words $[X^+]_i = 10[X^+]_o$. Suppose, in addition, that *initially* there is no voltage difference across the membrane (that is, $V_m = 0$). Furthermore there is no net charge on either side of the membrane, because the charges on X^+ and Y^- cancel each other. In the first instant after this condition is set up, there will be a tendency for X^+ to diffuse down its concentration gradient from inside to outside the cell. This will of course cause a redistribution of charge across the membrane; the inside of the membrane has lost some of its positive charge and the outside has gained some, so there will now be a voltage gradient across the membrane with the inside negative relative to the outside. This voltage gradient will tend to slow the diffusion of X^+, since the positive ion does not want to leave the region of negative charge. This continues over time with the flow of X^+ (the *ion current* across the membrane) becoming slower and slower, until eventually the voltage gradient becomes large enough to oppose the concentration gradient. At this point there is no longer any net flow of X^+, and hence the voltage is no longer changing (Fig. 3-1). It is important to note that the number of ions that flow across the membrane to give rise to the voltage difference is *very small* relative to the total number of ions in the intracellular and extracellular compartments. In other words, the volt-

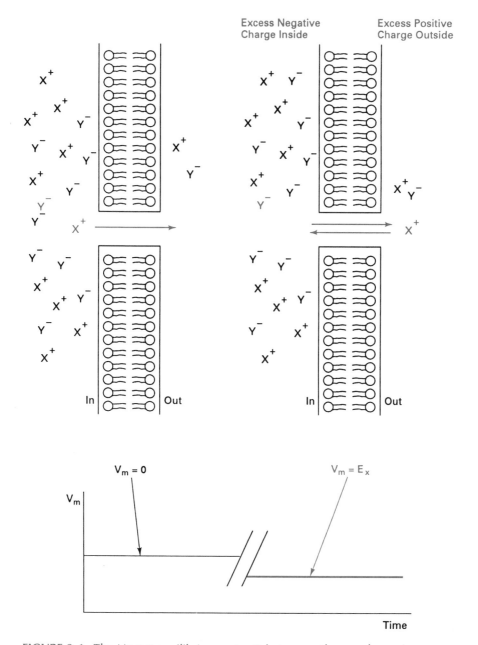

FIGURE 3-1. The Nernst equilibrium potential. Assume the membrane is permeant to a single charged ion, X^+, which is asymmetrically distributed on the two sides of the membrane. The counterion, Y^-, cannot cross the membrane. X^+ will flow across the membrane from the side of high concentration to that of low concentration, until the buildup of charge is sufficient to oppose net ion flow. The transmembrane voltage (V_m) is 0 before X^+ begins to flow. At equilibrium $V_m = E_x$.

age difference across the membrane is established without any significant change in the ion concentration gradient.

The voltage required to exactly oppose the flow of any given ion X is called the *equilibrium potential*, E_X, for that particular ion (Fig. 3-1). It will at once be evident that E_X is entirely dependent on the transmembrane concentration difference. Near the end of the nineteenth century the physical chemist Walther Nernst described the relationship between E_X (in millivolts), and $[X^+]_i$ and $[X^+]_o$, with the equation that bears his name:

$$E_X = \frac{RT}{zF} \log_e \frac{[X^+]_o}{[X^+]_i}$$

where R is the gas constant, T is the temperature in degrees Kelvin, z is the charge on the ion, and F is the Faraday (the amount of charge in coulombs carried by a mole of monovalent ions). For monovalent ions at room temperature (approximately 20°C), the Nernst equation reduces to

$$E_X = 58 \log_{10} \frac{[X^+]_o}{[X^+]_i}$$

For the 10-fold inside-to-outside concentration gradient we have been discussing here, it is evident that E_X will be -58 mV; in other words, when the inside of the cell is 58 mV more negative than the outside, the voltage gradient will balance the concentration gradient and there will be no net ion flow. Unless there is a change in the ability of X^+ to cross the membrane, the voltage across the membrane, V_m, will be equal to E_X. A change in the concentration gradient will of course lead to a shift in E_X and a corresponding shift in V_m.

A real membrane. Let us now return to the real world and consider what happens when there are several ions with different permeabilities and different concentration gradients across the membrane. The example we shall use is the membrane of the giant axon of a marine mollusc, the squid *Loligo,* which has been of enormous value in elucidating the physicochemical mechanisms underlying action potential generation and propagation. The extracellular medium for the squid giant axon contains ions at concentrations very similar to those in seawater, and the following relationships are *approximately* true for the ionic gradients across this membrane:

1. $[K^+]_i = 20[K^+]_o$
2. $[Na^+]_o = 10[Na^+]_i$

3. $[Cl^-]_o = [Cl^-]_i$ (This is not strictly correct since chloride ion concentration is somewhat higher outside than in; however, negative charges on intracellular organic macromolecules balance out the excess extracellular chloride.)

There are of course other ions both in seawater and inside the cell, but these are either present at low concentration or permeate the membrane poorly. Accordingly, they contribute very little to the transmembrane flow of ion current, and we may restrict our discussion to the major charge carriers potassium, sodium, and chloride.

For the squid axon (and most other axons) *at rest* the following relationships also hold:

4. The permeability of the membrane to chloride is essentially 0 (that is, $p_{Cl} = 0$).
5. The permeability to sodium (p_{Na}) is very low.
6. The permeability to potassium (p_K) is relatively high.

How then can we determine the membrane potential given these relationships? The Nernst equation can describe the membrane potential when only a single ion X^+ can flow, since under these conditions $V_m = E_X$. Almost 50 years ago David Goldman, and independently Alan Hodgkin and Bernard Katz, derived the following equation to describe the membrane potential in terms of the concentration gradient and permeation properties of several different permeant ions:

$$V_m = \frac{RT}{zF} \log_e \frac{K_o + [p_{Na}/p_K]Na_o + [p_{Cl}/p_K]Cl_i}{K_i + [p_{Na}/p_K]Na_i + [p_{Cl}/p_K]Cl_o}$$

When p_{Cl} and p_{Na} are zero and the membrane is permeable only to potassium, the permeability terms in both the top and bottom of the Goldman–Hodgkin–Katz equation are zero. Under these conditions the equation reduces to the Nernst equation and the membrane potential is determined only by the single permeant ion, potassium. When V_m is measured experimentally in the squid axon (and many other nerve cells), it is usually found to be very close to, but slightly less negative than, the equilibrium potential for potassium. In the particular case of the squid axon, for example, V_m is usually about -70 mV, whereas the E_K calculated from the potassium concentration gradient by the Nernst equation is -75 mV. The fact that the measured V_m is usually slightly less negative than E_K reflects the fact that the resting sodium permeability, although small, is not zero. Thus, to a limited extent, sodium also contributes to setting the resting membrane potential.

This resting potential, then, is exactly large enough to balance the ion flow caused by the various permeant ions with their different concentration gradients and permeabilities. At this voltage the *net* charge movement is zero. It is important to note that when only a single ion can cross the membrane the system comes to *equilibrium* and there is no net flow of that ion. As with the squid axon and other real cells, however, the V_m is not exactly equal to the equilibrium potential for any of the permeant ions. No individual ion is at equilibrium, and each will continually flow down its own concentration gradient. Thus there will be some current (I) carried by each ion. In this case the membrane is at a *steady state* rather than equilibrium. The total membrane current (I_m), which must be zero because the voltage is not changing, is the sum of the currents carried by the individual ions. In other words, I_m is given by

$$I_m = I_1 + I_2 + I_3 + \cdots + I_n = 0$$

where I_1, I_2, etc. are the currents carried by n different ions. It can be seen that for this sum to be equal to zero, different currents must have different signs. By convention the flow of positive ions across the membrane into the cell (inward current) is considered to be negative, and outward flow of positive ions (outward current) is positive. The opposite holds for the flow of negative ions.

If the total I_m is zero, and only sodium and potassium can flow, the currents carried by sodium and potassium (I_{Na} and I_K, respectively) must be equal and opposite. That is, $I_{Na} = -I_K$. How can this be when p_K is so much greater than p_{Na}? The answer is that I for any given ion is dependent on more than just its permeability. From Ohm's law, a fundamental law of physics, we know that the current flow between two points depends on the voltage difference (V) and resistance to current flow (R) between those points:

$$I = \frac{V}{R}$$

For the flow of an ion X^+ across a membrane, the relevant voltage difference is ($V_m - E_X$) and is called its *driving force*. We can consider that R is equivalent to the *inverse* of the permeability for that ion (intuitively we can see that permeability, a measure of the *ease* of ion flow, is inversely related to the *resistance* to ion flow). In reality, R is actually the inverse of the *electrical conductance* (G), a measure of the ease of ion flow that is similar to, but not identical with, permeability. For our purposes, however, conductance and permeability may be used interchangeably. Accordingly we can rewrite Ohm's law to describe the current carried by any ion as follows:

$$I_X = (V_m - E_X)G_X$$

Since $I_{Na} = -I_K$, it must follow that

$$(V_m - E_{Na})G_{Na} = -(V_m - E_K)G_K$$

Now remember that V_m is very close to E_K, so the driving force for potassium flow $(V_m - E_K)$ is very small. By contrast the driving force for sodium $(V_m - E_{Na})$ is large enough to generate an inward sodium current equal to the outward current carried by potassium, in spite of the very much lower sodium conductance. This point, that the current carried by a given ion is dependent on both the membrane conductance and the driving force for that ion, is a very important one to which we will return later.

Let us now suppose that G_{Na} *suddenly* becomes much higher than G_K and is maintained at this high level (Fig. 3-2a). Initially sodium ions will rush into the cell down their concentration gradient and the ion current carried by sodium (I_{Na}) will increase (Fig. 3-2b). At this time there is a net current, and the system is no longer at a steady state. As positively charged

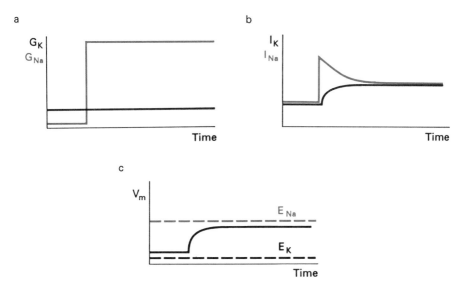

FIGURE 3-2. Conductance changes lead to voltage changes. Effect of a change in sodium conductance (a), on membrane currents carried by sodium and potassium (b), and on the membrane voltage (c).

sodium ions build up inside the cell, the membrane depolarizes (Fig. 3-2c). This depolarization brings the membrane potential V_m closer to E_{Na}, decreasing the driving force for sodium, and so I_{Na} will begin to decrease again (Fig. 3-2b). At the same time the V_m is farther from E_K, so I_K increases. These changes in the sodium and potassium currents combine to slow the rate of change of V_m (Fig. 3-2c). Eventually a new steady state is reached at a different voltage. I_K and I_{Na} are again equal and opposite, but are larger than they were before, reflecting an increase in total membrane conductance.

From these considerations we can see intuitively what the Goldman–Hodgkin–Katz equation tells us mathematically. When the ion concentrations are kept constant, V_m depends on the relative values of G_K and G_{Na}. Figure 3-3 describes the V_m when G_K is fixed and G_{Na} is varied from much smaller than to much greater than G_K. The two extremes are the limiting cases of the Goldman–Hodgkin–Katz equation, where it reduces to the Nernst equation and V_m is equal to either E_K or E_{Na}; intermediate conductance ratios lead to intermediate values for V_m.

Since V_m is a function only of ion conductances and concentration gradients, in theory a change in either or both of these parameters could be used to alter V_m. However, it seems likely that changes in concentration gradients sufficiently large to produce significant changes in V_m (1) would

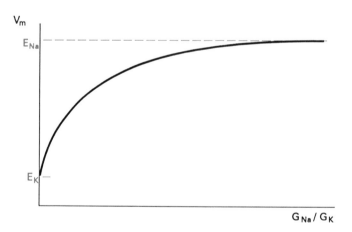

FIGURE 3-3. The membrane potential depends on the ratio of the membrane conductances for sodium and potassium. When the sodium conductance is very low, the membrane potential approaches the potassium equilibrium potential (E_K); in contrast, when the sodium conductance is very much higher than the potassium conductance, the membrane potential approaches the sodium equilibrium potential (E_{Na}).

be very slow and (2) would disrupt many other cellular functions. On the other hand the permeability properties of the plasma membrane can be modulated very rapidly by a variety of mechanisms. We shall see throughout this book that a *selective* change in the membrane conductance to one or another ion is used routinely by nerve cells as a means of changing V_m and rapidly producing meaningful electrical signals.

Ion Currents Flow Through Ion Channels

How does the plasma membrane selectively change its permeability to ions? At first this seems a remarkable feat. One might expect membrane permeability to be an inherent and unchanging entity, perhaps dependent on the lipid composition of the membrane, and as difficult to change as ion concentrations. But this picture of a sober and steady membrane permeability can be discarded when we recall that it is a class of proteins, the ion channels, that allows ion currents to flow. Proteins are high-spirited and debonair in comparison with most membrane lipids and their functional properties can be altered rapidly.

Single Ion Channels

The possibility that ion currents might flow through hydrophilic pores in the membrane was first suggested in the mid-1950s. Although this idea became widely accepted, more than 20 years passed before the activity of ion channels was measured directly. The breakthrough came with the advent of *single channel recordings,* methods for measuring the activity of individual ion channels either in their native membrane or after their insertion into artificial bilayer membranes constructed from phospholipids. The most important development was *patch clamp recording* (Fig. 3-4). This technique, developed by Erwin Neher, Bert Sakmann, and their co-workers, allows the current passing through single ion channels in the membrane of a cell to be measured directly. The information derived from this revolutionary approach has dramatically advanced our understanding of ion channel properties in neurons (and other cells).

To carry out a patch clamp recording a glass pipette, with an internal diameter of the order of a micron or so at its tip, is placed against the membrane of a cell. The application of suction to the inside of the pipette can lead to an electrical seal between the glass and the membrane. This seal becomes so tight that ions effectively are prevented from leaking out through it. Depending on the exact size of the patch of membrane under the pipette and the density of ion channels in the membrane, one or more

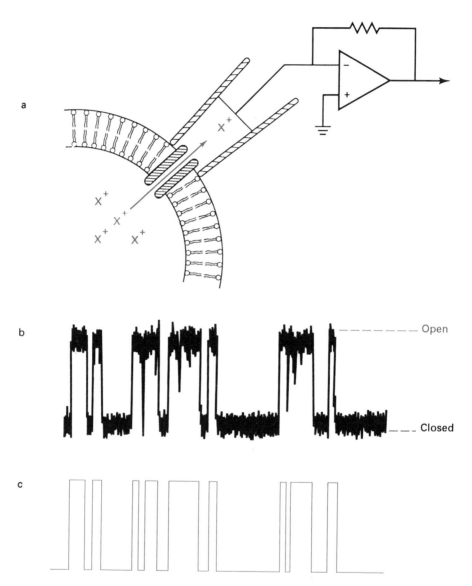

FIGURE 3-4. Patch clamp recording of single ion channel activity. *a*: Illustration of the cell-attached mode of patch clamp recording, with a current-to-voltage converter that is connected to the electrode (Hamill et al., 1981). *b*: An example of recordings of single channel activity obtained with this method. *c*: Simple computer programs can be used to produce idealized single channel records, which reproduce faithfully the openings and closings seen in the real record.

ion channel proteins may be isolated under the pipette. Current that is carried by ions flowing into or out of the cell through these channels can be detected by a sensitive current-to-voltage converter that is connected to the inside of the electrode (Fig. 3-4a).

For all of the ion channels whose activity has been recorded to date, the single channel patch clamp technique has revealed abrupt transitions between an open state, during which a detectable amount of current flows through the channel, and a closed state, during which no current flows. Figure 3-4 shows an example of a real channel recording (b) together with an idealized description of its opening and closing (c). The upward transitions are openings and the downward transitions are closings of the single ion channel. The computer-generated idealized record is free of noise and hence is suitable for quantitative analysis of channel activity.

For some purposes it may be desirable to have access to both the intracellular and extracellular sides of the patch membrane. In the cell-attached patch recording technique it is possible to alter the composition of the extracellular medium in the pipette, but there is no direct access to the inside of the patch. Fortunately, other configurations of single channel patch recording exist that do provide for manipulations of both the inside and the outside of the patch. Two such variants of patch clamping are termed *inside-out* and *outside-out* cell-free patch recording and these are illustrated in Figure 3-5. Both techniques rely on the fact that the seal between the glass pipette and the cell membrane is tight not only electrically but also mechanically. Accordingly, when a cell-attached patch pipette is pulled away rapidly from a cell, the patch of membrane frequently comes away with it. In the inside-out configuration the cytoplasmic membrane surface is exposed to the bathing medium, whereas the external membrane surface is accessible in the outside-out patch. Many ion channels can survive in such cell-free patches of membrane and a full characterization of the properties of the channels can therefore be carried out.

Ion Flow Through Ion Channels Is Fast

The fact that such measurements of single channel currents can be made at all should not be taken for granted—it really is rather astonishing. Ion channels are proteins, and when we measure the activity, the opening and closing, of a single ion channel, we are observing the activity of a single protein molecule! Compare this with the standard enzyme assay in a test tube, where one typically is measuring the sum of the activities of some 10^{10} or more protein molecules. Part of the reason that this measurement of single channel activity can be made lies in advances in modern electron-

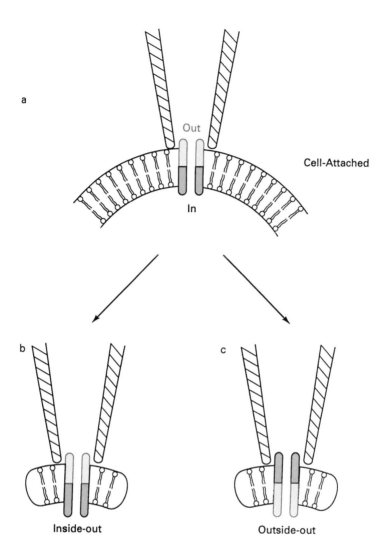

a

Out

Cell-Attached

In

b

Inside-out

c

Outside-out

FIGURE 3-5. Detached patches. Because the seal between the patch electrode and the plasma membrane is mechanically stable (a) it is possible to pull the electrode off the cell with the patch remaining attached to the electrode. Depending on the conditions under which this is done one can obtain either an inside-out patch (b), in which the former cytoplasmic portion of the channel (black) is exposed to the bathing medium, or an outside-out patch (c), in which the former extracellular portion of the channel (red) is exposed to the bathing medium (see Hamill et al., 1981).

ics; current-to-voltage converters capable of measuring as little as 10^{-13} A (0.1 pA) of current are available. In other words, there is a highly sensitive assay for ion channel activity. However, this assay would not be sufficiently sensitive to measure single channel currents if the rate of ion transport through channels were not remarkably fast.

The current flowing through a single ion channel, such as that illustrated in Figure 3-4, is typically in the 1–20 pA range. This corresponds to the movement of some 0.6–12×10^7 ions per second through the channel. If we think of an ion channel as an enzyme whose job it is to catalyze ion transport, then the turnover rate for this enzyme is of the order of 10^7–10^8 reactions per second. Turnover rates for most enzymes tend to be of the order of 10^2 per second, with the fastest being in the range of 10^5 per second. Active transport systems (see Fig. 4-9) also have turnover rates in the 10^2–10^4 per second range; indeed it can be shown that they have a theoretical limit of about 10^5 reactions per second because of the time it takes them to physically carry the ion across the membrane.

The uniquely high turnover rates for ion channels lead to the fundamental conclusion that the ion transport that they mediate *must* be via diffusion through a pore. The fact that we can measure single channel events at all makes this conclusion inevitable; single carrier currents could be no larger than about 10^{-3} pA and would not be detectable with presently available techniques. This in turn allows us to draw a picture of an ion channel (Fig. 3-6) as a membrane-spanning hydrophilic pore, which must be accurate in general outline if not in detail. The astonishing thing is that we can do this even though no high-resolution protein structural information is yet available for any ion channel. Low-resolution structural data have been generated recently for one kind of ion channel, the nicotinic acetylcholine receptor/channel described in Chapter 9. These data have confirmed the general outline that was inferred as long as 10 years ago from the electrophysiology.

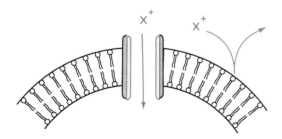

FIGURE 3-6. Our picture of ion channel structure. An ion channel, drawn as a hydrophilic pore (red) that spans the hydrophobic plasma membrane (black).

Different Kinds of Ion Channels

There are many different types of single channel activities, even in the membrane of a single neuron. These may be classified according to several different criteria, including

1. *single channel conductance,* a measure of the rate at which ions pass through the channel;
2. *ion selectivity,* the nature of the ions that are allowed to pass through the open channel;
3. *gating,* the opening and closing of the channel under the influence of such factors as the transmembrane voltage, the binding of neurotransmitters, hormones, and other agents to sites on the outside of the channel, and the actions of certain intracellular metabolites and enzymes; and
4. *pharmacology,* the susceptibility of the channel to various compounds that may block the pore or otherwise influence channel properties.

Single channel conductance. The voltage across the patch of membrane may be set to different levels, and the size of the current that flows through the open channels (Fig. 3-7a) can then be plotted against the voltage, as has been done in Figure 3-7b. For many channels, a straight line is obtained over a wide range of voltages. Such a plot provides two pieces of information, the *unitary conductance* of the channel and the *reversal potential* for the current that flows through the channel. The latter allows conclusions to be drawn about the ion(s) that can permeate the channel.

The conductance of a channel is a measure of the ease of flow of current through the channel. As previously mentioned, the electrical conductance G is the inverse of electrical resistance R. Conductance is closely related to permeability, which is the term we have been using to describe the ease with which an ion moves across the plasma membrane. The unitary or single channel conductance (g) is the slope of the open channel current versus voltage plot (Fig. 3-7b). It is given by the equation

$$g = \frac{\Delta i}{\Delta V}$$

This equation is simply Ohm's law. When i is given in amperes and V is in volts, the unit of conductance is siemens (S) or reciprocal ohms (note that i is used to denote the current passing through a single channel, and I the *macroscopic* or total membrane current; similarly g is the single channel conductance and G the macroscopic membrane conductance). Single channel conductances are usually given in picosiemens (pS) (10^{-12} S). The con-

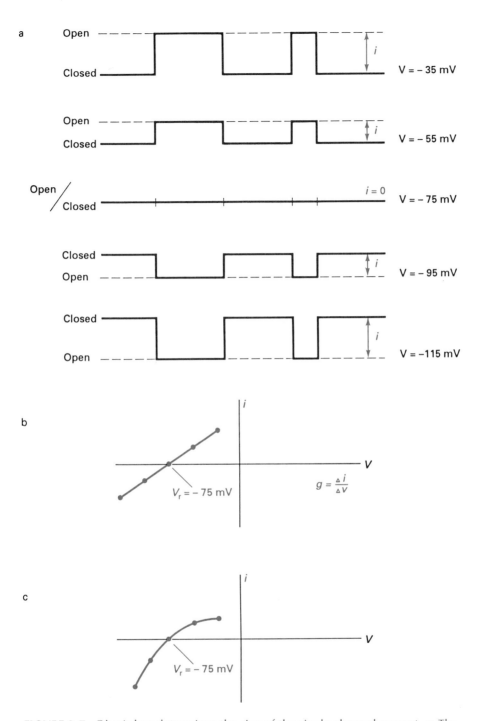

FIGURE 3-7. Ohm's law determines the size of the single channel current. *a*: The amplitude of the current passing through a single ion channel (i) varies as the voltage across the membrane is changed. *b*: A plot of the single channel current amplitude as a function of voltage. The reversal potential (V_r) is −75 mV. *c*: Some ion channels have nonlinear open channel current–voltage relationships.

ductances of channels in biological membranes that have been measured to date are generally in the 5–400 pS range.

Ion selectivity. Another important piece of information about an ion channel is the nature of the ions that normally flow through the open channel. An important clue to this information is provided by the reversal potential for the single channel current. The reversal potential is the voltage at which no current flows through the open channel. For the channel shown in Figure 3-7, this occurs at −75 mV. At this potential the channel may still open and close, but no net flow of ions occurs during the openings. From our previous discussion of Nernstian equilibrium potentials it will be evident that the reversal potential for the current through a channel is equal to the equilibrium potential for the ion that passes through the channel. Thus we may infer that current through the channel in Figure 3-7 is probably carried by potassium ions. Because E_K is very close to −75 mV, there will be no net flow of potassium at this potential even when the channel is open. It can be seen that these considerations for single channels are identical to those discussed earlier for total membrane currents, which result from the summed activity of large numbers of single channels.

As with the total membrane currents, the Nernst equation provides a ready test for the ion selectivity of a single channel. For example, for a potassium channel, increasing the extracellular concentration of potassium (that in the patch pipette) by a factor of 10 should shift the reversal potential by 58 mV to a more positive potential. Altering potassium concentration, however, should not alter the reversal potential of channels selective for other ions, such as sodium ions, if no change has been made in sodium concentration. Some channels, however, can allow more than one species of ion to cross the membrane. To determine the extent to which a potassium channel *selects* for potassium over other ions, these manipulations should be made in the presence of other potential permeant ions, for example, sodium ions. A channel that strongly prefers potassium over sodium will always exhibit a reversal potential at E_K, whatever the sodium concentration on either side of the membrane. If the reversal potential deviates from E_K, this indicates that the channel is not completely selective and allows some sodium to flow.

Again we may compare this with the total membrane current in a neuron at its resting potential. This current is equal to zero (i.e., it reverses its sign) *near* but not precisely *at* E_K because the membrane is permeable *mostly* but not *exclusively* to potassium. One possible interpretation of this finding is that there is one class of ion channels, permeable mostly to potassium and slightly to sodium, that is responsible for the total membrane current. As mentioned, such channels do exist. We know, however, that most channels in neurons are selectively permeable to one or another ion. The combination of a high potassium and low sodium permeability comes about because under resting conditions the potassium channels are sometimes open (and allow current to pass), whereas the sodium channels are closed virtually all of the time.

How do ion channels exhibit selectivity, often exquisite selectivity, for one ion over another. They can do this because they are far more than simple holes in the membrane. A detailed discussion of channel selectivity mechanisms is beyond the scope of this book, and in any event the final word on this topic will have to await high-resolution protein structure information. However, an ingenious series of measurements has provided insights that will almost certainly be confirmed by structural studies. By carefully measuring the permeabilities of a series of poorly permeable organic cations in the axonal sodium channel, the size of the narrowest place in the pore has been determined. Similar measurements have also been made for potassium channels. The fundamental conclusion from these studies is that an ion, together with its strongly bound shell of water molecules, must make a tight fit with this narrowest region, the so-called *selectivity filter,* to pass through it. According to this view, selectivity arises because evolution has designed the selectivity filters of different channels to just fit the diameter of one or another hydrated ion. Again this prediction from the electrophysiological measurements, that channels *must* have a narrowing somewhere in the pore, has been confirmed by the low-resolution structural map of the nicotinic acetylcholine receptor/channel.

Gating. By now it will be evident that ion channels are not simply inert pores in the membrane. Rather they are dynamic entities, which can undergo extremely rapid transitions between an open state, in which they conduct ions, and a closed state, which does not allow ions to pass. These open/closed transitions, which are readily apparent in single channel recordings such as those in Figure 3-4, must reflect conformational changes in the channel protein.

The process of opening and closing of a channel is often termed *gating,* because in our naivete we like to modify our picture of the ion channel as a pore (Fig. 3-6) by including a hinged gate, presumably an integral part of the channel protein, that can swing open to allow ion flow, or shut to prevent it (Fig. 3-8a). These two states of the protein are in dynamic equilibrium, and the amount of time the channel spends in each state will depend on the relative values of the free energies of the two states. These free energies in turn will be reflected by easily measured quantities, the *rate constants* for channel opening and closing (Fig. 3-8a).

Ion channel gating may be influenced by a variety of external conditions.

We often say that such conditions cause channels to "open" or "close," but we really mean that the relative free energies of the open and closed states have been changed, so that the channel is *more likely* to be open or closed than it was previously. This will be seen in the single channel records as a change in the rate constant for opening or closing, or sometimes in both (Fig. 3-8b). Such changes in single channel open probabilities for a large population of ion channels are responsible for changes in the membrane conductance, such as those described in Figure 3-2, and in more detail in the next chapter.

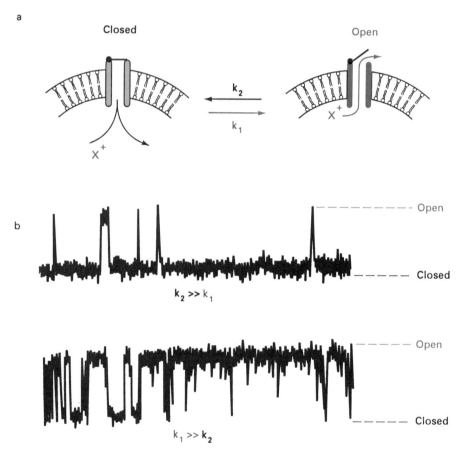

FIGURE 3-8. Ion channel gating. a: An ion channel drawn as a pore with a gate (bar). There is a dynamic equilibrium between the closed and open states, determined by the opening (k_1) and closing (k_2) rate constants. b: The proportion of time that the channel spends open depends on the relative values of k_1 and k_2.

Voltage-dependent gating. Many channels, particularly those that shape the ongoing electrical behavior of a neuron, are *voltage-dependent* channels. The frequency with which such channels open and close depends on the membrane potential. As we shall see in the next chapter, different types of channels may either increase or decrease the amount of time that they spend in the open state as the voltage across the membrane is made more positive. Figure 3-9a shows the behavior of a voltage-dependent potassium channel at different membrane potentials. At negative potentials such as the resting potential of the cell, the channel opens only infrequently or not at all. As the potential is made positive, the channel begins to open, until at potentials more positive than about $+20$ mV, the channel is fully activated. Note that the important point here is the amount of *time* that the channel spends in the open and closed states. The amplitude of the open channel current also changes with voltage, in the manner described in Figure 3-7, but this simply reflects the change in potassium driving force. Figure 3-9b is a graph of the open probability as a function of voltage. The data points fall on a sigmoid curve, the steepness of which reflects the channel's sensitivity to voltage.

What is the structural basis for this sensitivity of channel opening and closing to voltage. The hypothetical gate depicted in Figure 3-8 must act as a *voltage sensor,* detecting the strength of the electric field across the membrane. It is presumed that this part of the channel protein possesses some net charge and can move under the influence of the electric field, to unblock or block the channel pore and hence allow or prevent ion flow. In Chapter 5 we will consider the progress that has been made in identifying the voltage sensor in different kinds of voltage-dependent ion channels.

It is important to note that the open probability need not correlate with the amount of current that a channel passes. For example, for a voltage-dependent sodium channel, the reversal potential of the current flowing through the open channel is $+50$ mV. Thus no net flux of sodium occurs at $+50$ mV, even though the channel's voltage dependence is such that it is open all the time at this potential. In fact, the only way to demonstrate that the channel is indeed open at this voltage is by changing the sodium concentration on one or both sides of the membrane to shift E_{Na} away from $+50$ mV. The significance of this difference between open channel current and open probability will become evident in the next chapter when we consider the amount of membrane current carried by a *population* of ion channels.

Voltage dependence of current through the open channel. For some channels, when the current flowing through the channel is plotted against voltage, a straight line such as that in Figure 3-7b is not obtained. The open

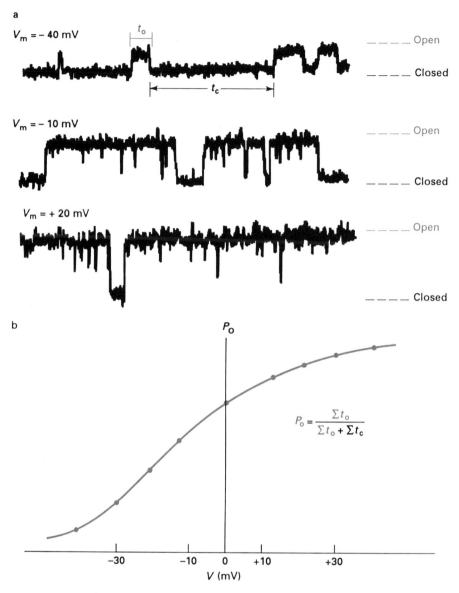

FIGURE 3-9. Gating of a voltage-dependent potassium channel. *a*: When the membrane voltage is varied, the amount of time that the channel spends open (t_o) and closed (t_c) changes. *b*: The channel open probability (P_o) can be plotted as a function of voltage. The data fit a sigmoid curve, with the open probability highest at the most depolarized voltages.

channel current–voltage relationship of such a channel is depicted in Figure 3-7c. In this example, a change of 20 mV in the region of depolarized membrane potentials produces a much smaller change in the amount of current flowing through the channel than does a change of 20 mV in the region of hyperpolarized potentials. Such a channel is said to display *rectification* in its single channel conductance. Rectification is a term from electronics that refers simply to a change in resistance (or conductance) as a function of voltage. Single channel rectification can depend on the concentrations of permeant ions on either side of the membrane. It can also result from blocking of current flow by nonpermeant ions confined to one side of the membrane, or can be an inherent property of the channel protein. It is important to note that this characterization of a channel refers to the amount of current that the channel passes *when it is open*. In the example in Figure 3-7c, the frequency with which the channel opens and closes at positive potentials may be the same as at negative potentials. In other words this single channel rectification is an example of voltage dependence, but it is not voltage-dependent gating. Since the word rectification is often *also* used to describe voltage-dependent gating, it is important to be aware that there are different uses of this term.

Single channel kinetics. Further characterization of an ion channel can be carried out by measuring the mean open time and mean closed time for a large number of transitions between the open and closed states. For a voltage-dependent channel these measurements must be made at several different voltages, and the voltage at which they were measured must be stated. Histograms can also be plotted of the number of openings or closings of a given duration. Such information enables simple models of the behavior of the channel to be made. Here we shall provide one example of the use of such information.

Some channels show "bursty" kinetic behavior (Fig. 3-10). Measurement of the closed times of this channel will show that periods during which the channel is closed fall into two groups—"short" closed times, which represent the brief closings of the channel *during* one burst of openings, and "long" closed times, which represent the times *between* the bursts of openings. A simple model for such bursty behavior consists of two closed states, C_1 and C_2. During a burst itself, the channel flips between the open state (O) and one of the closed states (C_2). Occasionally, however, the channel enters the other closed state (C_1). Return from this closed state is slow, accounting for the long periods between the bursts. A simple model of this type allows the calculation of rate constants for the transitions between each of the three states of the channel. The way that these rates are affected by membrane voltage, by neurotransmitters, by drugs, and by other parameters can then be analyzed.

Activation and inactivation. One important characteristic of the kinetic behavior of a voltage-dependent ion channel is its rate of *activation*. When the membrane potential is changed abruptly, the amount of current flowing through the open channel (i.e., the size of the single channel events) also changes abruptly in accord with Ohm's law, because the driving force has changed. The probability of the channel being open may change more slowly, however, until a new steady-state open probability is attained for

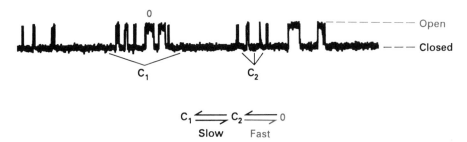

FIGURE 3-10. Single channel kinetics. The activity of channels cannot always be described in terms of a simple transition between one open and one closed state. Some channels exhibit bursty kinetic behavior. Such behavior can be explained if the channel has one open state (O) and two distinct closed states (C_1 and C_2).

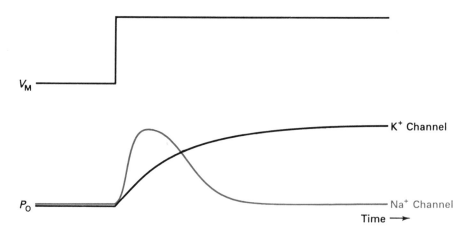

FIGURE 3-11. Activation and inactivation of ion channels. Sodium channels increase their opening probability rapidly after depolarization, but the open probability (P_o) then decreases again even during a sustained depolarization. In contrast, the open probability of some classes of potassium channels is increased more slowly by depolarization and remains increased throughout the duration of the depolarization.

the new voltage. This increase in the channel's open probability is termed its activation. The rate of activation of some voltage-dependent ion channels, for example, sodium channels, is very fast, reaching a maximum within a few milliseconds after the change in membrane voltage (Fig. 3-11). However, other channels, for example, a potassium channel that will be discussed in detail in the next chapter, exhibit very much slower rates of activation (Fig. 3-11).

As important for the electrical behavior of a neuron as the rate of activation of its different channels is their rate of *inactivation*. Some channels, once they have been induced to open by a change in voltage, maintain their new rate of opening for a prolonged period. This is the case for the slowly activating potassium channel (Fig. 3-11). Other channels, however, following their activation, undergo a progressive decrease in openings. This is illustrated in Figure 3-11 for our voltage-dependent sodium channel. The rate of loss of channel activity is termed the rate of inactivation. Some potassium channels also undergo inactivation. Thus channel activity can be not only voltage dependent but also *time dependent*. At early times, voltage-dependent activation causes sodium channels to open, but at later times inactivation begins to dominate and eventually the channels are never open, even though the depolarization is maintained. As we shall see elsewhere, inactivation of various ion channels plays an essential role in shaping action potentials and in determining the electrical characteristics of many neurons. For example, a delay in the response to external stimulation such as that shown in Figure 2-11c is produced by the progressive inactivation of a particular kind of potassium channel.

Gating controlled by neurotransmitters or intracellular messengers. The activity of many ion channels is tightly linked to the action of *neurotransmitters,* chemicals released from one neuron that influence the activity of other neurons. The gating of other channels may be influenced by intracellular *messenger* molecules or ions, for example, calcium ions. The way neurotransmitter binding or intracellular messengers influence the opening and closing of the channel gate will be discussed in detail in Chapters 9 and 10.

A phenomenon that is analogous to the inactivation of voltage-dependent ion channels also occurs with some neurotransmitter-gated ion channels. After they have been induced to open by application of the neurotransmitter, a progressive loss of channel openings may be observed if the neurotransmitter remains present. Such loss of activity is often termed *desensitization.* As will be discussed in Chapters 10 and 11, the mechanism of desensitization of some receptors is very different from that of voltage-dependent channel inactivation.

It will thus be evident that full characterization of an ion channel may require the examination of how its activity is influenced by the application of neurotransmitters, and by various biochemical events at the internal face of the channel, in addition to (or in place of) voltage. These different modes of channel modulation are not mutually exclusive, and it will be particularly interesting when the structural work does catch up with the electrophysiology and we are able to understand the structural basis for these regulatory phenomena. Changes induced by neurotransmitters in the properties of ion channels contribute in an important way to regulation of short-term and long-term changes in the electrical activity of neurons. Further chapters will cover this topic in considerably more detail.

Summary

The flow of ion currents across the plasma membrane gives rise to electrical signaling in nerve cells. These ion currents arise from the activity of individual ion channels, specialized plasma membrane proteins that provide a hydrophilic pore through which particular ions can diffuse down their electrochemical gradients. Because ion channels provide for extremely rapid ion transport, the amount of current flowing through an individual ion channel can be measured. Such measurements have demonstrated that ion channels can exist in an open state, which allows ions to flow, and a closed state, during which no ions can pass. Gating between these states can be influenced by factors such as the membrane voltage and the actions of neurotransmitters or intracellular messengers.

Each species of channel generates a current that either depolarizes or hyperpolarizes the cell, at a characteristic rate that depends on its sensitivity to voltage (or other influences) and its rate of activation and inactivation. These depolarizations or hyperpolarizations in turn influence the activity of this and other species of voltage-dependent ion channels. Thus the normal pattern of firing of a neuron, and its response to stimulation, can be seen as a play of interactions among the currents generated by the different kinds of ion channels. The next chapter will provide more details of the different types of ion channels in a neuron, and the ways in which their different activities interact to produce neuronal electrical activity.

Combinations of ion currents give rise to complex patterns of neuronal electrical activity

A great deal of what we know about ion channels, including information about selectivity and kinetics of different types of channels, was already understood before the invention of single channel recording. This understanding developed because for almost 50 years it has been possible to record the *macroscopic membrane current* in some cells. The macroscopic current is the combined current flowing simultaneously through all the active ion channels in the cell. Before discussing just how such measurements are made, let us examine how the microscopic currents flowing through a population of ion channels combine to generate the much larger macroscopic current recorded from the whole cell.

Macroscopic Ion Currents Result from the Activity of Populations of Ion Channels

The voltage-dependent sodium channels that open rapidly in response to a change from a negative to a more positive membrane voltage provide an excellent example (Fig. 4-1). When one channel is present in a patch (Fig. 4-1a), the response to depolarization is an increase in channel open probability, followed by a decrease again as the channel inactivates. When several sodium channels are present in the patch (Fig. 4-1b,c), the current record begins to resemble the *whole cell sodium current* (Fig. 4-1d), measured (with all other currents blocked) by the techniques we will describe below. In other words, the whole cell sodium current is the sum of the currents passing through all of the sodium channels in the plasma membrane. In the whole cell, the rapid change in the probability of opening of the individual channels is manifest as a rapid increase in the total current.

Note also in Figure 4-1 that the total current returns to zero even though the depolarization is maintained, as a consequence of voltage- and time-dependent inactivation of the individual ion channels. This may be con-

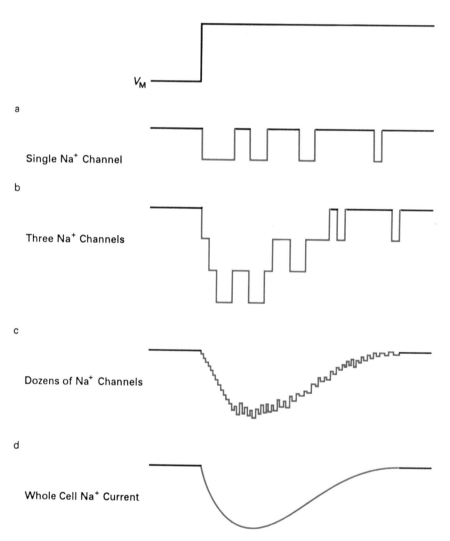

FIGURE 4-1. Macroscopic membrane currents result from the summed activity of individual ion channels. *a:* During a sustained depolarizing voltage pulse, single sodium channels open and then close again as they inactivate. The more channels there are in the patch (*b* and *c*), the more closely the patch current resembles the sodium current recorded from the whole cell (*d*).

trasted with the slowly activating potassium channels we introduced at the end of Chapter 3. In this case the gradual change in the probability of opening of the individual channels (Fig. 3-11) is manifest as a slow increase in the whole cell potassium current to some maintained steady-state level. These current traces emphasize once again the fundamental fact that the activity of many ion channels is both voltage and time dependent.

These considerations can be expressed in a more quantitative way by means of a simple yet useful equation. The macroscopic current I carried by one type of ion channel is given by

$$I = Np_o i$$

where N is the number of functional channels of that type present in the membrane, i is the current carried through a single channel when it is open, and p_o is the probability of a channel being open. As we have seen, i depends on the voltage across the membrane, and we shall see that p_o (and sometimes N) may be a function of voltage and time, and may be modulated by neurotransmitters and/or intracellular metabolic events.

One difficulty in the interpretation of whole cell macroscopic current recordings is that the membrane of a neuron has many different types of ion channels, which are selective for different ions and whose gating is influenced in different ways by voltage and neurotransmitters. Is it indeed possible to measure *selectively* the whole cell sodium current, as we have shown in Figure 4-1? Although it is not always easy, by a careful choice of voltages and by the use of drugs that block the opening of specific classes of channels, one can often record currents that represent the opening and closing of a single class of ion channel. This will become evident in our discussion of the ion currents responsible for the action potential and for patterns of neuronal firing.

Ionic Mechanisms of the Action Potential

Changes in ion permeability such as those described above are exactly what happens during the nerve impulse. We will begin by summarizing the sequence of changes during an action potential (Fig. 4-2), and then discuss in depth the experimental evidence for this sequence of events.

When an axon is depolarized beyond the action potential threshold, the depolarization itself causes large numbers of voltage-dependent sodium channels to open. This is seen as a rapid increase in G_{Na} (Fig. 4-2a), which quickly rises to a level very much higher than G_K (as in Fig. 3-2). As a result inward sodium current increases (Fig. 4-2b), the membrane depolarizes further, and V_m approaches E_{Na} (Fig. 4-2c). In contrast to the situation described in Figure 3-2, however, a further series of changes occurs as a *result of the depolarization*. By the peak of the action potential there is (1) sodium channel inactivation, and hence a rapid decrease in G_{Na} back to its resting level, and (2) a slower increase in G_K. As a result the inward I_{Na}, which had been transiently very large, begins to drop, and, somewhat more slowly, I_K begins to increase. As the outward I_K becomes larger than the inward I_{Na}, the net current flow is now outward (hyperpolarizing), and it begins to drive V_m back toward the resting level. Notice that the increase in G_K is prolonged (Fig. 4-2a), and thus for some period after the spike the V_m will actually be *more negative* than the normal resting potential (Fig. 4-2c), that is, closer to E_K. This phenomenon is called the spike *afterhyperpolarization*. Finally G_K begins to decrease again, and at some time soon after the end of the action potential the membrane permeabilities have returned to their normal resting levels. The time course of these changes (Fig. 4-2) can vary from cell to cell, but in general axonal action potentials

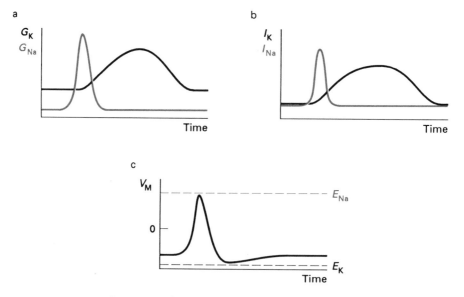

FIGURE 4-2. Membrane conductances, currents, and voltage during an action potential. *a:* The first change during an action potential is an increase in the sodium conductance (G_{Na}) to a level much greater than at rest. This is followed by a slower and longer lasting increase in the potassium conductance (G_K). *b:* The sodium (I_{Na}) and potassium (I_K) currents change in accordance with the changes in conductance. *c:* The result of these changes is the characteristically shaped action potential. See Hodgkin and Huxley (1952a,b,c) and Hodgkin et al. (1952). A sophisticated quantitative analysis is presented by Hodgkin and Huxley (1952d).

are very fast; in the squid giant axon, for example, the entire sequence of events described above is over in a few milliseconds. In other cells, for example pituitary cells of vertebrates, action potentials may last for tens of milliseconds.

Voltage clamp studies of the squid giant axon. The studies that led to the elucidation of the sequence of events described represented a partnership between several talented investigators on the one hand, and a magnificently well-suited experimental preparation on the other. The work of J. Z. Young, K. C. Cole, Alan Hodgkin, and Andrew Huxley on the *squid giant axon* is a great success story in twentieth-century science. Hodgkin and Huxley received the Nobel Prize for their studies, and some have remarked that it is unfortunate that the Atlantic squid *Loligo* cannot be similarly honored.

Young, a British zoologist, found in the mid-1930s that the mantle of the squid is innervated by giant nerve fibers up to 1 mm in diameter. The giant axon arises from the fusion of a large number of smaller neurons. It can be removed from the animal, and the axoplasm can be extruded and replaced by saline solutions of defined ionic composition; in other words the transmembrane ion gradients can be manipulated by the experimenter. The axonal plasma membrane is surprisingly robust and survives this maltreatment with its electrical properties intact. The large size of the axon makes it easy to place electrodes both inside and outside the membrane to measure (and control) the transmembrane voltage. The importance of *controlling* the voltage arises from the fact that, as discussed, the sodium and potassium conductances themselves change as a function of voltage and the membrane is not at steady state during the action potential. It will be evident that depolarization produces an increase in G_{Na}, which then produces further depolarization. This in turn further increases G_{Na}, and an unstable positive feedback loop results that gives rise to the *regenerative* all-or-none action potential. Accordingly, the only way to study the regulation of membrane conductance effectively is to measure the conductance properties at fixed voltages.

To deal with this problem, K. C. Cole and his colleagues invented an electronic feedback system called the *voltage clamp* to hold the membrane potential constant at a voltage chosen by the investigator. In its simplest form (Fig. 4-3a) the voltage clamp consists of two separate electrodes, one connected to a voltage measuring amplifier to measure the transmembrane voltage and the other connected to a current-passing amplifier. A negative feedback loop is created by adding a feedback amplifier, which compares a voltage set by the experimenter ($V_{command}$) with the measured V_m. The difference between these two voltages is known as the error signal, and the

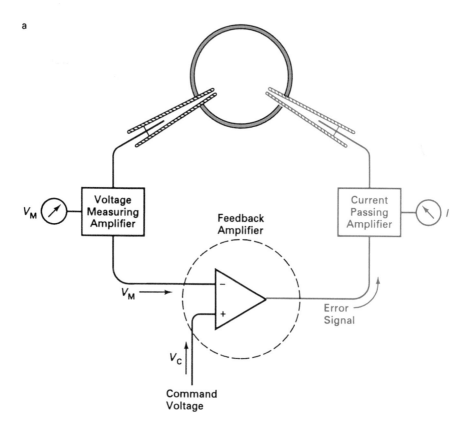

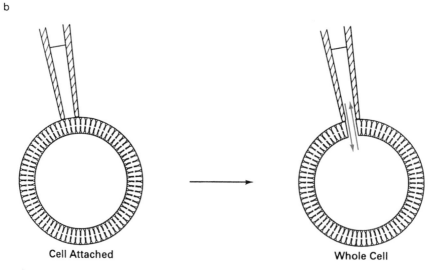

Cell Attached   Whole Cell

FIGURE 4-3. How is the whole cell current measured? *a*: Schematic drawing of a two electrode voltage clamp, which uses three separate amplifiers to control the membrane voltage. *b*: Whole cell mode of the patch recording technique (see Hamill et al., 1981).

feedback system injects current through the current-passing electrode to maintain the error signal as close as possible to 0. By this means V_m is forced to be equal to $V_{command}$; in other words the membrane voltage is controlled by the experimenter, and the signal that is measured is the amount of current required to maintain that particular voltage. This current is, in fact, equal to the ionic current flowing across the membrane at that voltage.

A more recent method that has been developed to allow the measurement of currents in whole cells is a variant of the patch clamp technique known as *whole cell patch recording* (Fig. 4-3b). In this method a conventional patch electrode is sealed to a cell as described in Chapter 3, and the membrane under the patch is then destroyed either by a pulse of suction or by a large abrupt change in voltage. The solution in the pipette can then exchange freely with the cytoplasm of the cell, and the cell can be voltage clamped with appropriate electronics connected to the inside of the pipette. This configuration can be considered analogous to a very large outside-out patch, consisting of most of the cell's plasma membrane, with a very large number of ion channels contributing to the current flow across the membrane.

One problem that sometimes arises with the whole cell patch clamp technique is that the exchange of molecules and ions between the pipette solution and the cell's cytoplasm can prove disruptive to the normal function of the cell (Fig. 4-3b). Although this exchange may also be an asset for some types of experiments, many interesting regulatory phenomena may simply disappear during whole cell patch recordings, and under these circumstances the less invasive intracellular microelectrode voltage clamp is the method of choice. For cells that are too small to tolerate the insertion of two independent microelectrodes, a single microelectrode can be used to measure voltage and inject current. The electrode is "switched" electronically between voltage measuring and current injecting modes to voltage clamp the cell. To provide an effective clamp the switching must occur at a rapid rate so that membrane voltage does not change significantly during the brief periods of voltage measurement (when no current is being injected and the cell is briefly out of clamp). For technical reasons this technique is less effective for recording rapid and large changes in membrane currents, but it is extremely useful with small cells or cells within intact areas of the nervous system that cannot be visualized easily for penetration with two microelectrodes.

Voltage- and time-dependent ion currents. Hodgkin and Huxley (as well as Cole) immediately recognized the importance of controlling the membrane voltage, and carried out a series of seminal experiments (interrupted by

World War II) on voltage clamped squid giant axon. Keep in mind that of the three parameters involved in regulating the transmembrane ion current—the ionic gradients, the voltage, and the ionic conductances—the first two can be manipulated by the experimenter in this preparation, and, accordingly, the regulation of the membrane conductance can be investigated in a rigorous way. The results of these studies were published in a classic series of papers by Hodgkin and Huxley (one of them in collaboration with Bernard Katz) in 1952. They are not easy reading, but are essential for the serious student of neurophysiology and membrane biophysics.

Hodgkin and Huxley asked what happens when you voltage clamp the axon near the resting potential, and either hyperpolarize it or depolarize it before returning to the original voltage (the *holding* potential). As shown in Figure 4-4a, small hyperpolarizing or depolarizing pulses of the same size produce small inward or outward currents, respectively. These *leak* currents are not time dependent (they reach their maximum amplitude in a time that is too short to be resolved by the recording instrumentation), and they are the same size in the inward and outward directions for equal amplitude hyperpolarizations and depolarizations. When the current amplitude is plotted against the voltage during the pulse (the *pulse* or *command* potential) for these small hyperpolarizations and depolarizations, the resulting *current–voltage* (I–V) *relationship* is a straight line (Fig. 4-4b). Such a straight line I–V relationship is also seen for a linear resistor in a nonbiological electrical circuit (Fig. 4-4c). Unidentified time- and voltage-*independent* ion channels contribute to the leak current. It can be seen that the I–V curve intersects the zero current axis at V_r. This is not surprising, since V_r is defined as the voltage at which the net ion current flow is 0.

This linear leak current is all that is seen for hyperpolarizing voltage clamp pulses, whatever their amplitude. What happens when a larger pulse, one that normally exceeds the threshold for action potential generation, is given in the depolarizing direction? Remember that the membrane voltage is held constant by the voltage clamp, so no action potential is permitted to occur. It is immediately obvious that the membrane current is *not* at steady state during these larger depolarizing pulses, and time-dependent currents flow (Fig. 4-5a). There is an inward-going (negative by convention) current during the first millisecond or two after the beginning of the depolarizing pulse, and then the current reverses sign and becomes outward or positive during the remainder of the pulse. When a series of such pulses is given to different command voltages, a family of curves is generated as shown in Figure 4-5b. Note that as the amplitude of the voltage pulse increases, the early inward component of the current (red) first increases, and then begins to decrease until it reverses sign and becomes outward at very large depolarizations. In contrast the late outward component of the

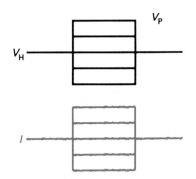

a

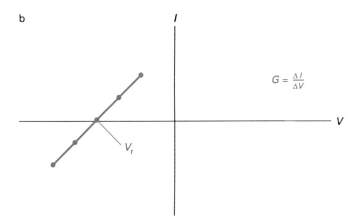

b

$$G = \frac{\Delta I}{\Delta V}$$

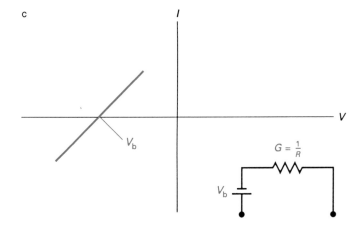

c

$$G = \frac{1}{R}$$

FIGURE 4-4. Leak currents. *a*: When small depolarizing or hyperpolarizing voltage clamp steps are made from the holding potential (V_H) to the pulse potential (V_P), small time-independent currents (I) are seen. *b*: A plot of these currents as a function of voltage. *c*: A similar current–voltage relationship is seen for a linear resistor in a nonbiological electrical circuit, containing a battery of voltage V_b.

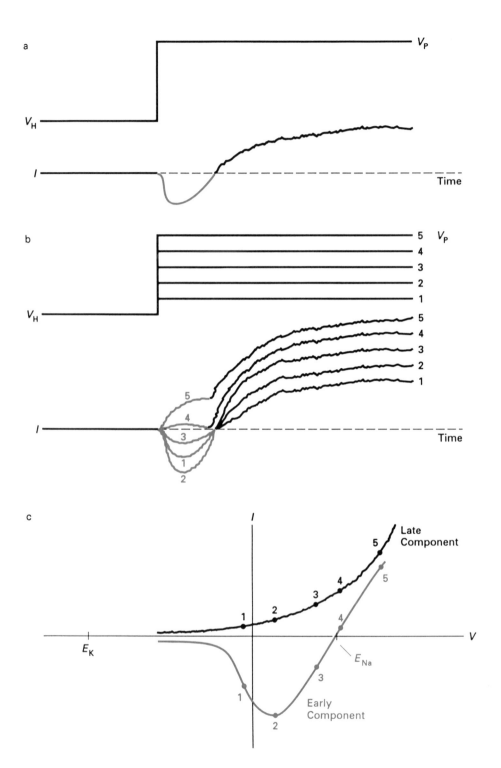

a

V_P

V_H

I

Time

b

5 V_P
4
3
2
1

V_H

5
4
3
2
1

5
4
3
2
1

I

Time

c

I

Late
Component

5

5

4

4

3

3

1 2

E_K

E_{Na}

V

1

3

2

Early
Component

current (black) remains outward and continues to increase in amplitude with larger depolarizations.

The relationship between the imposed membrane voltage and the membrane current that flows at that voltage can be investigated in more detail by constructing I–V curves. The early and late components of the current can be examined separately by measuring the current at different times. The peak inward current (usually after about 1–2 msec) is taken as a measure of the early component, and the current near the end of the pulse is the late component (also called by Hodgkin and Huxley the *delayed* outward current). As we shall see below, the early and late components can also be separated on the basis of other criteria, confirming that it is appropriate to make this distinction on the basis of their kinetic properties. The I–V curves in Figure 4-5c confirm the conclusions drawn from an inspection of the current traces themselves. Both components of the current exhibit markedly nonlinear I–V curves, the early inward component first increasing and then decreasing in amplitude, then reversing in sign. The late outward component continues to increase (with increasing slope) over the voltage range examined. For these large depolarizations, then, the membrane is no longer behaving like a linear resistor; rather it rectifies, and the membrane conductance exhibits *voltage dependence* in this voltage range. As we know from our consideration of voltage-dependent gating in Chapter 3, these conductance changes result from voltage-dependent changes in the open probability of the ion channels responsible for the membrane currents.

Sodium and potassium carry the inward and outward membrane currents. Hodgkin and Huxley noted that the early inward current has the right sign, amplitude, voltage dependence, and kinetics to be responsible for the upstroke of the action potential. For example, the fact that the inward current turns on only at voltages depolarized from rest provides an explanation for the phenomenon of a threshold for action potential generation. Similarly the delayed outward current has the right characteristics to be responsible for the repolarization. They subsequently went on to ask what ions are the charge carriers for

FIGURE 4-5. Nonlinear voltage-dependent currents. a: When the depolarizing pulses are made very much larger than those in Figure 4-4, time-dependent currents are seen to flow. During a sustained depolarization, there is an early component of the current that is inward (red) and a later component that is outward (black). b: When a series of pulses are given to different depolarizing pulse potentials, the currents change in a characteristic way. c: a plot of the early and late peak currents as a function of voltage (see Hodgkin and Huxley, 1952a,b,c; Hodgkin et al., 1952).

the inward and outward currents, using a series of *ion substitution* experiments. When the sodium was removed from the extracellular medium and replaced by an equivalent amount of some nonpermeant monovalent cation (for example choline), the late component of the current was not affected, but the early component was *outward* over the entire voltage range examined. This is because the sodium concentration gradient is now reversed, and when the sodium channels open, sodium *leaves* rather than enters through the channels. When the extracellular sodium concentration is varied over a wide range, so as to systematically vary E_{Na}, it can be seen that the V_r for the early current is always equal to E_{Na}. This confirms that this component of the membrane current is carried entirely by sodium and there is no significant contribution by any other ion.

The delayed current is not affected by the extracellular sodium concentration. Hodgkin and Huxley suspected that the delayed current was carried by the outward flow of potassium ions, but were unable to confirm this directly because they had difficulty changing the intracellular potassium concentration without damaging the axons. However, subsequent ion substitution experiments, on squid axon and other cell types, confirmed that V_r for the delayed current is always equal to E_K, demonstrating that potassium, and only potassium, is the charge carrier for this current component. Because of its kinetics, and the voltage-dependent gating that gives rise to rectification in the I–V curve, this current is often called the *delayed rectifier potassium current*.

Pharmacological tools have also proven to be extremely useful in separating the two components of axonal current. A number of creatures that use toxins either for self-defense or to subdue their prey have evolved toxins targeted against sodium channels. Probably the most useful of these naturally occurring toxins is *tetrodotoxin,* which is found in the Japanese puffer fish, and which, at nanomolar concentrations in the extracellular medium, can block axonal action potentials by blocking selectively the inward flow of sodium. It is thought that tetrodotoxin enters the mouth of the sodium channel and physically occludes the pore. In the presence of tetrodotoxin (TTX) the early sodium component of the current is eliminated and the delayed potassium component can be studied in isolation (Fig. 4-6).

It is interesting that the puffer fish is a delicacy in Japan, and chefs are specially trained to remove the TTX-containing organs for the preparation of this dish; however, these organs are not removed entirely, because the tingle one gets from a small dose of tetrodotoxin is apparently one of the major reasons this dish is so popular. In spite of the undoubted skill of these highly trained chefs, mistakes are made and there are still several deaths per year in Japan from tetrodotoxin poisoning.

The story of potassium channel blockers has (until very recently) been less colorful. Several organic compounds with quaternary ammonium groups, the most useful of which is *tetraethylammonium* (TEA), are selective blockers of the delayed rectifier potassium current in squid axon, allowing the early sodium current to be examined in isolation (Fig. 4-6). As we shall see later in this chapter, there can be many kinds of potassium currents in nerve cells, and these blockers do not affect all potassium currents to the same extent. More recently the pharmacology of potassium channels has expanded with the discovery that certain toxins, for example the bee venom toxin *apamin,* the scorpion venom component *charybdotoxin,* and the snake venom toxin *dendrotoxin,* can selectively block certain classes of potassium channels.

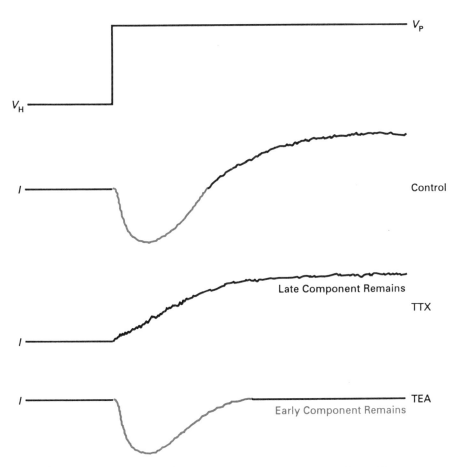

FIGURE 4-6. Pharmacological agents selectively block the early and late components of the current. Tetrodotoxin (TTX) blocks the early component of the current, whereas tetraethylammonium (TEA) blocks only the late component.

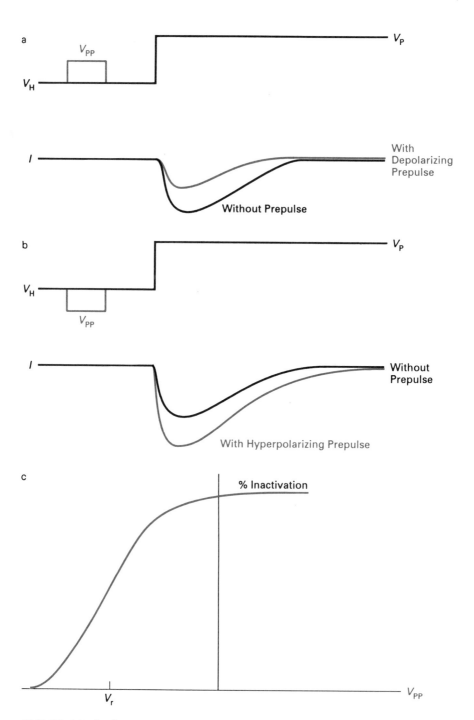

FIGURE 4-7. Sodium current inactivation. When the late component of the current is blocked, for example, with TEA, it is possible to examine the kinetics of the sodium current in isolation. Effect of a depolarizing (a) or hyperpolarizing (b) prepulse (V_{pp}) on the amplitude of the sodium current during a subsequent depolarizing pulse (V_p). c: Degree of inactivation as a function of voltage.

Sodium current inactivation. An essential feature of an action potential is the activation of the sodium current by membrane depolarization. But as we have seen, depolarization not only makes the sodium current turn on, it also causes it to turn off very soon thereafter. This *inactivation* occurs during, and results from, the depolarizing pulse. It is distinct from *deactivation* (or reversal of activation), which occurs after the end of the pulse as a result of the return of the membrane voltage to the hyperpolarized holding potential. Inactivation of the sodium current of course reflects the inactivation of individual sodium channels, as discussed in Chapter 3. A comparison of Figures 4-5 and 4-6 makes it clear that the switch from net inward to net outward current during a depolarizing pulse is due not only to the slow turning on of the delayed outward potassium current, but also to the fact that the opposing inward sodium current has inactivated and gone to sleep.

The voltage-dependent inactivation of the sodium current can be investigated by using *conditioning prepulses,* prior to the voltage clamp test pulse during which the sodium current is measured. If the membrane is depolarized briefly (prepulsed) immediately before the test pulse, it is found that the amplitude of the sodium current is *smaller* than in the absence of a prepulse (Fig. 4-7a). This is because the prepulse depolarization has caused inactivation of a portion of the current, and it has not yet recovered from this inactivation by the time the test pulse is given. In contrast, with a *hyperpolarizing* prepulse, the sodium current during the test pulse can be larger than the control (Fig. 4-7b). This shows that at the holding potential, which often is close to the cell's resting potential, the level of depolarization is sufficient to cause some resting or steady-state inactivation, which is removed by the hyperpolarizing prepulse. This is illustrated in a plot of the relationship between membrane voltage and extent of inactivation (Fig. 4-7c).

We have seen that sodium channel inactivation is a time-dependent process that contributes to the switch from inward to outward current during an action potential. The recovery of sodium current from inactivation is also time dependent. The time course of recovery from inactivation can be investigated by varying the time between a depolarizing prepulse and a test pulse. The prepulse produces inactivation, and recovery to the normal amplitude sodium current takes several tens of milliseconds. Typically recovery is half complete after about 15 msec in the squid axon. This of course means that sodium current inactivation will long outlast the action potential that produces it.

Figure 4-8 summarizes the sequential activation, inactivation, and recovery from inactivation of sodium channels. Under resting conditions (upper left) the black activation gate is closed and no current flows. When the membrane is depolarized, this activation gate undergoes voltage-dependent opening. As shown on the upper right, the channel is now open and current can flow. The same depolarization that opens the activation gate also causes slower closing of the red inactivation gate (lower right). Although the activation gate is still open, no current can flow because the inactivation gate is closed. After the end of the depolarizing pulse, deactivation, the reversal of activation, occurs and the activation gate closes. Again no current can flow (lower left). At this time a depolarization cannot evoke any current, because even though the activation gate will open as a result of the depolarization, the inactivation gate remains closed for some time following the pulse. Only after inactivation is removed is the channel back in its resting state (upper left) and available to be opened by depolarization. In

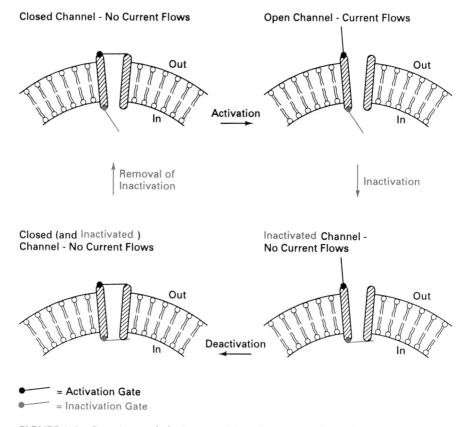

FIGURE 4-8. Opening and closing transitions in voltage-dependent sodium channels. The opening and closing transitions in sodium channels during and following a depolarizing pulse are summarized. See text for details.

the next chapter we shall discuss the biochemical and molecular approaches that provide information about the structures of the voltage-dependent activation and inactivation gates.

The functional implications of the long time course of recovery from sodium channel inactivation are profound. Since the peak of the action potential is a voltage at which inactivation will be complete (Fig. 4-7c), sodium current will be inactivated for some period of time, and the cell will be unable to fire a second action potential until there is sufficient recovery from this inactivation. It will be evident, then, that *sodium inactivation is the ionic mechanism underlying the refractory period*. The elevated threshold during the refractory period reflects sodium current inactivation, and the return of the threshold to normal (Fig. 2-5) corresponds to removal of inactivation. Effectively, sodium inactivation sets the upper limit of action potential frequency in an axon. We shall see, however, that a variety of potassium currents also participate in determining the actual rate at which a neuron fires.

Ion pumps maintain the ion concentration gradients. It is evident that neurons cannot continue to fire action potentials forever. When membrane currents flow, ions are moving down their concentration gradients. The currents are relatively small at steady state, but of course are much larger during action potentials. If action potential firing continues, eventually the sodium and potassium concentrations on the two sides of the membrane will be equal, the membrane potential will be zero, and the state of the cell can be well described by the word "dead." It may take a long time to run down the ionic concentration gradients in an axon as large as the squid giant axon, but in smaller cells significant changes may occur after relatively few action potentials. Fortunately the energy driven pumps come to the rescue before any damage is done.

A particular active ion transporter, the *sodium–potassium-ATPase* or *sodium–potassium pump,* mediates the pumping of sodium out of and potassium into the cell to maintain the ion concentration gradients. Another way to think of this is to say that the pump is responsible for *charging up the membrane battery.* The pump is an enzyme that hydrolyzes ATP and uses the energy to move each ion against its concentration gradient (Fig. 4-9). This transport activity may involve some sort of movement or rotation of the pump in the membrane. The stoichiometry of the sodium–potassium-ATPase, the transport ratio for sodium and potassium, is not 1:1 but instead is 3:2. In other words, three sodium ions are transported outward for every two potassium ions transported inward, and, as a result, the pump produces a net outward current of one ion per ATP hydrolyzed. This kind of pump is said to be *electrogenic,* because its activity causes the cell to hyperpolarize, and it contributes (although usually only to a limited extent) to the setting of the resting potential.

Although Chapter 3 deemphasized their contribution to neuronal excitability, we can see that ion pumps do indeed play an essential role. They are not quite as flashy as the channels that allow for such rapid flow of ions, but they are always there in the background, working away quietly to ensure that the ion concentration gradients that are essential for electrical signaling are maintained.

Action potentials jump along the axon. We have mentioned that the speed of action potential propagation along the axon is determined in part by myelination. This comes about because the myelin sheath, which is a large number of layers of glial plasma membrane wrapped about the axon, acts as an excellent electrical insulator. The space between the axon and the

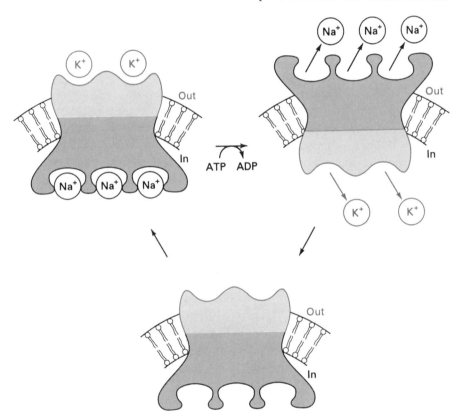

FIGURE 4-9. The sodium–potassium-ATPase. Depicted is the ATPase, a plasma membrane pump that uses the energy of hydrolysis of ATP to move three sodium ions from the inside of the cell to the outside, and, at the same time, move two potassium ions from the outside to the inside.

myelin is *not* an ion-containing extracellular solution, and *no* current can flow across the axonal membrane in the regions of myelination. Remember, however, that the myelin sheath is interrupted periodically at the nodes of Ranvier (Fig. 1-5), and at these nodes the ionic conduction pathways we have been discussing are indeed present. Thus action potentials can be generated at nodes, and the passive spread of the action potential depolarization can bring the adjacent node above threshold, allowing the action potential to "jump" along the axon from node to node. Nature has designed the myelinated axon with the nodes spaced some 1–2 mm apart, so that the depolarization produced at one node is still well above threshold by the time it reaches the next node. To put this another way, the space constant of the axonal membrane and the spacing between nodes are designed to ensure that the action potential is propagated. The high resistance of myelin forces the current to move down the axon rather than leak out across the axonal membrane, and thus the myelin sheath itself contributes to the space constant. In addition, the voltage-dependent sodium channels are not spread evenly throughout the axonal plasma membrane, but are packed together at a very high density at the nodes and are sparse in the intervening membrane under the myelin. All of these factors together allow the action potential to jump from node to node, achieving conduction with the minimum use of ion channels or energy-consuming pumps.

This *saltatory conduction* (from the Latin *saltare* meaning to leap or dance) permits conduction at speeds many times faster than in nonmyelinated axons of the same diameter. Certain diseases of the nervous system, the best known of which is *multiple sclerosis,* are characterized by loss of myelin from some myelinated axons. These demyelinating diseases can have severe consequences, because they result in the slowing (and sometimes the complete block) of axonal conduction, with devastating effects on the neuronal pathways in which the demyelinated axons participate.

Diversity of Ion Currents and Ion Channels

Thus far we have been discussing the axonal membrane as if it had only two ion currents, a voltage-dependent potassium current that is responsible for the axon's electrical activity (actually its lack of activity) at rest and its repolarization following an action potential, and a voltage-dependent sodium current that underlies the large and rapid depolarization during the action potential. In the squid giant axon, and perhaps in many other axons, this picture is essentially correct.

In contrast to axons, the electrical behavior of neuronal cell bodies and dendrites is more varied. Some neuronal cell bodies do *not* fire action potentials, and are said to be *electrically inexcitable* because they lack the rapid voltage-dependent sodium current. Other neurons produce pacing, bursting, and the wide range of responses to external stimuli described in Chapter 2. The cell bodies of such cells may exhibit many other ion currents in addition to or in place of the classical action potential currents. These other currents, which may be regulated by (1) voltage, (2) neurotransmitters that bind to receptor sites on the extracellular side of the channel, (3) intracellular calcium, (4) intracellular metabolic modulators, or (5) some combination of these factors, interact to give rise to complex patterns of neuronal electrical activity such as that exhibited by *Aplysia* neuron R15 (see Fig. 2–10c).

Since ion currents reflect the activity of ion channels, the diversity in ion currents must be matched by an equivalent diversity in the ion channels that underlie these currents. In fact single channel recording techniques have revealed that the diversity is even greater than had been imagined; in many cases these techniques reveal that currents that had been thought to be carried by a single population of ion channels, in fact reflect the activity of several different kinds of ion channels. In addition channels are appearing for which no macroscopic current equivalent had been observed. Only through a knowledge of the properties of the ion channel can experimental strategies be devised to investigate the macroscopic currents, and their role in the electrical behavior of the neuron.

Voltage-Dependent Ion Channels

The rest of this chapter will focus on the major classes of *voltage-dependent* ion channels. We do not mean to give short shrift to the *neurotransmitter-gated* ion channels, which are essential for chemical synaptic transmission and for the modulation of neuronal electrical properties. This group, including a variety of potassium, calcium, and chloride channels, will be dealt with in detail in subsequent chapters. Some of the neurotransmitter-gated channels are also voltage dependent, and what was formerly thought to be an absolute dichotomy between these two classes is now blurring. Thus, the choice of channels to be discussed here, rather than in later chapters, is by necessity somewhat arbitrary.

Channels that are voltage dependent can be classified into two main groups. In the first group are those channels whose open probability is strongly influenced by voltage only at rather depolarized membrane potentials. Such channels include, among others, the voltage-dependent sodium and delayed rectifier potassium channels already encountered in the squid axon. The second group includes channels whose opening and closing are influenced by changes in membrane potential close to, or negative to, the normal resting potential. This latter group is more likely to contribute to the subthreshold electrical behavior of a neuron, such as whether it fires spontaneous bursts and whether its firing rate adapts to a maintained external stimulus. It is particularly interesting that the activity of many of these subthreshold channels is subject to modulation by neurotransmitters and intracellular metabolites. Table 4-1 lists the names of some voltage-dependent ion channels and the approximate voltage range over which the channels gate.

We reemphasize here a very important concept concerning the amount of current carried through a population of voltage-dependent ion channels. Remember from Ohm's law that the current depends on *both* the driving force and the membrane conductance for the ion in question. The driving force for any ion always depends on the membrane voltage (relative to the ion's equilibrium potential), and for a *population* of voltage-dependent channels the conductance will also depend on the voltage. However, these dependencies may be very different. For example, the axonal sodium channel will be fully open at a V_m of $+50$ mV (at least until it inactivates), but no current will flow because this happens to be E_{Na} and hence the driving force $(V_m - E_{Na})$ is zero. Accordingly it is essential to take into account *both* the channel open probability and the amount of current that flows through the channel when it is open in determining the total amount of membrane current contributed by a particular type of voltage-dependent ion channel at any given voltage.

Channels That Carry Inward Current

In axons, the most important channels for the generation of inward currents are the voltage-dependent sodium channels whose properties have

TABLE 4-1 Examples of Voltage-Dependent Ion Channels[a]

Channel Type	Activation Voltage Range	Physiological Function
Axonal sodium channel	-30 to $+20$ mV	Upstroke of action potential
Calcium channels	Variable (see Table 4-2)	Calcium action potentials; calcium-mediated intracellular events (see Fig. 10-13)
Delayed rectifier potassium channel	-20 to $+30$ mV	Action potential repolarization
Transient potassium channel	-60 to -10 mV	Spacing of action potentials
Anomalous rectifier potassium channel	-50 to -100 mV	Regulation of resting level of neuronal activity

[a]The activation voltage range may be somewhat different in different cell types.

already been described in detail. However there are several other classes of channels that contribute substantially to inward current flow in many neuronal cell bodies (and even in some axons).

Calcium channels. In most neurons, a depolarizing voltage clamp step elicits an inward current with kinetics very different from those seen in the squid axon. The current may rise to its peak more slowly, and inactivate only partially and far more slowly than in the squid axon (Fig. 4-10a). When pharmacological treatments and/or ion replacement are used to eliminate any sodium current, an inward current that rises slowly to its peak and inactivates only partially (if at all) can still be elicited by the depolarizing pulse (Fig. 4-10a). In fact when the neuron is released from voltage clamp under these conditions, it is often found that it can still fire action potentials *even in the complete absence of sodium,* although the shape and duration of these action potentials can be very different from those observed in the presence of sodium (Fig. 4-10b). Further ion substitution experiments reveal that most neurons exhibit a substantial *voltage-dependent calcium current* (I_{Ca}). In some cases this is responsible for much or all of the regenerative depolarization during the rising phase of the action potentials.

It is interesting that the very brilliance of the studies of Hodgkin and Huxley, which defined the sodium and potassium currents as both necessary and sufficient to account for action potentials in the squid axon, led to skepticism in assessing the work of early pioneers in the calcium current field. Although the calcium current experiments stood up to critical scrutiny, there was a reluctance on the part of many neurophysiologists to complicate, with another ion current, what had been a simple and satisfying picture of membrane excitability. The neurophysiology community did not suspect in the 1960s just how drastically this simple picture was to be modified in the years to come.

Diversity of calcium channels. In some neurons, a plot of the peak calcium current as a function of voltage (Fig. 4-10c) closely resembles that for the sodium current (Fig. 4-5c), except that the current approaches 0 at very depolarized voltages, reflecting the more depolarized value for the calcium reversal potential, E_{Ca}. In other cells this curve appears more complex indicating that more than a single population of calcium channels gives rise to the current–voltage relationship. In fact, on closer inspection with single channel recordings, even the simpler I–V relationship can turn out to be generated by more than one species of calcium channel. It now appears that there are at least three distinct categories of calcium channels in many mammalian neurons, which can be distinguished on the basis of their voltage dependence of activation and inactivation, kinetics, single channel con-

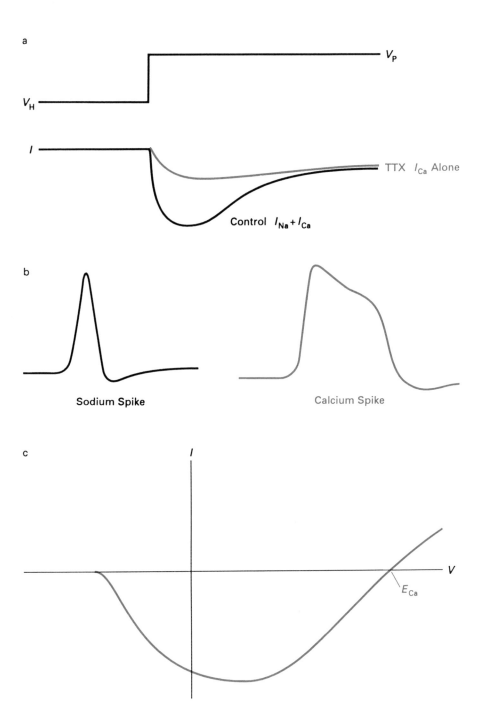

FIGURE 4-10. Currents carried by calcium ions. a: When TTX is used to block the sodium component (I_{Na}) of inward current, a calcium current component (I_{Ca}) remains. b: The sodium action potential is contrasted with the action potential in cells in which the inward current is carried predominantly by calcium. c: Current–voltage relationship for the calcium current.

ductance, and pharmacology (Table 4-2). Some invertebrate neurons may also contain several different kinds of calcium channels (e.g., see the discussion of *Aplysia* bag cell neurons in Chapter 11).

Calcium channels are of particular interest because calcium is far more than simply a charge carrier across the plasma membrane. As essential as calcium ions are in contributing to action potentials and other aspects of neuronal electrical activity, this role may be secondary to the intracellular messenger actions of calcium. Calcium that enters the cell interacts with calcium binding proteins to regulate a variety of intracellular enzymes. Furthermore, intracellular calcium ion regulates the gating of several types of ion channel, and can even feed back and participate in the inactivation of one of its own channels. In addition an essential characteristic of neuronal signaling, the release of chemical neurotransmitters at synapses, is controlled directly by intracellular calcium. In this sense calcium can be thought of as the *transducer* of an electrical signal, depolarization, into chemical signals inside the cell.

All of these features set calcium apart. It will thus come as no surprise that the activity of calcium channels themselves is subject to intricate modulatory influences, in cardiac and skeletal muscle as well as in neurons. We shall be hearing much more about the regulation and consequences of calcium channel activity throughout this book.

Calcium-dependent cation channels. One of the many things that intracellular calcium does is to activate yet another inward current, which is carried by a nonselective cation channel that allows the flow of both sodium and potassium. Because the driving force for sodium is so much greater than that for potassium near the resting potential, the net current carried by this channel is inward in a resting cell, but may be outward during the

TABLE 4-2 Properties of Three Classes of Calcium Channels (T, N, L) Described in Neurons and Other Cell Types

Property	Channel Type		
	T	N	L
Activation voltage range	More positive than −70 mV	More positive than −10 mV	More positive than −10 mV
Inactivation voltage range	−100 to −60 mV	−100 to −40 mV	Little inactivation
Single channel conductance	8–10 pS	11–15 pS	23–27 pS
Blocked by cadmium	No	Yes	Yes
Blocked by cobalt	Strongly	Weakly	Weakly

Modified from Tsien (1987).

depolarization that occurs during an action potential. This current helps to provide a depolarizing drive toward the action potential threshold, and it may be important in the generation of repetitive firing, including bursting behavior as in neuron R15. Other than the fact that intracellular calcium is required for its activation, its properties have not been investigated very thoroughly in neurons, and we will not discuss it further here.

Channels That Carry Outward Current: The Potassium Channels

The palette of calcium channel diversity seems pale in comparsion with that exhibited by the potassium channels. Some half dozen or more voltage-dependent potassium currents were first identified on the basis of voltage clamp experiments. This number has expanded dramatically as single channel and molecular biological approaches have been brought into play on this question. As in the case of other channels, kinetics, voltage dependence, pharmacology, and more recently single channel properties have been used to distinguish between and characterize the various potassium channels. We will summarize briefly here the properties of some of these channels, and again will defer a detailed description of their activity and regulation to later chapters. We will find that some of these potassium channels, like the voltage-dependent calcium channels, are subject to modulatory influences that allow for the regulation of neuronal properties.

Calcium-dependent potassium currents. In most cells, for example, the large cell bodies of many molluscan neurons that have been widely used for voltage clamp studies of membrane currents, the total outward current carried by potassium exhibits a steady-state I–V relationship very different from that seen in the squid axon (we are using the term "steady state" to refer to the sustained noninactivating current, measured many tens or hundreds of milliseconds after the onset of a depolarizing pulse). The current–voltage curve has a characteristic "N" shape in the range of depolarized voltages (Fig. 4-11a), because it is the sum of at least two distinct current components (Fig. 4-11b,c). When the cell is injected with a calcium chelating agent such as EGTA, or calcium entry is prevented during the depolarizations by pharmacological block of calcium channels, the resulting I–V curve (black in Fig. 4-11b) looks identical to that of the delayed rectifier potassium current in squid axon (compare with Fig. 4-5c). When the kinetics, voltage dependence, and pharmacology of this current are examined it can be seen to exhibit the properties of a classic delayed rectifier, which we have already discussed in detail above.

The other outward current component that contributes to the shape of the steady-state I–V curve at positive voltages is the one that was blocked

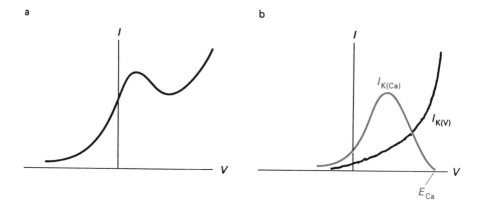

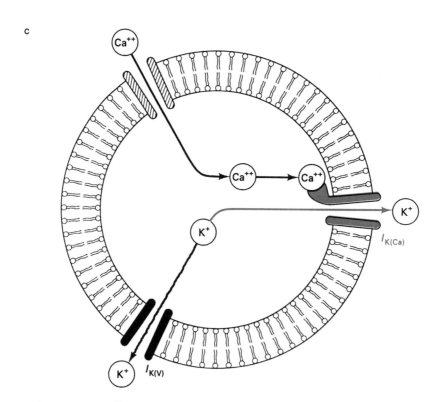

FIGURE 4-11. Different components of outward current. *a*: In most neuronal cell bodies, when the inward current is blocked, the current–voltage relationship for the peak outward current has a characteristic N shape (see Meech and Standen, 1975). *b*: Two components of outward current. *c*: The activation of two classes of potassium channels is illustrated.

by preventing calcium entry or chelating intracellular calcium. It is a *cal-cium-dependent potassium current*. Its I–V curve (red in Fig. 4-11b) is obtained by subtracting the delayed rectifier component from the total out-ward current. This current is activated in part by the depolarization per se, but also by the calcium that enters during the depolarizing voltage pulse (Fig. 4-11c). A comparison of the I–V curve in Figure 4-11b with that in Figure 4-10c reveals that the voltage dependence of the calcium-activated potassium current mirrors that of the calcium current; this of course arises from the requirement for calcium entry, through voltage-dependent cal-cium channels, to contribute to the activation of this potassium current. As the voltage approaches E_{Ca}, the driving force for calcium entry decreases, and hence there is less activation of the calcium-dependent potassium cur-rent (other calcium-dependent intracellular processes may exhibit a similar voltage dependence). This requirement for intracellular calcium also explains why the current is eliminated by blocking calcium entry during the depolarization. However this current, and other calcium-dependent intracellular events, may still be evoked by intracellular injection of cal-cium, or physiological treatments that cause the release of calcium from intracellular stores.

Kinetic and pharmacological studies of voltage clamp currents suggested that there might be some heterogeneity of the calcium-dependent potas-sium current, but the extent of this heterogeneity became apparent only from single channel experiments. Many cell types contain a large conduc-tance calcium-dependent potassium channel (a so-called *maxi* channel), but there are smaller ones as well, and there is even heterogeneity *within* the maxi class. For example, in rat brain plasma membrane preparations, there are at least two separate maxi channels that can be distinguished on the basis of their gating kinetics (Fig. 4-12) and pharmacology. It appears that more than one type of calcium-dependent potassium channel can be present in a single cell, but the functional significance of this heterogeneity remains to be determined. Since calcium-dependent potassium currents contribute (with the delayed rectifier) to action potential repolarization, and also to interspike currents that help to control the frequency of repetitive firing, it is possible that the different kinetic properties of different calcium-depen-dent potassium channels permit them to undertake distinct functional roles.

Transient potassium current. Thus far we have been talking about steady-state potassium currents that inactivate very little if at all during a long depolarizing pulse. There is also a transient potassium current, often known as *A-current,* that activates rapidly and then inactivates in a manner analogous to the sodium current. To measure this current the membrane potential must first be set to a very negative holding potential for several hundred milliseconds to remove the voltage-dependent steady-state inactivation (this is reminiscent of the prepulse protocol for examining sodium current inactivation and removal of inactivation—see Fig. 4-7). When the membrane is depolarized from this very negative holding potential, an outward A-current is seen (Fig. 4-13a) that mirrors the inward sodium current (Fig. 4-1 and 4-6), albeit with a more prolonged time course. The voltage dependence of A-current inactivation is such that in neurons that have a relatively positive resting potential (more positive than about -45 mV),

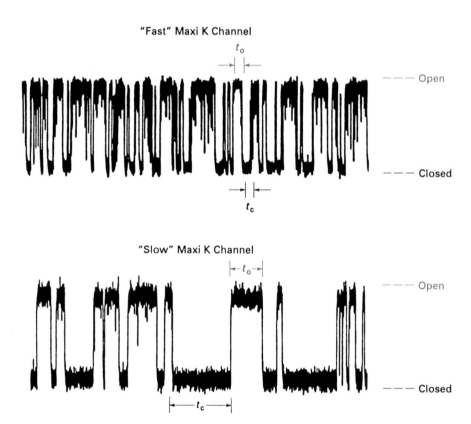

FIGURE 4-12. Different classes of calcium-dependent potassium channels. At least four distinct calcium-dependent potassium channels can be seen in plasma membrane fractions from mammalian brain. Two of them are shown here. They have very similar single channel conductances (240 pS), but differ in their gating kinetics (note the differences in t_o and t_c). These records were obtained following reconstitution of calcium-dependent potassium channels into artificial phospholipid bilayer membranes (see Reinhart et al., 1989).

inactivation is more or less complete at the resting potential (Fig. 4-13b; compare with Fig. 4-7c). That is why, in such cells, the steady-state inactivation must first be removed to examine this current.

The A-current is active in the subthreshold region of membrane potential and helps to determine the frequency of repetitive firing in neurons. Although it is largely inactivated near the resting potential and completely inactivated during action potentials, some portion of the inactivation is removed by the afterhyperpolarization that normally follows an action potential. Hence the A-current is active for a short while after an action potential and slows the return of the membrane potential toward the spike threshold. This in turn slows the firing frequency in a repetitively firing neuron.

Another role for A-current is to allow a delay to occur between an excit-

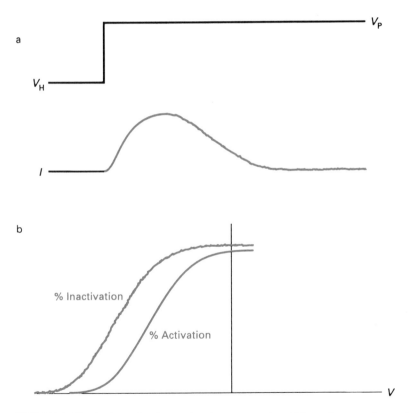

FIGURE 4-13. An inactivating potassium current. a: When the membrane is depolarized from very negative holding potentials, a rapidly inactivating potassium current can be observed. This potassium current has been termed the A-current. b: Extent of activation and inactivation of the A-current as a function of voltage.

atory stimulus and the onset of action potentials. This occurs in neurons with a relatively negative resting potential, in which there is little steady-state inactivation. When such a neuron is depolarized, the A-current is activated and tends to oppose the change in membrane potential toward the threshold. As the A-current inactivates during the depolarization, however, the neuron begins to depolarize more rapidly. Following a delay set by the kinetics of inactivation, the neuron finally reaches threshold. An example of this is found in the *Aplysia* ink gland motor neurons, which were shown in Figure 2-11. The prolonged noxious stimulus causes a depolarization that leads to progressive inactivation of A-current. Only when the A-current has undergone inactivation do the neurons fire the action potentials that trigger ink release.

Again voltage clamp experiments have shown that there are multiple types of A-currents in different cells. Although this has not been investigated thoroughly at the single channel level, molecular biological approaches have provided evidence (to be discussed in detail in the next chapter) that these are due to distinct species of A-channel proteins. These distinct A-current channels, perhaps with different kinetic properties or voltage dependence, probably contribute in different ways to the regulation of neuronal firing rates.

A potassium current activated by hyperpolarization. All the inward and outward currents we have discussed thus far are activated by depolarization, either because they are gated directly by voltage, or because depolarization-induced calcium entry is required for their activation. This gives rise to rectification in the I–V relationship, an increase in the slope of the curve with depolarization. However, in many cells, there also exists a potassium current that is activated by *hyperpolarization*. This causes a *decrease* in the slope of the I–V curve with depolarization (Fig. 4-14), a phenomenon known as *anomalous* or *inward* rectification.

This all seems rather strange. Why would a cell bother with a channel that passes only inward but not outward potassium current? This question becomes particularly pressing when we remember that under normal conditions a neuron's V_m can never be more negative than E_K, since potassium is the charge carrier with the most negative reversal potential. Thus only under the artificial hyperpolarizations imposed by the voltage clamp will the inward flow of potassium ever occur. One clue to the role of the anomalously rectifying potassium channels comes when we realize that they are not perfect rectifiers, and can pass some outward current in the voltage range up to about 30 mV depolarized from E_K (Fig. 4-14). Although the amount of current is not large, very few other membrane currents are active in this voltage range; accordingly it may play an important role in regulating the resting level of neuronal activity (see Chapter 11).

Other potassium channels. We have by no means exhausted the complement of potassium channels with the description of the major classes of voltage-dependent channels above. There are others that are gated by ligands or intracellular metabolites, and we have emphasized that some of these may exhibit voltage dependence as well. These other channels include the so-called M-current and S-current potassium channels, whose activity can be modulated by the neurotransmitters muscarine (M-current) and serotonin (S-current), respectively. We will describe in subsequent chapters the ways in which these currents contribute to important physiological phenomena. One important question is why neuronal membranes contain so many distinct conductance pathways for a single ion, potassium. It seems likely that a series of potassium currents with different properties is the best way to achieve an extraordinary diversity and flexibility in the types of firing patterns that neurons generate.

Summary

The flow of ions through populations of ion channels in the neuronal plasma membrane gives rise to transmembrane ion currents. It is the sum of the various currents flowing at any point in time that determines the neuron's membrane potential. The activities of the sodium and potassium channels responsible for axonal action potentials are dependent on voltage. Voltage clamp studies, which allow the measurement of the current flowing through these channels at fixed voltage, have provided a detailed understanding of the sequence of membrane permeability changes that gives rise to action potentials.

Many neuronal cell bodies, and even some axons, contain a large number

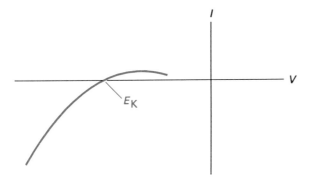

FIGURE 4-14. A potassium current that activates with hyperpolarization. Current–voltage relationship for the inwardly rectifying potassium current, which displays a decrease in slope conductance with depolarization.

of permeability pathways in place of, or in addition to, the action potential sodium and potassium channels. We know about several different kinds of calcium-selective ion channels, and a large and ever-expanding number of potassium channels. Many of these channels are regulated by voltage, by neurotransmitters, or sometimes by both. These multiple ion channels, with their diverse and often interacting regulatory mechanisms, allow the neuron to modulate its electrical properties in complex and subtle ways.

Structure and function
of ion channels

It will be evident from the discussion thus far that biophysicists and electrophysiologists have made major contributions to our knowledge about ion channels and neuronal excitability. More recently, chemists and molecular biologists have stepped into the picture, providing visions of the molecular structures of some of the ion channels described in Chapter 4. Together, these different approaches are now beginning to explain, at a molecular level, the very different patterns of electrical activity in neurons controlling diverse physiological functions. In this chapter we will describe some of the insights that are emerging from these combined approaches.

Structure of the Sodium Channel

The first voltage-dependent ion channel to have yielded its biochemical structure to scientists is the workhorse of axonal conduction, the sodium channel. This is in large part because of the relatively large number of pharmacological agents that interact specifically with the sodium channel protein. These include tetrodotoxin and a related toxin from marine dinoflagellates, *saxitoxin* (STX), which block the channel when applied to the outside of the membrane. Other agents that activate the channel, or affect its gating in other ways, are listed in Table 5-1. Using these specific toxins to label sodium channel proteins, several groups in the early 1980s were able to extract and purify the channel from membranes of skeletal muscle, rat brain, and the electric organ of electric eels. The latter is a particularly rich source of the channel, which the eel uses to generate sufficient current to stun its prey. In all cases a very large glycoprotein (about 260 kDa) was purified and termed the α-subunit. In addition, two smaller proteins were found in rat brain that copurify with the α-subunit and are termed the $\beta 1$ (39 kDa) and $\beta 2$ (33 kDa) subunits.

The full amino acid sequence of the α-subunit of the eel sodium channel has been deduced from the sequence of nucleotides in a piece of DNA (*complementary* DNA or cDNA) whose sequence is complementary to that of the messenger RNA, which encodes the protein. This is a molecular biological technique that is rapidly becoming a standard way of determining the amino acid sequence of proteins. The cDNA is synthesized in a test tube from the messenger RNA template, using the enzyme *reverse transcriptase.* Messenger RNA usually is available only in very small amounts, and its susceptibility to breakdown by the degradative enzyme ribonuclease makes it difficult to manipulate. However, these drawbacks do not apply to DNA. The amount of cDNA can be amplified by molecular cloning, and its sequence can be determined readily. The protein predicted from the α-subunit cDNA sequence contains about 2000 amino acids and accounts for 208,000 of the total molecular weight, the remainder apparently being accounted for by carbohydrates attached to the channel protein.

What can an unwieldy string of 2000 amino acids tell us about the structure of the channel? Luckily there are a few general rules relating amino acid sequence to protein structure, that can be used to provide clues as to which regions of the protein may lie within the lipid membrane itself. One of the more important of these rules is that a string of 23 or so hydrophobic amino acids can span a normal cell membrane, and that such a string is likely to be organized in the form of an α-helix of amino acids. Such strings of hydrophobic amino acids in a protein can readily be picked out from a

TABLE 5-1 Agents That Bind to Sodium Channels and Alter Their Properties

Agents	Actions
Tetrodotoxin (TTX) Saxitoxin (STX) μ-Conotoxins	Block channel from outside
Batrachotoxin Aconitine Veratridene Grayanotoxin	Alkaloid neurotoxins that activate the channel
α-Scorpion toxins Sea anemone toxins	Slow the rate of inactivation
β-Scorpion toxins	Shift voltage dependence of inactivation to more negative potentials
Local anesthetics	Block channel, mostly from inside

hydrophobicity plot. Each amino acid is assigned a hydrophobicity value, which reflects its ability to interact with water. Amino acids with nonpolar side chains interact poorly with water—are strongly hydrophobic—and are given a high positive value. In contrast polar amino acids are given a negative value. A running average of these values over several amino acids is then calculated around each amino acid in the protein sequence, and plotted as a function of position along the protein chain. Such a hydrophobicity plot for the α-subunit of the sodium channel is shown in Figure 5-1a. Twenty-four possible transmembrane stretches of amino acids can be found.

The complexity of the structure of the sodium channel can be substantially reduced when it is realized that the sequence contains four internal regions each of which strongly resembles the others. These are marked I, II, III, and IV in Figure 5-1. Within each of these homologous regions there are six possible transmembrane segments labeled S1–S6. With this information it is possible to construct a model for the arrangement of different parts of the protein across the plasma membrane (Fig. 5-1b). It should be emphasized that such a structure remains only a hypothetical model until direct structural determinations are made on the protein itself.

The Use of *Xenopus* Oocytes to Study Ion Channels

Does the α-subunit alone constitute a complete functional sodium channel or are other proteins, such as the $\beta 1$- and $\beta 2$-subunits, required for conduction, gating, and selectivity? One way to attack this question is to take the purified protein and to reconstitute it into synthetic lipid membranes, an approach that will be described in more detail for the acetylcholine receptor/channel in Chapter 9. Another approach is to express the α-subunit in a cell that does not normally contain voltage-dependent sodium channels. Ooctyes of the South African toad *Xenopus laevis* have been particularly useful for this purpose.

A heterologous expression system. To an electrophysiologist, *Xenopus* oocytes are dull cells. They bear few of the interesting channels that give neurons and other excitable cells their distinctive electrical characteristics (Fig. 5-2). On the other hand, they are very large and their electrical properties can easily be measured under voltage clamp after penetration with several electrodes. Early in the 1970s it was found that when messenger RNA encoding a protein from some other cell type is injected into the oocyte, it can be translated faithfully into the protein. Moreover, the oocytes carry out many of the normal posttranslational modifications of the proteins, such as the addition of carbohydrates or phosphate groups,

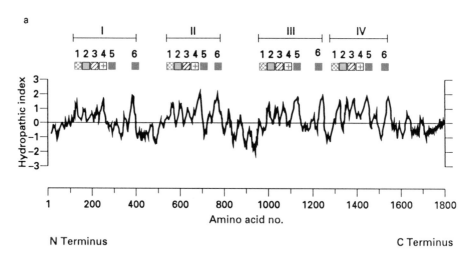

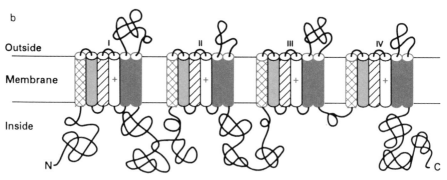

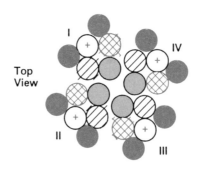

FIGURE 5-1. The sodium channel. *a*: A hydrophobicity plot for the α-subunit of the sodium channel. *b*: A model for the arrangement of different parts of the protein across the plasma membrane (from Noda et al., 1984).

and are able to insert the protein into the appropriate compartment of the cell. So it is with ion channels. When messenger RNA prepared from an excitable tissue is injected into the oocytes, and several days are allowed to elapse for the synthesis of proteins, new currents can be recorded across the oocyte membrane. As shown in Figure 5-2, the exact pattern of channels expressed depends on the source of the RNA.

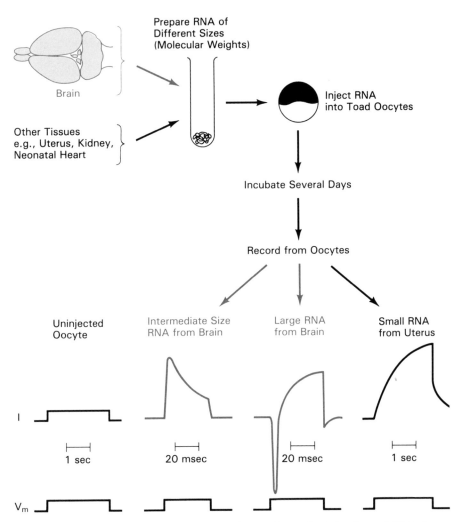

FIGURE 5-2. Heterologous expression of ion channels. Injection of RNA from brain and other tissues into *Xenopus* oocytes results in the appearance of new channels in the membrane of the oocyte.

Expression of mutant channels. Such *heterologous expression* of messenger RNA coding for ion channels (and other proteins) can also be used to systematically investigate the relation between structure and function in the proteins that are expressed. Pieces of DNA can be constructed in which the sequence coding for the protein is modified or *mutated* at specific sites. This DNA can then be transcribed into messenger RNA, and the mutated messenger RNA can then be expressed in oocytes. Such *site-directed mutagenesis* experiments can be used to ask what role particular amino acids or structural features play in the function of the protein. For example, as we shall discuss below, mutations in certain regions of the sodium channel α-subunit have provided insights about those portions of the protein that are involved in voltage-dependent activation and inactivation. Other examples of this approach will be given in our discussion of neurotransmitter receptors and transduction mechanisms in Chapters 9 and 10.

Expression of sodium channels. When purified messenger RNA that codes only for the α-subunit of a rat brain sodium channel is injected into an oocyte, functional sodium channels are formed. These results, coupled with artificial lipid membrane reconstitution experiments, suggest that the α-subunit is all that is required to generate sodium currents that both activate and inactivate in response to depolarization. It is worth pointing out, however, that some of the electrical properties of the sodium channels expressed from pure α-subunit messenger RNA are not identical to those observed when total messenger RNA from rat brain is used. In particular, the rate at which the channel inactivates appears slower when the pure α-subunit RNA is expressed. We will come back to this point in our discussion of the structure of the inactivation gate later in this chapter.

Gating Currents and the Voltage Sensor

Given the amino acid sequence of a channel protein, for example, the α-subunit of the sodium channel, there are many things that a biophysicist would dearly love to know. For example, which part of the molecule forms the ion pore? Where is the site of selectivity of the channel for sodium ions? Such questions are as yet unanswered. Although these issues are usually the province of biophysicists rather than neurobiologists, one aspect of the structure of a channel will be addressed briefly here. That is the question of what structure in these proteins renders them sensitive to changes in voltage.

In Chapter 3 we saw that for a channel to respond to changes in voltage across the cell membrane, it must contain some charged structure that acts as a voltage sensor. As a result of a change in the voltage across the membrane, this charged structure may undergo a physical movement, which in turn, may trigger the rearrangement of the protein that allows the opening or closing of the ion channel pore. The charge on the voltage sensor is known as the *gating charge.*

Immediately after a change in voltage, the movement of a gating charge within an ion channel protein should register as a flow of current across the membrane. This *gating current* is very much smaller than that due to the flow of ions through the channel. In addition, it is very brief. Gating current flows only while the channel protein is undergoing the movement to a new conformation and this can occur very rapidly. The ionic current, on the other hand, starts to flow only after the new conformation of the protein is achieved (Fig. 5-3a). It has been possible to measure the gating currents for sodium and potassium channels in the squid giant axon and some other preparations (Fig. 5-3b). To do this it is first necessary to eliminate, either by the use of drugs or by electronic wizardry, the very much larger contaminating currents flowing across the lipid membrane through the ion channels. The movements of charge that can be recorded under these conditions generally match those expected for the movement of a physical "gate" that controls the entry and exit of ions through the channel. A perfect match, however, is not obtained because not all of the charge movements within an ion channel protein lead directly to the opening or closing of the channel.

The S4 region. The identity of the voltage sensor is not yet known with certainty. A leading contender is found in the S4 segments of the sodium channel. Figure 5-4a shows that this stretch of amino acids contains repeated basic residues, either arginines or lysines, in every third position. A similar S4 region exists in other known ion channels (Fig. 5-4a). If the S4 region were to form an α-helix within the membrane, these positively charged residues would come to be arranged in a spiral form around the helix, as is shown in Figure 5-4b. Such an arrangement of charged residues in the membrane is indeed a prime candidate for a voltage sensor. The positively charged residues are likely to be stabilized by negatively charged amino acids, such as aspartate or glutamate, situated on adjacent helices. It has been proposed that a change in the electric field across the α-helix, such as would occur when the membrane is depolarized, leads to an uncoupling of the positive residues from their partners, followed by a displacement or rotation of the helix. This would result in the movement of charge in the direction of the imposed electric field and the establishment of a new equilibrium conformation.

It is possible to produce mutated forms of the α-subunit of the sodium channel, in which the basic amino acids in the S4 region are replaced by neutral or negative charges. When messenger RNA coding for such mutant

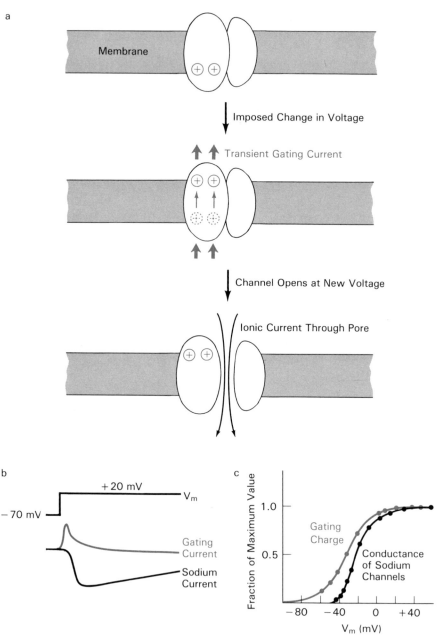

FIGURE 5-3. Gating currents. *a*: Generation of gating currents by charges in an ion channel protein after a change in voltage. *b*: After a step in voltage, the gating current precedes the ionic current. *c*: A comparison of the voltage dependence of gating charge movement with that of the conductance of sodium channels in an axon (this topic is covered further in a review by Armstrong, 1981).

a

Sodium I

Val-Ser-Ala-Leu-Arg -Thr-Phe-Arg -Val -Leu -Arg-Ala-Leu-Lys -Thr-Ile - Ser-Val-Ile -

Sodium II

Leu-Ser-Val -Leu-Arg-Ser-Phe -Arg-Leu-Leu-Arg -Val -Phe-Lys -Leu-Ala-Lys -Ser-Trp-

Calcium I

Val-Lys -Ala-Leu-Arg-Ala-Phe-Arg-Val-Leu-Arg-Pro-Leu-Arg-Leu-Val-Ser- Gly-Val -

Calcium II

Ile - Ser-Val-Leu-Arg -Cys-Ile -Arg-Leu-Leu-Arg-Leu-Phe-Lys - Ile - Thr-Lys -Tyr-Trp-

Shaker

Leu-Arg-Val- Ile -Arg-Leu-Val- Arg-Val -Phe-Arg - Ile- Phe-Lys-Leu-Ser-Arg-His -Ser-

Rat Rck1

Leu-Arg -Val- Ile - Arg-Leu-Val-Arg -Val -Phe -Arg -Ile - Phe-Lys-Leu-Ser-Arg-His -Ser-

b

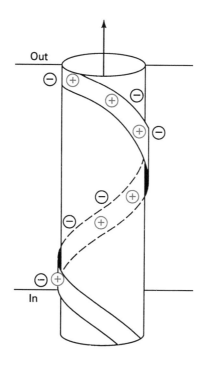

FIGURE 5-4. The S4 region. *a*: The amino acid sequence of S4 regions in domains I and II of sodium and calcium channels and in two potassium channels, *Shaker* A and *RCK1*. *b*: Possible arrangement of basic residues in S4 regions in a helix that spans the membrane (modified from Catterall, 1986).

channels is expressed in oocytes, the dependence of the sodium current on voltage is shifted in ways that are generally consistent with the idea that the S4 segment is a sensor of the membrane voltage.

The inactivation gate. If the S4 segment is important for sodium channel activation, does it also contribute to inactivation? It is known that kinetics and voltage dependence are not the only ways to distinguish between sodium current activation, deactivation, and inactivation. In a classic experiment in the early 1970s, the inside of the squid giant axon was perfused with pronase, a heterogeneous mixture of proteolytic enzymes. This treatment removes inactivation but does not affect activation or deactivation of the sodium current. This experiment showed unequivocally that the ion channel component responsible for inactivation—the so-called *inactivation gate*—is a protein domain that must be accessible from the cytoplasmic face of the membrane. It also demonstrated that the inactivation gate is distinct from the activation/deactivation gate, and suggested that the latter is located on the extracellular side of the membrane (or at the very least is not accessible to pronase from the cytoplasmic side). Recall that deactivation is simply the reverse of activation, whereas inactivation is a separate process. These findings have been instrumental in shaping our ideas about how sodium channels work, as illustrated in Figure 4-8.

Studies with mutated channels and with antibodies have confirmed that sodium channel inactivation is mediated by structural components distinct from the S4 region. For example, we mentioned earlier that sodium channels, expressed in oocytes from pure α-subunit messenger RNA, inactivate at an abnormally slow rate. Interestingly, this rate of inactivation can be increased if a subfraction of messenger RNA coding for much smaller proteins in rat brain is coinjected with the large α-subunit RNA (Fig. 5-5a). This finding suggests that some small protein, possibly the β1- or β2-subunit, may also be required to "fine tune" the normal kinetics of the channel.

Other evidence indicates that the intracellular loop between the third and fourth membrane-spanning domains plays an important role in the inactivation of the channel. Mutants in which this loop has been cut suffer a severe loss of the inactivation process (Fig. 5-5b). Similar loss of inactivation results when antibodies binding to this region are applied to the internal face of the channel protein. This suggests that pronase may eliminate inactivation in the squid axon by cutting the α-subunit at some site or sites within this loop. We emphasize again, however, that such studies are in their infancy, and many more direct structural determinations will be required before discrete elements within ion channels can be assigned specific functions.

Biological insights from sodium channel cloning. In addition to questions of the relation of structure to function, the molecular cloning of sodium channels has raised a series of more specifically biological questions. It has been found that there are several species of sodium channel, each with similar, but not identical, protein sequences. At least three, and perhaps many more, sodium channels, encoded by different genes, may be expressed in the nervous system. For example, the sodium currents shown in Figure 5-5 result from the expression of the so-called type II channel from rat brain. Thus even sodium channels, which display far less heterogeneity in their electrical properties than do calcium or potassium channels, constitute a family of diverse proteins. It remains to be determined whether different members of the family are expressed in different neurons or in different locations within a neuron, and, if so, what their relative contributions are to different aspects of neuronal firing.

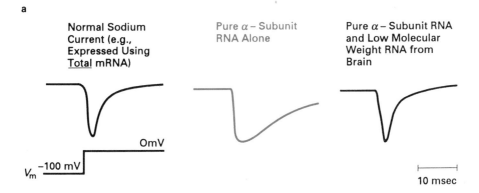

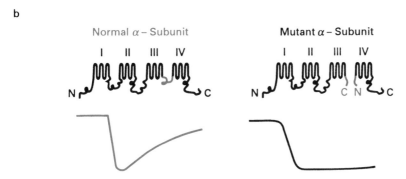

FIGURE 5-5. Inactivation of sodium currents. a: Expression in *Xenopus* oocytes of RNA for α-subunit of sodium channel. Inactivation of current from α-subunit is slower than that from total brain RNA. Coinjection of α-subunit RNA with a subfraction of small RNAs from brain (which alone induce no channels) restores normal inactivation (Auld et al., 1988). b: Loss of inactivation after cutting the cytoplasmic loop between domains III and IV (Stuhmer et al., 1989).

Structure of Calcium Channels

Chapter 4 described the existence of a number of classes of voltage-dependent calcium channels. As is the case for sodium channels, drugs and toxins that selectively bind to some of these channels can be found. Prominent among these are the *dihydropyridines,* which bind to the L-type calcium channel whose properties were described in Table 4–2. This has allowed the purification, reconstitution, and molecular cloning of a calcium channel, the so-called dihydropyridine receptor, from skeletal muscle. The protein that binds dihydropyridines is shaped much like the α-subunit of the sodium channel, and has been termed the α1-subunit. It contains four related regions, each consisting of six hydrophobic (and thus presumed membrane-spanning) segments. The S4 regions are homologous to those in the sodium channel (Fig. 5-4). Within the muscle membrane, this protein is associated with several additional proteins of unknown function. One of these, the α2-subunit (approximately 140 kDa), has three presumed transmembrane segments, but bears no structural resemblance to the α1-subunit.

As with the α-subunit of sodium channels it has been possible to obtain functional calcium channel expression following injection of the calcium channel α1-subunit messenger RNA into oocytes. Another particularly elegant demonstration that the dihydropyridine receptor protein does comprise a component of the calcium channel has come from experiments on muscle cells of mice suffering from *muscular dysgenesis,* a genetic disorder in which the gene encoding the dihydropyridine receptor protein is defective. Such cells lack an L-type calcium current when depolarized (Fig. 5-6). All that is required to restore this current to normal is to inject the cells with exogenous DNA encoding a normal receptor protein (Fig. 5-6).

The dihydropyridine receptor calcium channel in skeletal muscle differs from the calcium channels of neurons. In addition to allowing calcium to enter the cell, the muscle channel also plays an important role in coupling depolarization of the membrane to calcium release from internal membrane stores. The pursuit of the identities of calcium channels in neurons, and the factors that regulate these channels, is an active field to which we shall return in Chapter 11 when we discuss modulation of the properties of neuronal ion channels.

Mutant Flies and the Quest for Potassium Channels

Unlike blockers of sodium and calcium channels, the agents that bind to potassium channels have not, as yet, proven useful in their purification. The beginning steps in identifying the molecular basis for the large array of potassium channels have come from a very different direction, the study of mutant flies.

The fruit fly *Drosophila melanogaster* has been an invaluable resource for genetic studies for many years. Later in this book, we shall see examples of how it is currently providing insights into neuronal development and mechanisms of learning. To examine one of these creatures closely, it must be anesthetized with ether (otherwise it just flies away). Occasionally, instead of going to sleep as a good fly should, a fly will shake its legs, wings, and abdomen when exposed to the anesthetic (Fig. 5-7). This turns out to

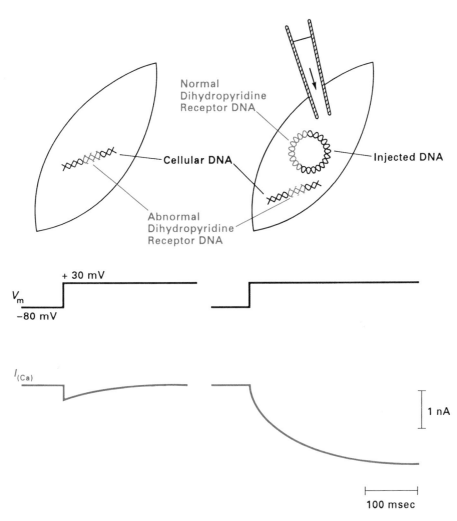

FIGURE 5-6. Injection of dihydropyridine receptor DNA restores calcium current in muscle cells of mice with muscular dysgenesis. Cells with this genetic mutation lack a noninactivating L-type calcium current and possess only a transient inactivating calcium current (Tanabe et al., 1988).

be due to mutations in specific genes, and a number of strains of mutant flies that exhibit this behavior have now been bred. Among these strains of mutant flies are several that have defects in a particular gene, a piece of DNA that has been termed the *Shaker* locus.

The **Shaker** *locus codes for a potassium channel.* The defect in some of the *Shaker* flies was found to result from the total loss of the potassium A-current (I_A) in a number of nerve fibers and muscle cells (Fig. 5-7). In these cells, the A-current contributes substantially to the repolarization of action potentials. Thus, abnormally long action potentials are recorded in giant axon fibers of *Shaker* flies (Fig. 5-7). It should be noted that the adjective "giant" is only a relative term; an entire fly could fit with only a little squeezing into a squid giant axon. When the prolonged action potentials in the giant fibers arrive at the neuromuscular junction, they cause unusu-

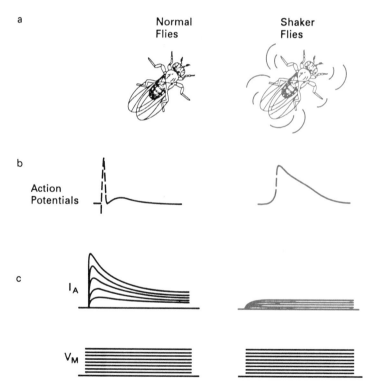

FIGURE 5-7. The *Shaker* mutation. Action potentials (*b*) and potassium currents (c) in normal and *Shaker* flies (a) (from the work of Mark Tanouye, Larry Salkoff, Bob Wyman, and their colleagues).

ally prolonged transmitter release, prolonged muscular contraction, and shaking. Not all mutations within the *Shaker* locus cause the entire loss of A-current. Some mutations alter only the amplitude or the kinetics of the current. Such findings suggested that the *Shaker* locus may contain a gene encoding the A-current channel.

An advantage of *Drosophila* is that the position at which a mutation (such as *Shaker*) occurs can be mapped precisely on one of the chromosomes, and the DNA at this position in normal and mutant flies can then be cloned and sequenced. Identifying the sequence of the protein encoded by the *Shaker* locus proved, however, to be a difficult task, which took several years of characterization of the DNA around the locus. When the long-sought sequence finally became known, it was found to possess some rather interesting features.

The first arresting feature of the proteins encoded by the *Shaker* locus is that they are only about one-quarter the size of the sodium or calcium channel proteins we described above. Only one domain consisting of six putative transmembrane regions, with one S4 region closely resembling that in the other voltage-dependent channels (Fig. 5-4), is found in the protein. Evidence that these proteins are indeed functional potassium channels comes from the finding that normal A-currents can be expressed in oocytes injected with RNA coding for *Shaker* proteins.

Alternative splicing gives rise to different **Shaker** *products.* The second remarkable feature is that more than one type of channel can be made from the RNA produced from the *Shaker* locus (Fig. 5-8). The DNA coding for the channel does not run continuously through the locus. Rather, regions coding for the protein (termed *exons*) are separated by stretches of noncoding DNA (termed *introns*). Because RNA is synthesized from this DNA, it initially includes these noncoding introns. During the production of the mature messenger RNA that encodes the protein, the introns are removed and the exons are *spliced* together. In the *Shaker* locus there exist several different regions that can code for *alternative* carboxyl-terminal and amino-terminal regions of the channel protein (Fig. 5-8). Different patterns of cutting and splicing of RNA can therefore produce channels that possess the same sequence in the central region, which contains most of the hydrophobic segments and the S4 segment, but which have different sequences at their carboxyl or amino termini. Figure 5-8 demonstrates the production of two *Shaker* proteins, termed ShA and ShB. These two proteins have identical sequences at their amino termini but differ after the fifth hydrophobic segment.

The different proteins that can be constructed from the *Shaker* locus do in fact give rise to channels with different electrical properties. Figure 5-8 shows currents, recorded in oocytes, after injection of messenger RNA cod-

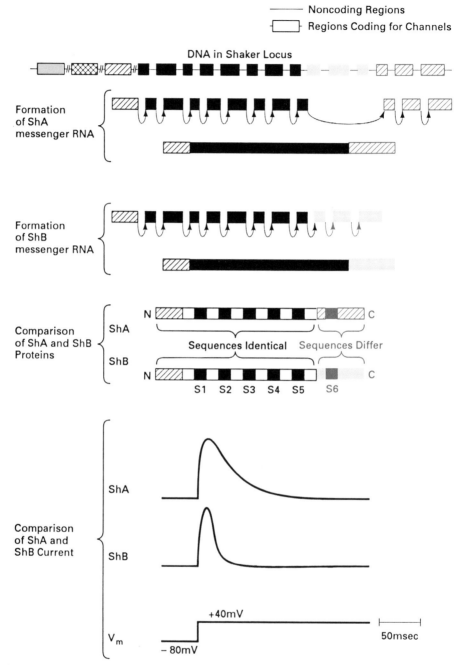

FIGURE 5-8. Alternative splicing in the *Shaker* locus. Lily and Yuh Nung Jan and their co-workers cloned the *Shaker* locus and showed that RNA transcribed from this locus can be processed to encode one of several different *Shaker* proteins, including *Shaker* A *(ShA)* and *Shaker* B *(ShB)* (Schwarz et al., 1988).

ing for ShA and ShB. In response to depolarization of the membrane, both the ShA and ShB channels activate and then inactivate as the depolarization is maintained. The rate at which ShB inactivates, however, is more rapid than for ShA.

In addition to ShA and ShB, several other *Shaker* channels, with different carboxyl- or amino-terminal regions, exist. Although all of these give rise to transient potassium currents of the A-current type, the rates of activation, inactivation, and recovery from inactivation vary substantially. In addition to *Shaker,* several other genes, encoding proteins similar to the *Shaker* protein, have now also been found in fruit flies, as well as in species such as rat and mouse. Some of these closely related proteins give rise to delayed rectifier potassium currents rather than the inactivating A-currents. In fact it appears that there is a graded spectrum of potassium channel types, all of which belong to this close-knit family. Not all of the *Shaker*-like genes give rise to multiple protein products in the way the *Shaker* locus does. However, the rather astonishing number of putative potassium channel genes that are being discovered provides evidence that the diversity of potassium channels is likely to be far greater than even that suggested by electrophysiological studies.

A number of important questions still remain regarding the structure of *Shaker*-like potassium channels. Do other subunits, as yet uncharacterized, influence the properties of the channels in intact cells? Although it is possible that a single *Shaker* polypeptide chain represents one channel, it seems more likely that several—perhaps four—such chains come together in the membrane to form a structure that resembles the sodium channel α-subunit. Do the proteins exist as *homo-oligomers,* an aggregate of four or more identical proteins, or do *hetero-oligomers* exist, in which different proteins such as the ShA and ShB proteins interact to form a mixed complex? Do the other classes of potassium channels listed in Chapter 4 resemble the *Shaker* proteins? The answers to these questions are actively being sought, and there is no dearth of possibilities for how the mix of potassium channels in a cell comes to be tailored so uniquely to the biological role of that cell.

A nonconformist voltage-dependent ion channel. We may have given the impression that all voltage-dependent ion channels possess one or more homologous sequences containing six membrane-spanning domains. However, there is evidence that the voltage-dependent channel responsible for the very slowly activating potassium currents shown in the last trace of Figure 5-2 has a radically different structure. It is a small protein of 130 amino acids, with only one transmembrane segment, and with an amino acid sequence and predicted structure unrelated to those of the voltage-dependent sodium, calcium, and potassium channels described in this chapter. Thus generalizations about structural features common to *all* voltage-dependent ion channels will have to await the elucidation of structures of many more members of this class of channel.

Summary

The hydrophobic nature and low density of ion channel proteins have made it difficult to investigate their properties biochemically. However, advances in molecular cloning techniques have allowed the discovery of the amino acid sequences of several voltage-dependent ion channels. Sodium channels, calcium channels, and several kinds of potassium channels share certain structural features. Among these are the presence of several membrane-spanning regions, and a region of charged amino acids that may constitute the voltage sensor. Experiments using site-directed mutagenesis have begun to provide insights into the structural basis for certain channel functions. But the discovery of one kind of voltage-dependent potassium channel with a completely different structure highlights the limitations to our knowledge of the relation between structure and function in ion channel proteins.

Intercellular communication

The first section of this book described the membrane specializations that permit the transfer of information from one part of a neuron to another. The next seven chapters address another fundamental aspect of nervous system function, *intercellular signaling*. This includes the mechanisms that neurons use to communicate with one another and with the outside world. Chapter 6 compares two fundamentally different modes of interneuronal communication. The first is the direct transfer of ions and small molecules from one neuron to another via *electrical synapses (gap junctions)*. The second involves the release or secretion from one cell of some chemical, a *neurohormone* or *neurotransmitter,* that diffuses to and affects the activity of a target cell. Often the membranes of the secreting and target cells are immediately adjacent to one another at the highly specialized structure known as the *chemical synapse.* As described in Chapter 7, mechanisms of release of neurotransmitters at chemical synapses are best understood from the study of two highly specialized synapses, the *nerve-muscle synapse* in vertebrates and the so-called *giant synapse* in the stellate ganglion of the squid. The various classes of neurotransmitters and neurohormones that have been found in nervous systems, and details of their synthesis and metabolism, are presented in Chapter 8. The following two chapters describe the neurotransmitter and neurohormone *receptors.* These are specialized membrane proteins that recognize and bind signaling molecules and *transduce* the extracellular chemical signal into an electrical response in the target cell. Two distinct classes of transduction mechanism are considered. Chapter 9 discusses receptors in which the binding site is part of the same molecule or macromolecular complex as the ion channel whose activity is regulated by the neurotransmitter—the *directly coupled* receptor/ion channel systems. These are contrasted in Chapter 10 with the *indirectly coupled* systems, in which the occupation of the receptor by neurotransmitter sets in motion a chain of biochemical events. These events lead ultimately to a change in the activity of an ion channel that is *not* intimately associated with the receptor. Chapter 11 addresses the concept of *neuromodulation*, the long-term alteration of neuronal electrical properties as a result of neurotransmitter or neurohormone action, and the intracellular biochemical mechanisms that are responsible for these alterations. Finally, Chapter 12 describes neurons that act as *sensory receptors* by converting information from the outside world into electrical signals that can be passed on to other neurons in the brain. Sensory receptor neurons use the mechanisms discussed in Chapters 9 through 11 to convert sensory information into a change in the properties of membrane ion channels.

How neurons communicate: gap junctions and neurosecretion

Within any organism, cells must be able to communicate. This is of course important in tissues other than the brain, but is *essential* for proper nervous system function. There are three general ways in which cells talk to each other:

1. Direct transfer of molecules and ions from the cytoplasm of one cell into that of another. As we mentioned in Chapter 1, this is mediated by *gap junctions.*
2. The release of a chemical that diffuses to, and acts on, another cell. This release process is termed *secretion.*
3. Direct physical contact. A cell can be influenced profoundly by events triggered when molecules in its plasma membrane interact with the membranes of adjacent cells.

This chapter will deal with the first two of these modes of communication, modes that neurons use on a day-to-day basis to generate specific behaviors. The third pattern of communication plays a very important role in the development of neurons and their connections, and we shall discuss it in more detail in Chapters 13 through 15.

Gap Junctions, Connexins, and Electrical Synapses

Intracellular communication through gap junctions is conceptually the very simplest form of cell-to-cell interaction. Small molecules and ions in one cell diffuse through pores in the plasma membrane directly into the cytoplasm of a neighboring cell (Fig. 6-1). These pores can be visualized in the electron microscope and are found in clusters that sometimes have a crystalline appearance. Such an array was captured in the electron micrograph of Figure 6-2 by the technique of *freeze fracture*. In the preparation of tissues for freeze fracture, plasma membrane that has been frozen is allowed to break within the plane of the membrane itself. The exposed halves of the lipid bilayers are then coated with platinum and carbon to produce a form of "bas-relief" view of the inner plane of the membrane. These methods permit ready visualization of intramembranous particles, which represent integral membrane proteins such as receptors and ion channels. Figure 6-2 shows flattened sheets of membranes with arrays of *gap junction particles*. These particles contain the proteins that form the cell-to-cell pore.

Gap junctions are abundant in tissues such as the liver and the lens, which have been used as sources for the purification of gap junction pro-

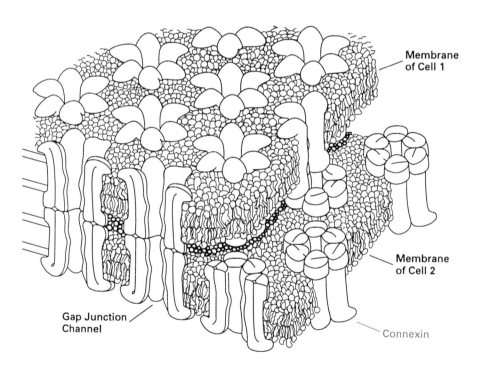

Membrane of Cell 1

Membrane of Cell 2

Gap Junction Channel

Connexin

FIGURE 6-1. Gap junctions. Pores spanning two cell membranes are made of connexin proteins. This picture evolved from the X-ray diffraction work of Makowski et al. (1977).

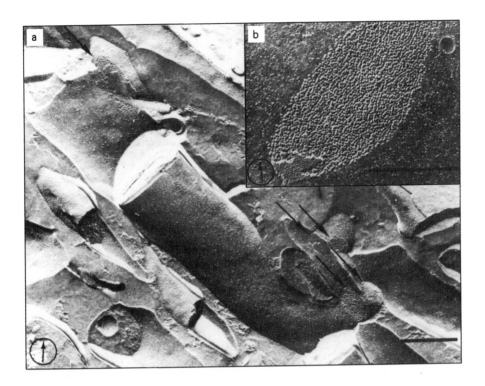

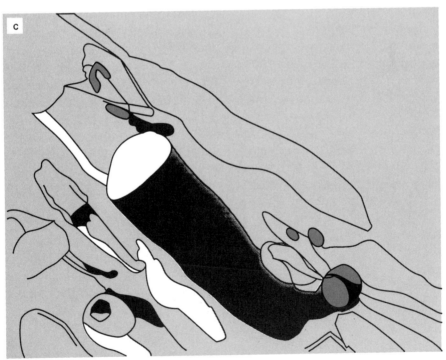

FIGURE 6-2. Freeze fracture replica of gap junction particles. *a*: Electron micrograph showing a bundle of processes of *Aplysia* neurons connected by gap junctions. *b*: Magnification of a single array of gap junction particles. Some of the arrays, marked with arrows in *a*, are shown in the drawing (c) (from Kaczmarek et al., 1979).

teins. These proteins have been termed *connexins*. A pore is believed to be formed by a complex of six connexin proteins in the membrane. To link the pore to the cytoplasm of an adjacent cell the connexins bind to another hexameric complex of connexins in that cell (Fig. 6-1). In liver, the major connexin protein has a molecular weight of 27,000. In other tissues, connexin proteins with different molecular weights may be found. This diversity may reflect the fact that the properties of gap junctions, and the way these properties are altered by neurotransmitters or hormones (see Chapter 11), vary in different cells.

Perhaps surprisingly, the electrical properties of gap junctions appear similar in many respects to those of the ion channels we have encountered in previous chapters. Despite the fact that gap junctions allow the passage of relatively large molecules, with molecular weights up to about 1000, recordings of the opening and closing of gap junctions resemble the gating of channels that allow specific ions to cross the plasma membrane. This can be measured by whole cell patch clamp of a pair of cells that are coupled by only a small number of pores (Fig. 6-3). When the membrane potential of one cell is maintained more negative than that of its partner, current flows across the pore only while the cell-to-cell channel is open. The conductance of the open channel, for example in pairs of heart cells, is about 50 pS.

Despite the ubiquity of gap junctions, their biological role in most tissues is not well understood. Gap junctions are likely to play a role in the embryonic development of many tissues, a time when many gap junctions are formed and then broken again. One of their functions may be to allow the transfer between cells of small molecules that are important for development, and of second messenger molecules involved in intracellular signaling (see Chapter 10). This, in turn, allows a group of cells to act as one functional unit. In the nervous system and other excitable tissues, gap junctions take on a special significance. Connections between nerve cells via gap junctions are often called *electrical synapses*, in recognition of the fact that they are involved in rapid electrical signaling and information transfer. For example, they allow groups of neurons to synchronize their electrical activity. However, electrical synapses are also found between pairs of neurons that do not always fire in synchrony. In such cases, their role may be to allow synaptic inputs into one neuron to be registered in a neighboring neuron. Later in the book (see Chapter 16), we shall encounter specific examples of neurons that couple electrically through gap junctions.

Neurosecretion

Although gap junctions constitute a very basic and rapid form of communication, they are limited to bidirectional interactions between neighboring

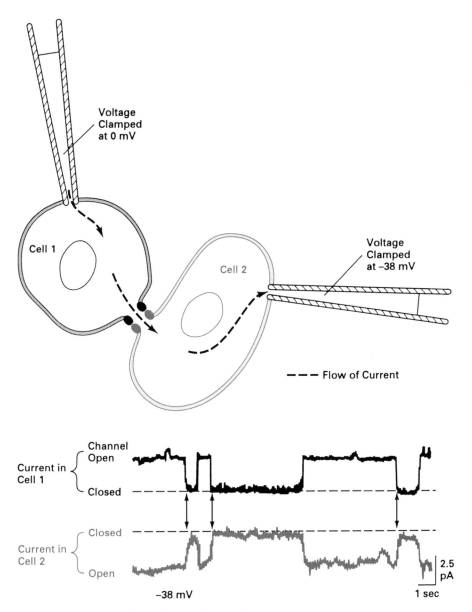

FIGURE 6-3. Single channel recordings of gap junctions. In an experiment carried out by Burt and Spray (1988), two cells are voltage clamped at different potentials. When the gap junction channel opens, the current that flows between the cells is recorded as outward current in cell 1 and inward current in cell 2.

cells. When information has to be transferred from one cell to a distant cell, the usual mechanism is for the first cell to release a chemical into the extracellular space. This chemical then either diffuses to, or is transported to, the target cell. A large number of different chemicals may act as extracellular signals. These include amino acids and other small organic compounds, lipids, small peptides, and large protein complexes. Chemicals that are released from a neuron, and alter the excitability of another neuron or a muscle cell, are termed *neurotransmitters*.

Figure 6-4 shows some of the mechanisms that allow molecules to leave a cell. Lipid molecules diffuse readily across a cell membrane. Once synthesized, they therefore can leave the cell readily without specialized machinery (Fig. 6-4a). For certain other, more hydrophilic, molecules, carriers or pores exist in the plasma membrane to allow them to cross the membrane (Fig. 6-4b). The major pathway for the release of neurotransmitters, however, appears to be *vesicular secretion*. At the synaptic terminals of neurons, many *secretory vesicles* can normally be found (see Chapter 1). When a neuron is stimulated, these transmitter-containing vesicles fuse with the plasma membrane to release their contents into the synaptic cleft (Fig. 6-4c). This pathway probably evolved from the requirement of many cells to release large peptides and proteins that would not normally be able to cross the lipid membrane, and also from the need to insert proteins such as ion channels and other integral membrane proteins into the plasma membrane. Thus it is not surprising that many insights into the release of neurotransmitters have come from work with nonneuronal cells. In the remainder of this chapter we shall discuss the general mechanisms for secretion of proteins and other transmitters in neurons and exocrine cells. In the next chapter we shall consider in more detail the physiology of transmitter release at two well-studied chemical synapses.

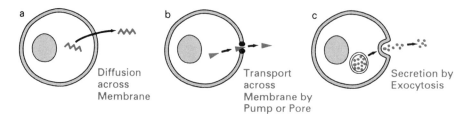

FIGURE 6-4. Three pathways (a–c) by which cells release substances into the external medium.

Constitutive and Regulated Secretion of Protein

In all eukaryotic cells, proteins that are destined for release into the extracellular space are synthesized in the *rough endoplasmic reticulum* (Fig. 6-5).
Newly synthesized proteins enter the *Golgi apparatus*, a tightly packed
stack of intracellular membranes in which the proteins undergo a variety
of posttranslational modifications. The proteins are then transferred to vesicles that bud off the Golgi apparatus and move to the cell surface. Here
the vesicles release their contents into the extracellular space by fusion with
the plasma membrane, a process known as *exocytosis*. In some cases, the
arrival of the vesicles at the cell surface is followed immediately by exocy-

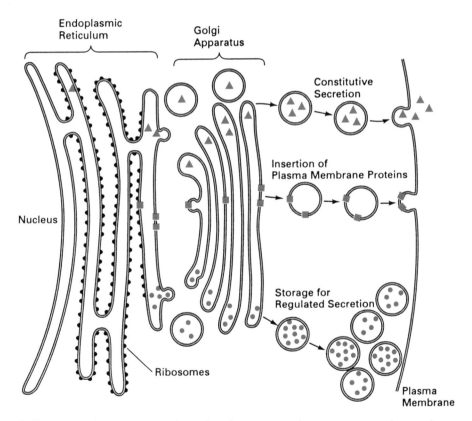

FIGURE 6-5. Constitutive and regulated secretion. The upper two pathways show
the movement of proteins from ribosomes to the plasma membrane for constitutive release into the external medium or insertion into the plasma membrane.
The lower pathway shows the buildup of protein-containing vesicles for regulated
secretion.

tosis. This form of secretion has been termed *constitutive secretion*. The insertion of integral membrane proteins, including ion channels and receptor molecules, into appropriate regions of the plasma membrane also occurs by constitutive secretion (Fig. 6-5). Recall from Chapter 1 that the organelles involved in these processes are not restricted to neurons, and that the neuronal cell body has many features in common with other cells.

Neurotransmitters and hormones, in contrast, are not released immediately following their synthesis. They are packaged into vesicles, and the vesicles are transported along microtubules to the release sites, such as those at axon terminals. They are then stored at these sites until an appropriate stimulus is received by the cell. In neurons, this stimulus is usually a depolarization that causes calcium to enter the cell through voltage-dependent calcium channels. A large amount of neurotransmitter can then be released by the cell when the vesicles fuse with the plasma membrane. Such secretion, which is regulated acutely by external stimulation, has been termed *regulated secretion* (Fig. 6-5).

Even within these general classes of constitutive and regulated secretion there must be a large number of variants. Different proteins are inserted at different locations in the plasma membrane of a cell by constitutive pathways. Neurons contain at least two kinds of secretory vesicles that participate in regulated secretion. So-called *large dense core vesicles* contain primarily peptide neurotransmitters. *Small synaptic vesicles* contain nonpeptide transmitters and the enzymes required for their synthesis. This diversity of intracellular traffic implies that newly synthesized proteins must somehow be assigned to the correct vesicle pathway. At least part of the information for the correct assignment may be found in the sequence of the protein itself. We shall now list some of the signals that may exist in a newly synthesized protein, that determine whether part or all of the protein will be secreted, and by which secretory pathway. The focus in this section will be on peptide neurotransmitters.

Signals on Secreted Proteins

The signal sequence. The first part of a protein to be synthesized is the amino terminus. Proteins that are to enter one of the secretory pathways contain a stretch of hydrophobic amino acids at the amino terminus known as the *signal sequence* (Fig. 6-6). Although the exact signal sequence differs in different proteins, this hydrophobic stretch of amino acids is recognized by cytoplasmic factors (the *signal recognition particle*) and by components of the membrane of the rough endoplasmic reticulum. These factors then assist the protein to cross this membrane. Without crossing into the lumen of the endoplasmic reticulum the protein could not eventually be packaged into membrane vesicles. Proteins that lack a signal sequence do not cross the membrane, and thus become cytoplasmic proteins. The signal sequence is usually cleaved from the remainder of the protein by proteolytic enzymes shortly after it has crossed the membrane.

a   Vasopressin Precursor

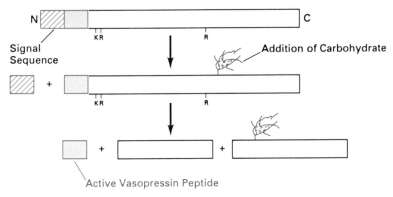

b   ELH Precursor

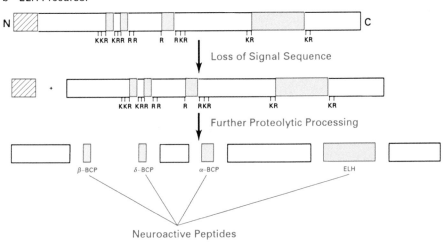

K = Lysine ⎫
R = Arginine ⎬ at cleavage sites

FIGURE 6-6. Processing of neuropeptide precursors. *a*: Formation of the hypothalamic peptide vasopressin. *b*: Formation of the *Aplysia* peptides α-, β-, and γ-BCP and ELH. K, lysine; R, arginine.

Sorting signals. Sorting of newly synthesized proteins must take place, so that different proteins are assigned to the appropriate vesicles and thus to the appropriate secretory pathway. In large part, this sorting appears to occur within the Golgi apparatus. The nature of the *sorting signal* is not understood. It appears, however, that the signal is intrinsic to the protein sequence itself. For example, DNA that encodes the protein precursor for insulin (proinsulin) can be introduced into a pituitary cell line that does not normally make insulin. When this is done, not only is the DNA faithfully transcribed and translated into proinsulin, but this protein is also packaged into vesicles of the regulated secretion pathway, and insulin can be secreted from the cell in response to an appropriate stimulus. The sorting of peptide transmitters and hormones into vesicles frequently is associated with a condensation of the protein into a dense aggregate. Such aggregates presumably serve to increase the concentration of peptide in a vesicle. It is because they appear electron opaque when viewed in an electron microscope that peptide-containing vesicles are frequently described as *dense core vesicles* or granules (Fig. 6-7).

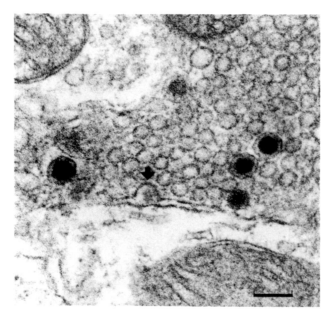

FIGURE 6-7. Dense core granules and clear vesicles. An electron micrograph of part of the terminal of a rat neuron. Several larger dense core granules coexist with smaller, clear synaptic vesicles. The arrow marks a vesicle that is undergoing exocytosis. The scale bar represents 100 μm (courtesy of Dr. Asa Thureson Klein).

Proteolytic processing. Most neuroactive peptides are not synthesized in the form in which they are eventually secreted. Instead, they are synthesized as part of a larger, inactive precursor protein or *prohormone* (Fig. 6-6). A good example of this is the precursor of insulin, proinsulin, which we introduced above. Two others are the precursors for *vasopressin,* which is synthesized in certain neurons in the hypothalamus, and for *egg laying hormone* (ELH), which is synthesized in *Aplysia* bag cell neurons. Proteolytic cleavage of the precursors to smaller fragments, including the active peptides, occurs in the secretory vesicles and also in the Golgi. In some cases, several different neuroactive peptides may be generated from a single precursor, for example, the production of the transmitters, α, β, and γ bag cell peptides (BCPs), together with ELH, from the ELH precursor (Fig. 6-6b). A sequence of two basic amino acids, either lysine or arginine, is frequently the site for proteolytic cleavage within the precursor (Fig. 6-6). Cleavage sites also exist, however, that are not marked by two basic amino acids.

Posttranslational modifications. A newly synthesized protein, and the peptide fragments generated from such a protein, can be modified further by the action of enzymes in the Golgi or the secretory vesicles themselves. Table 6-1 lists the more common covalent modifications, some of which may be essential for the biological activity of a peptide. For example, one common modification is the amidation of the carboxyl terminal. This is carried out by the enzyme *peptidyl glycine (alpha)-amidating mono-oxygenase,* which is located within secretory granules. The enzyme acts on peptides that have a glycine residue at their carboxyl terminal by removing the glycine and amidating the penultimate amino acid residue. Figure 6-8 illustrates the amidation of the invertebrate neuropeptide FMRFamide and of vasopressin. It has been found that the biological activity of neuropeptides such as vasopressin is very greatly reduced if the peptide is not amidated.

TABLE 6-1 Some Covalent Modifications of Peptide Transmitters and Hormones

Modification	Possible Function	Example
Conversion of N-terminal glutamate to pyroglutamate	Increase stability to proteases	Neurotensin
Amidation of C-terminal amino acid	Increase stability to proteases	Substance P
Glycosylation (usually addition of sugar residues to asparagine)	Targeting to appropriate location?	Thyrotropin—also commonly found in integral membrane proteins
Sulfation	Unknown	Cholecystokinin

This and other modifications may also influence the stability of the peptide once it has been released. Some posttranslational modifications of newly synthesized proteins may participate in the sorting of the protein into the appropriate vesicles. A precedent for this is found in proteins of the lysosomes. Newly synthesized lysosomal proteins contain asparagine residues to which mannose phosphate groups are added. These mannose phosphate

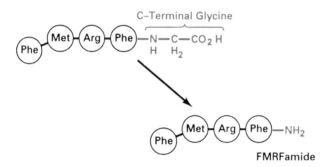

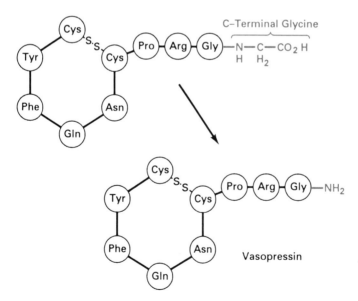

FIGURE 6-8. C-terminal amidation. Formation of the neuropeptides FMRFamide and vasopressin by removal of a C-terminal glycine.

groups are then recognized by receptors in the Golgi, causing these sugar-linked proteins to be targeted selectively to lysosomes.

Secretion of Nonpeptide Neurotransmitters

In contrast to peptide transmitters, small neurotransmitter molecules including acetylcholine, the amino acid transmitters such as GABA, glycine, and glutamate, and the amines such as dopamine, norepinephrine, histamine, and serotonin, are not synthesized only at the soma (see Chapter 8). Instead, it is the enzymes that catalyze their synthesis that are synthesized in the soma, then transported to the release sites at axon terminals. Depletion of neurotransmitter by electrical activity can therefore be followed by rapid resynthesis within the terminal. Vesicles that contain these neurotransmitters are small and very homogeneous in size (typically 50 nm in diameter). It is these small synaptic vesicles that are associated with the very rapid release of transmitter that occurs at specialized synaptic junctions (see Chapter 7). Because they do not generally contain a dense aggregate of neuropeptides, they are not electron dense and therefore appear as clear vesicles in electron micrographs (Fig. 6-7). We shall deal with the enzymes for the synthesis of nonpeptide neurotransmitters, and their location within the terminals, in Chapter 8. Below we discuss mechanisms of transmitter release, from vesicles as well as from nonvesicular stores.

Exocytosis of Neurotransmitter-Containing Vesicles

There is abundant evidence that regulated secretion in many nonneural cells occurs through *exocytosis*. For example, in a chromaffin cell of the adrenal medulla, stimulation by a transmitter causes the secretory granules to fuse with the plasma membrane, releasing their contents to the extracellular space. The vesicle membrane that has been added to the plasma membrane is then rapidly resorbed by *endocytosis*. In many such cells, the process of exocytosis can be observed by light microscopy of living cells. Release of neurotransmitter from small synaptic vesicles at synaptic junctions also occurs through exocytosis of transmitter-containing vesicles, but visualization of this process has been harder to achieve, largely because of the technical difficulties of working with small synaptic terminals. A discussion of the evidence for exocytosis at synaptic junctions is deferred to the next chapter.

Detection of exocytosis by capacitance measurements. There are still many unanswered questions about the dynamics of exocytosis, even in cells in which the process can be visualized readily in the microscope. For example, must the membrane of a vesicle undergo complete fusion into the plasma membrane to release transmitter? Alternatively, can a vesicle fuse with the plasma membrane transiently, release its neurotransmitter through a pore that spans the two membranes, then reseal without losing its integrity? The latter form of hypothetical release mechanism has been termed a "kiss and run mechanism" (Fig. 6-9a). A variant of the patch clamp technique, devised by Erwin Neher and his colleagues, is providing answers to these sorts of questions. The technique is based on the fact that a cell's capacitance, which can be measured using the "whole cell" configuration of the patch clamp, is directly proportional to the surface area of the cell membrane. To provide a continuous measure of cell capacitance, a high frequency ($\sim$800 Hz) sinusoidal voltage command is applied to the cell. The current that flows across the membrane is measured with an amplifier known as a *lock-in amplifier*. This determines the magnitude of the currents that are in phase with the voltage oscillation and those that are 90° out of phase. The latter can be related directly to the capacitance of the cell. When a vesicle fuses with the plasma membrane, there is a small increase in the size of the surface membrane (Fig. 6-9b). This is registered as an increase in cell membrane capacitance, allowing the whole cell patch clamp apparatus to detect directly the exocytosis of single secretory vesicles.

Figure 6-9c shows measurements of capacitance in a mast cell from the peritoneum of a mouse. When stimulated, mast cells release massive amounts of histamine from their secretory granules and their activity is responsible for many of the symptoms of an allergic attack. In whole cell recordings, introduction of a nonhydrolyzable analog of GTP into the patch pipette is sufficient to trigger this release, suggesting that a GTP-binding protein (see Chapter 10) may contribute to the onset of secretion. A stepwise increase in capacitance is recorded as each vesicle fuses with the plasma membrane (Fig. 6-9c, left). Stepwise *decreases* in membrane capacitance can also be observed (Figure 6-9c, right); these represent membrane retrieval by endocytosis.

For this technique to work the secretory vesicles must be large, as is the case in chromaffin cells and mast cells, so the amount of added membrane will be detectable as an increase in capacitance. Although it has not yet been possible to apply the technique to the exocytosis of synaptic vesicles in intact nerve terminals, the approach has provided several insights into the process of exocytosis in nonneuronal cells. An early step in exocytosis in mast cells is the formation of a pore between the inside of the vesicle and the external medium. The initial conductance of the pore is about 230 pS. With time the conductance increases as the pore dilates and the vesicle fuses with the plasma membrane. The initial stages of pore formation, however, appear to be reversible. This is observed as the occurrence of "flicker," in

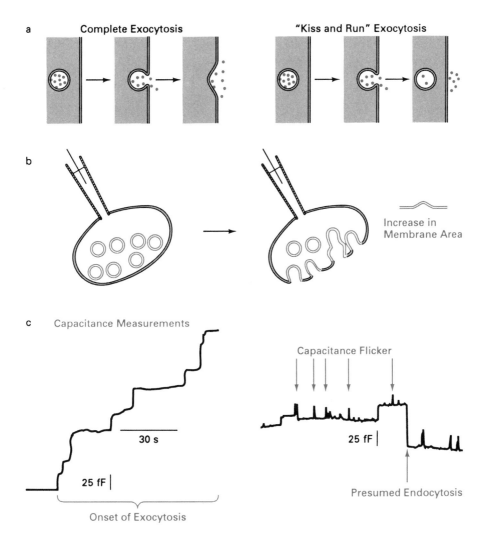

FIGURE 6-9. Exocytosis and capacitance measurements. *a*: Diagram contrasting total exocytosis with a "kiss-and-run" mode of exocytosis. *b*: Exocytosis leads to an increase in the area of plasma membrane. *c*: Capacitance measurements in a mast cell demonstrating stepwise increases *(left)* in capacitance at the onset of exocytosis, capacitance "flicker," and stepwise decreases *(right)* in capacitance. Experiments of this sort have been carried out by Erwin Neher, Wolf Almers, and their colleagues.

which the capacitance fluctuates between two levels (Fig. 6-9c). This indicates that vesicles can fuse with the plasma membrane and dilate partially, but then return to a "closed" state without undergoing full exocytosis. It is not yet known if the contents of secretory vesicles are released during such partial fusions. Nevertheless, such findings lend some support to the hypothesis that release can occur during a "kiss and run" mode of exocytosis.

Proteins of Synaptic Vesicle Membranes and the Mechanism of Exocytosis

Despite the relatively detailed phenomenological description of exocytosis we now have, little is known as yet about its molecular mechanism. In many cells, however, it is abundantly clear that the immediate trigger for release is an elevation of intracellular calcium. The experiments that have demonstrated this fact for neurons are discussed in the next chapter. There are two general hypotheses to explain how an elevation of calcium ions leads to the fusion of vesicles with the plasma membrane. In the first hypothesis, exocytosis simply requires vesicles to come into contact with the appropriate region of the plasma membrane. The vesicles, however, are normally inhibited from fusing with the membrane by the cytoskeleton. The action of calcium is to release this inhibition by "dissolving" some component of the cytoskeleton. According to the second hypothesis, there are specific proteins on the membrane of the vesicle that interact with sites on the inner surface of the plasma membrane in a calcium-dependent manner. In this view, an elevation of intracellular calcium promotes directly the vesicle–plasma membrane interaction that leads to exocytosis.

At present we have no way of differentiating between these hypothetical mechanisms. The answer may eventually come from the study of the proteins that are in, or attached to, the membrane of synaptic vesicles. Relatively pure preparations of synaptic vesicles from the nervous system may be prepared by the technique of *subcellular fractionation*. In this technique, nervous tissue is homogenized in a medium that allows subcellular organelles, such as mitochondria and synaptic vesicles, to remain intact. Because different organelles have different sizes, they can then be separated by centrifugation through layers of sucrose of different densities. In this way it is possible to obtain relatively pure preparations of vesicles from neuronal as well as other secretory tissues.

Table 6-2 lists several of the proteins that are associated with secretory vesicles. It should be pointed out that the small synaptic vesicles, which appear clear in electron micrographs (Fig. 6-7) and which contain small neurotransmitter molecules such as acetylcholine, glutamic acid, GABA, or glycine, have a somewhat different composition than do the slightly larger dense core vesicles that are predominantly peptide containing. In particular, one set of proteins appears to be specific to the small vesicles. This set includes *synaptophysin,* an integral membrane protein with a cytoplasmic tail that binds calcium ions. Also specific to the small vesicles are *synapsin I* and *synapsin II,* proteins that do not reside within the membrane but bind tightly to the surface of the vesicles. Synapsin I is also able to bind actin, a component of the cytoskeleton at nerve terminals, suggesting that it might serve as a link between the vesicles and the cytoskeleton. As we shall see in Chapter 7, the addition of phosphate groups to synapsin I may regulate the amount of neurotransmitter that can be released by a nerve impulse.

Nonvesicular Release of Neurotransmitters

Most neuroscientists believe that exocytosis is the major form of transmitter release at most synaptic junctions, although this has been very difficult to prove unequivocally. Nonetheless, there are several examples of the

TABLE 6-2 Components of Secretory Vesicle Membranes

Component	Function
Proteins associated with the membrane of peptide-containing granules	
Cytochrome b_{561}	Keeps ascorbic acid in a reduced state—required for enzyme that amidates the C-terminal of released peptides
Peptide α-amidase	
Proton pump	Maintains acidic pH in vesicles
Calpactins	A set of proteins that bind to outside of vesicle membrane in the presence of calcium ions
Proteins associated with the membrane of small clear synaptic vesicles	
Proton pump	Maintains acidic pH in vesicles
Neurotransmitter transporters	Required for loading of neurotransmitters into the vesicles
Synaptophysin	Integral membrane protein that crosses the membrane of the vesicle four times—has a cytoplasmic tail that may bind calcium—may be phosphorylated at a tyrosine residue
Synaptobrevin	Integral membrane protein that crosses the membrane only once—long cytoplasmic tail—very short segment in lumen of vesicle
Synapsin I	Protein that binds to surface of vesicles—also binds actin— can be phosphorylated at three different sites
Synapsin II	Binds to surface of vesicles—can be phosphorylated at one site
p29, p65	As yet uncharacterized components of small synaptic vesicles

release of transmitter by neurons that are likely to result from mechanisms *other* than calcium-dependent exocytosis. One such example is found in the retina. Some of the synapses made by the *photoreceptors,* and also synapses made by a retinal cell termed the *horizontal cell,* lack synaptic vesicles. Moreover, in contrast to most synapses, these synapses do not require extracellular calcium to release their neurotransmitter. Using one type of horizontal cell from the retina of the catfish, it has been shown that graded release of the neurotransmitter GABA occurs in response to depolarization of the presynaptic membrane, with no change in intracellular calcium concentration. It has been proposed that a carrier protein exists in the plasma membrane and that, in response to depolarization, this protein transports GABA from the cytoplasm to the extracellular space. Other examples of apparent nonvesicular release of transmitter also exist.

Summary

Two ways neurons communicate with one another are by direct electrical coupling and by the secretion of neurotransmitters. Electrical coupling arises from the existence of proteins, known as connexins, that form pores linking the cytoplasm of adjacent cells. Ions (as well as small molecules) can carry signals from one cell to another through these pores. Neurosecretion is a more complex process in which different categories of molecules are sorted into vesicles in the cytoplasm. A variety of chemical processes within these vesicles ensures that they contain biologically active transmitters or hormones. Stimulation of cells, which usually leads to an elevation of intracellular calcium, allows the vesicles to fuse with the plasma membrane and to release their contents in the extracellular space. Elucidation of the molecular mechanism by which this occurs remains a major challenge for cell biologists.

Synaptic release
of neurotransmitters

In the previous chapter we saw that secretion of proteins occurs through the exocytosis of membranous vesicles. This is a process that has been elaborated throughout evolution in endocrine, exocrine, and neuronal cells. It allows cells to send specific chemical signals that diffuse to, and act on, recipient cells. In many cases, the properties of neurons are very much like those of endocrine cells, whose business is chemical communication. Some neurons release peptides directly into the blood, just as an endocrine cell does. Other neurons release their transmitters locally into the extracellular space, where the transmitter diffuses slowly over some distance and influences many other neurons. Many neurons, however, differ from these other cell types in that they have been under evolutionary pressure to develop very rapid chemical communication with specific target neurons and muscle cells. To this end they have developed long axons that bring the source of the messenger substance right up to the membrane of their target cell. In addition, neurons often use small chemical transmitters that can be synthesized directly at the terminal rather than being transported from the soma, as is the case with peptide transmitters. The characteristics of the release of such transmitters at synaptic terminals differ from those of many other secretory cells. In this chapter we give an account of transmitter release at two thoroughly studied synaptic junctions, the *vertebrate neuromuscular junction* and the *giant synapse of the squid*.

Transmitter Release Is Quantized

We have seen that in many cells secretion occurs through the exocytosis of packets of peptide or hormone that are stored inside vesicles. The release of neurotransmitter at most chemical synapses also occurs in small packets or *quanta*. This was first demonstrated through electrophysiological studies at the neuromuscular junction by Bernard Katz and his colleagues. They placed electrodes in the postsynaptic muscle cell and measured the extent of depolarization of the muscle in response to synaptic stimulation. This depolarization provides a direct measure of the amount of synaptic transmitter released under various experimental conditions.

The vertebrate neuromuscular junction. Figure 7-1 is a drawing of the frog neuromuscular synapse, which uses acetylcholine as its transmitter. Large

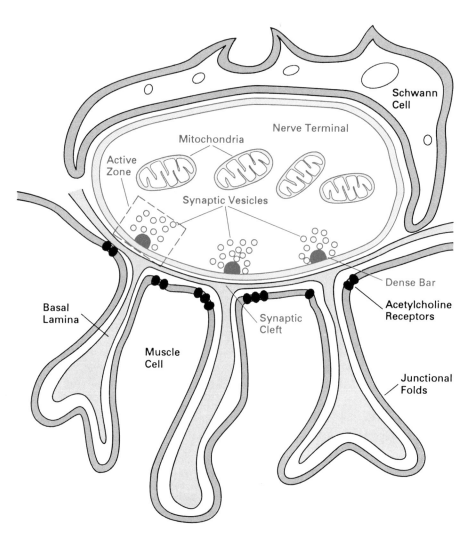

FIGURE 7-1. Diagram of a vertebrate neuromuscular junction.

numbers of synaptic vesicles are associated with specialized areas of the pre-synaptic membrane that have been termed *active zones*. These are located close to structures called *dense bars*, which are located at the cytoplasmic side of the presynaptic membrane, but whose nature is not well understood. The dense bars may serve to align the synaptic vesicles at the sites of neu-rotransmitter release. On the muscle cell, clusters of *receptor molecules*, to which the released acetylcholine binds, are located in the areas of mem-brane closest to the active zones of the presynaptic terminals.

EPPs and MEPPs. When the presynaptic nerve is stimulated, an action potential travels along the axon to the presynaptic terminal. The depolar-ization of the terminal causes the release of acetylcholine. This in turn acts on the receptors in the postsynaptic membrane to depolarize the muscle, by mechanisms that will be discussed in Chapter 9. This depolarization is termed an *end-plate potential* (EPP) (Fig. 7-2). Under normal conditions the end-plate potential is many tens of millivolts in size, sufficient to trigger an action potential and the subsequent contraction of the muscle. However,

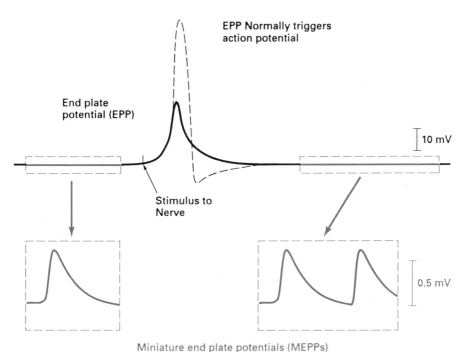

FIGURE 7-2. EPPs and MEPPs. Membrane potential recordings in a muscle cell. An EPP evoked by a nerve stimulus normally triggers an action potential. When, however, its amplitude is decreased or the postsynaptic action potential is blocked, its time course matches that of MEPPs.

even in the absence of nerve stimulation, small depolarizations can be recorded with an electrode in the muscle (Fig. 7-2). These spontaneously occurring depolarizations are only about 0.5 mV in amplitude, but in most other respects they are very similar to the larger end-plate potential evoked by nerve stimulation. In particular, the time course of the small depolarizations matches that of the end-plate potential. In addition, the spontaneously occurring potentials, like the end-plate potential, can be blocked by antagonists of the acetylcholine receptor such as curare (see Chapter 9). The potentials can also be prolonged by agents that prevent the hydrolysis of acetylcholine by the enzyme acetylcholinesterase in the synaptic cleft. Finally, the frequency of occurrence of these small depolarizations increases on depolarization of the presynaptic terminal. These and other findings indicated that these small potentials are due to the spontaneous release, at random intervals, of a small amount of acetylcholine from the presynaptic terminal. The small potentials were therefore named *miniature end-plate potentials* (MEPPs).

EPPs are made up of multiple MEPPs. Katz then showed that the stimulus-evoked synaptic potential, that is, the end-plate potential, was caused by the simultaneous occurrence of a large number of individual potentials each of which appeared identical to a MEPP. To analyze the synaptic potentials in detail, the neuromuscular junction was bathed in a medium that contained lower than normal levels of calcium ions and a higher concentration of magnesium ions. As we shall see, such a medium reduces the amount of transmitter that is released on stimulation. In these particular experiments, release was reduced to such an extent that the average amplitude of the postsynaptic potential following nerve stimulation was a few millivolts, only a few times larger than a spontaneously occurring MEPP. Under these conditions the evoked end-plate potential did *not* have a fixed amplitude with each stimulus to the nerve. Instead, some stimuli failed to generate an end-plate potential, some generated end-plate potentials that were equal in amplitude and duration to a single MEPP, while others generated end-plate potentials that were equal in amplitude to two or more individual MEPPs (Fig. 7-3). When a count is made of the number of times that end-plate potentials of different amplitudes occur in such an experiment, a histogram such as that in Figure 7-3 is generated. Although there is some variability in the size of individual MEPPs, the peaks in the histogram of evoked responses clearly correspond to the amplitudes of integral numbers of MEPPs.

A simple calculation shows that a single MEPP arises from the simultaneous action of a large number of acetylcholine molecules in the synaptic cleft. We now know that the conductance of the channel that is opened by

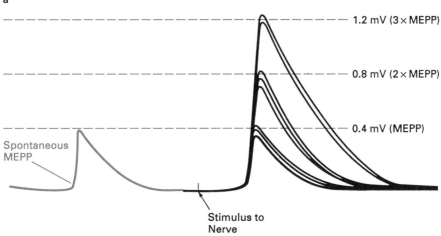

a

Spontaneous MEPP

1.2 mV (3 × MEPP)

0.8 mV (2 × MEPP)

0.4 mV (MEPP)

Stimulus to Nerve

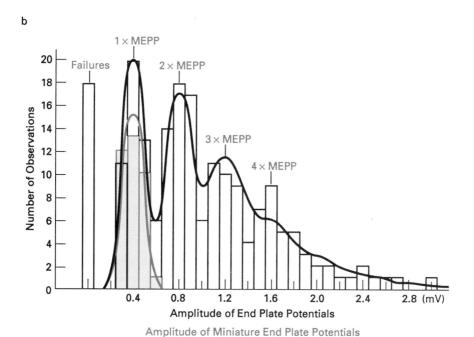

b

Number of Observations

Failures

1 × MEPP

2 × MEPP

3 × MEPP

4 × MEPP

0.4   0.8   1.2   1.6   2.0   2.4   2.8 (mV)

Amplitude of End Plate Potentials

Amplitude of Miniature End Plate Potentials

FIGURE 7-3. EPPs are made of multiple MEPPs. *a*: Under conditions of low transmitter release, nerve-evoked EPPs have amplitudes that correspond to a unit number of MEPPs. *b*: A histogram showing the relation between the size of EPPs and MEPPs in an experiment by Boyd and Martin (1956) on the cat neuromuscular junction.

acetylcholine is about 40 pS, and that the current flowing though the open channel is in the range of 5 pA near the muscle resting potential. The resistance of the muscle membrane is such that the opening of a single acetyl-choline-gated channel in the postsynaptic membrane can produce a depolarization of less than 1 μV. As much of the acetylcholine released into the synaptic cleft is hydrolyzed before it is able to interact with the receptor, and two molecules of acetylcholine are required to open a single acetylcho-line-gated channel (see Chapter 9), it can be estimated that approximately 5000 acetylcholine molecules are released synchronously into the synaptic cleft to generate a single MEPP.

We shall see later that many factors can alter the strength of synaptic transmission. The fact that transmitter release occurs in quanta can sometimes make it possible to determine whether an alteration in the strength of synaptic transmission results from a change in the amount of transmitter that is released (rather than from a change in the sensitivity of the postsynaptic cell to the transmitter). In particular, if a change in number of quanta released on stimulation is detected, this immediately implies that some alteration has occurred in the presynaptic terminal. This *quantal analysis* is therefore a highly useful technique for investigating the properties of those relatively few synapses where such measurements can be made.

Morphological Evidence for Exocytosis during Synaptic Transmission

In the 1950s, Katz and his colleagues suggested that the packets or quanta of acetylcholine released from the presynaptic terminals at the neuromuscular junction correspond to the content of acetylcholine within single synaptic vesicles. They suggested further that, even at rest, vesicles occasionally fuse with the presynaptic membrane, resulting in exocytosis of the contents of a single vesicle, and the generation of a MEPP in the postsynaptic muscle cell. Depolarization by a presynaptic action potential greatly increases the probability of exocytosis of vesicles, leading to the simultaneous release of the contents of many vesicles. A normal end-plate potential would result from the exocytosis of several hundred synaptic vesicles. However, direct visualization of exocytosis at the neuromuscular junction was achieved only 20 years later, by using electron microscopy combined with the freeze-fracture technique (Fig. 7-4).

In the last chapter, we saw the use of freeze fracture to detect arrays of gap junction particles. When applied to presynaptic membranes at the neuromuscular junction of a frog, this technique reveals deformations of the plasma membrane, due to the presence of the dense bars under the presynaptic membrane. Aligned in two rows on either side of the dense bar are

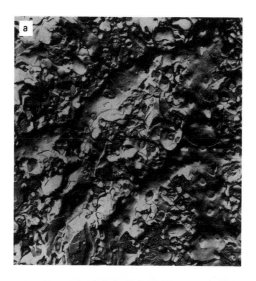

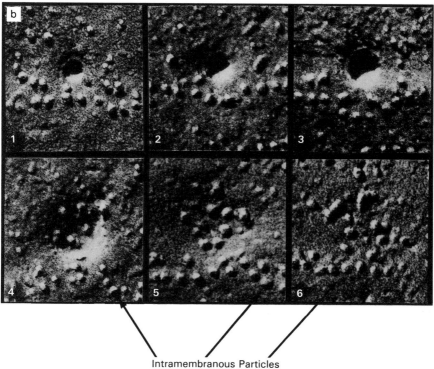

Intramembranous Particles

FIGURE 7-4. Freeze fracture evidence for exocytosis at the neuromuscular junction. *a*: Deformations of the plane of the plasma membrane are made by intramembranous particles, underlying structures and fusing vesicles. *b*: Images of freeze fracture replicas of the presynaptic membrane, made by Tom Reese and his colleagues, showing progressive stages (1–6) in the exocytosis of a single vesicle (courtesy of Tom Reese).

strings of intramembranous particles (Fig. 7-4). It is widely believed that these particles represent calcium channel proteins, although definitive evidence for this is yet to come. The electron micrographs of Figure 7-4 were made using freeze fracture coupled with a technique that allowed the terminals to be frozen very rapidly, and at precise times, following the stimulation of the motor nerve. In a terminal at rest, or in one that has been stimulated but that has not yet begun to release transmitter, the dense bar and the two rows of particles can be detected. At the very time that acetycholine release occurs, however, further deformations of the membrane can be seen. These newly formed "pits" are thought to represent fusion of synaptic vesicles with the plasma membrane. The presence of these pits suggests strongly that release of transmitter occurs through exocytosis.

For technical reasons it has not been possible to stimulate nerves when a small number of quanta are released, and then to compare directly the number of quanta with the number of pits formed by exocytosis of synaptic vesicles. However, the neuromuscular junction has been stimulated in the presence of the potassium channel blocking agent 4-aminopyridine, which prolongs the action potential and greatly increases the amount of transmitter released. In this condition the amount of acetylcholine released does match the number of pits observed in freeze-fracture replicas.

Synaptic Transmitter Release Is Dependent on Calcium

Although the detailed molecular events that lead to the exocytosis of neurotransmitter-containing vesicles are obscure, it has been known since the work of Sidney Ringer in the nineteenth century that calcium ions in the extracellular medium are required for the normal function of the neuromuscular junction and other synapses. At most chemical synapses, eliminating calcium from the external medium prevents the release of transmitter evoked by nerve stimulation. In contrast, an increase in the external concentration of calcium frequently enhances the amount of neurotransmitter that is released.

The squid giant synapse. A major drawback to the investigation of neurotransmitter release is the small size of most vertebrate presynaptic terminals, including those of the neuromuscular junction. This precludes normal electrophysiological recordings from the terminals. A synapse at which this is not a problem is the giant synapse in the stellate ganglion of the squid (Fig. 7-5). This synapse is used in escape behavior. When stimulated, this syn-

FIGURE 7-5. The squid giant synapse. *a*: Position of the giant synapse within the stellate ganglion. *b,c,d,e*: Presynaptic and postsynaptic responses recorded in experiments by Bernard Katz and Ricardo Miledi, and by Rodolfo Llinas and coworkers. *b*: The normal response. *c*: The postsynaptic axon is voltage clamped and the postsynaptic current is recorded. *d,e*: Presynaptic sodium and potassium cur-

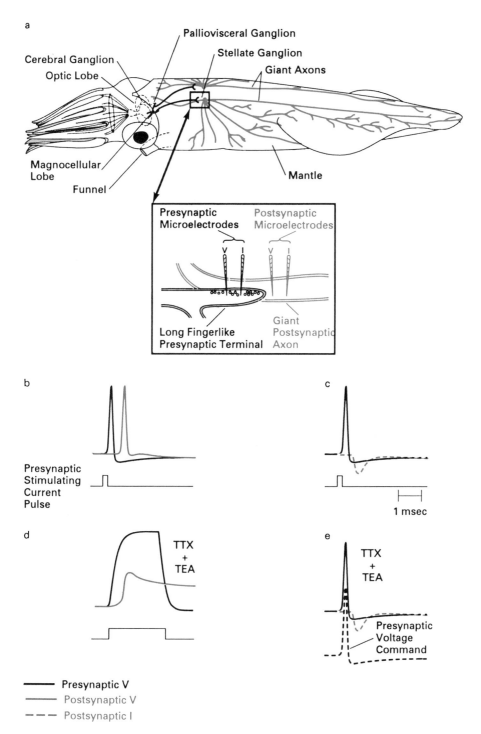

a

Cerebral Ganglion
Optic Lobe

Pallioviseral Ganglion

Stellate Ganglion

Giant Axons

Magnocellular
Lobe

Funnel

Mantle

Presynaptic
Microelectrodes

Postsynaptic
Microelectrodes

V   I

V   I

Long Fingerlike
Presynaptic Terminal

Giant
Postsynaptic
Axon

b

Presynaptic
Stimulating
Current
Pulse

c

1 msec

d

TTX
+
TEA

e

TTX
+
TEA

Presynaptic
Voltage
Command

—— Presynaptic V
—— Postsynaptic V
– – – Postsynaptic I

rents have been blocked by TTX and TEA. *d*: The prolonged postsynaptic potential evoked by a long presynaptic depolarization. *e*: The presynaptic voltage is driven to follow that of a normal action potential, resulting in a normal postsynaptic current.

aptic pathway triggers the ejection of water through the mantle of the squid, propelling the animal rapidly away from a source of danger. Several cell bodies of the average molluscan neuron could be comfortably enclosed in the presynaptic terminal of the giant synapse (the terminal is about 50 μm in diameter and 700 μm in length). Accordingly, two or more independent microelectrodes can be placed in the presynaptic terminal as well as in the postsynaptic cell (Fig. 7-5a). As in the case of the neuromuscular junction, the amplitude of the postsynaptic voltage change is taken as a direct measure of the amount of transmitter released.

Work with the squid giant synapse confirmed and extended the findings of Katz on the neuromuscular junction, and definitively established a role for calcium and for voltage-dependent calcium channels in the release of transmitter following a presynaptic action potential. For example, it is possible to evoke transmitter release at this synapse by direct injection of calcium ions into the presynaptic terminal. As in the somata of many neurons, voltage-dependent calcium currents can be recorded in the presynaptic terminal of this synapse (together with the much larger, voltage-dependent sodium and potassium currents that shape the action potential). Stimulation of an action potential in the terminal usually liberates sufficient transmitter to depolarize the postsynaptic axon to threshold and to trigger a postsynaptic action potential (Fig. 7-5b). If, however, the voltage of the postsynaptic cell is clamped near its resting potential, then the stimulation of the synapse generates an inward current in the postsynaptic cell (Fig. 7-5c). This inward current represents the current flowing through the transmitter-activated ion channels in the postsynaptic membrane.

It is possible to block the sodium and potassium channels in the presynaptic membrane, leaving calcium flux as the only voltage-dependent ion current. This is accomplished by applying the sodium channel blocker TTX to the external medium, and by injecting the potassium channel blocking agent TEA into the terminal (see Chapter 4). In this condition, postsynaptic responses can still be recorded when the presynaptic terminal is depolarized (Fig. 7-5d). Of course, in the presence of TTX and TEA, normal action potentials no longer occur in the presynaptic terminal. However, when the voltage in the terminal is driven by the voltage clamp to follow the normal shape of an action potential, the postsynaptic current exactly matches the normal postsynaptic response (Fig. 7-5e). This indicates that the normal presynaptic sodium and potassium fluxes are not required for release of neurotransmitter, but that calcium entry alone is sufficient.

Calcium channels are clustered near release sites. The spatial distribution of calcium channels in the squid terminal closely matches the sites of transmitter release. This was first suggested by the fact that the calcium current recorded *in* the presynaptic terminal is very much greater than that in regions of the axon *leading to* the terminal. Direct measurement of calcium levels in the terminal, however, required the use of substances that could be introduced into the terminal to measure changes in calcium concentration. The first of these to be used was a protein known as *aequorin*, which emits light when it binds calcium. This protein was used to provide direct evidence that the calcium concentration in the terminal is raised during transmission. Subsequently, other calcium-indicator dyes have been used for this purpose, including the dye *Arsenazo III*, and the now widely used calcium indicator *fura-2*. In its structure fura-2 resembles the common chelator of calcium ions, EGTA (Fig. 7-6a). It is, however, a fluorescent compound, whose excitation spectrum changes when it binds calcium (Fig. 7-6b).

Using a fluorescence microscope coupled to a computer, digital images can be made of changes in calcium concentration occurring during transmitter release. Figure 7-6c illustrates the application of fura-2 imaging to the squid giant synapse. During a brief burst of action potentials, calcium levels within the terminal rise. This elevation occurs first at the active zones that are closely apposed to the postsynaptic fiber. During the burst, a steep gradient of calcium concentration develops across the terminal. After stimulation ceases, the gradient dissipates. Such images further illustrate that calcium channels are clustered near the release sites.

Three important features of calcium-dependent transmitter release. Electrical measurements, coupled with measurements of intracellular calcium, have provided three further important conclusions about the mechanism of transmitter release: (1) transmitter release depends on a high power of the calcium concentration; (2) transmitter release is more sensitive to calcium ions than to other divalent cations; and (3) release occurs very rapidly following calcium entry. We shall now discuss briefly the evidence for each of these conclusions.

Transmitter release requires binding of several calcium ions. As described above, transmitter release depends on the presence of extracellular calcium. At low concentrations of extracellular calcium, release of transmitter at the frog neuromuscular junction depends on the fourth power of the external calcium concentration. Experiments at the squid giant synapse, in which calcium currents and increases in calcium concentration can be measured directly, indicate that calcium *entry* into the terminal does not depend on the fourth power of the external calcium concentration, but rather that the *release process* itself depends on a high power of intracellular calcium (Fig. 7-7a). The simplest interpretation of such findings is that some reaction in the terminal requires the binding of several calcium ions before release of transmitter occurs.

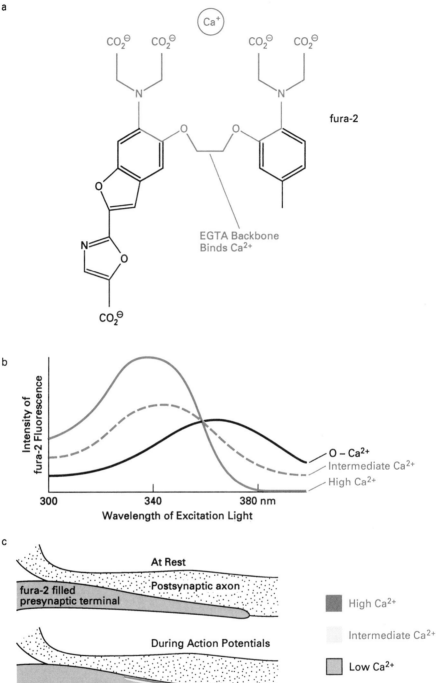

a

$CO_2^{\ominus}$ $CO_2^{\ominus}$ $CO_2^{\ominus}$ $CO_2^{\ominus}$

Ca^+

N   N

O   O

fura-2

O

N   O

EGTA Backbone
Binds Ca^{2+}

$CO_2^{\ominus}$

b

Intensity of
fura-2 Fluorescence

O – Ca^{2+}
Intermediate Ca^{2+}
High Ca^{2+}

300   340   380 nm

Wavelength of Excitation Light

c

At Rest

fura-2 filled
presynaptic terminal

Postsynaptic axon

During Action Potentials

High Ca^{2+}

Intermediate Ca^{2+}

Low Ca^{2+}

FIGURE 7-6. Measurement of intracellular calcium. a: Structure of the calcium indicator dye fura-2, which was synthesized by Roger Tsien. b: The intensity of fura-2 fluorescence varies with calcium concentration (Grynkiewicz et al., 1985). c: Diagram of how calcium levels in the squid presynaptic terminal, measured using fura-2 fluorescence, change during stimulation (Smith and Augustine, 1988).

The release mechanism prefers calcium over other divalent cations. When the extracellular calcium at a synapse is replaced by other divalent cations, transmitter release is usually reduced or abolished. Ions such as nickel, cadmium, manganese, cobalt, and the trivalent ion lanthanum act as calcium channel blockers and would not be expected to evoke release. However, the ions barium and strontium readily enter the cell through voltage-dependent calcium channels. Even for these ions, release of transmitter at the squid synapse is very much lower than for an equivalent amount of calcium influx. Such experiments suggest that the ion-sensitive step in the release process prefers calcium strongly over the other ions in the order Ca>Sr> Ba.

Transmitter release occurs very rapidly following calcium entry. When the presynaptic terminal of the squid is depolarized to allow calcium channels to open, release generally increases after a delay of a few milliseconds. This is, however, not a good measure of the rate at which the release process itself is activated by calcium, because the presynaptic calcium current activates over a period of milliseconds. A more direct measurement of the response time of release can be obtained during measurements of *calcium tail currents.*

If, in a voltage-clamp experiment, the presynaptic terminal is stepped from its resting potential to a very positive potential close to the equilibrium potential for calcium ions, then calcium channels open in response to the depolarization. Nevertheless, at this potential calcium ions do not enter the terminal, because there is little or no driving force for calcium entry (Fig. 7-7b). Thus no transmitter release occurs with such large depolarizations. The voltage-clamp circuitry can then be used to step the membrane potential very rapidly back to the resting potential. Immediately after this step, the calcium channels remain open for a finite time at the resting potential, before they eventually close. Because the driving force for calcium entry is large at the resting potential, calcium ions enter through the still open calcium channels, resulting in a rapid but transient calcium tail current, measured after the end of the depolarizing pulse. This tail current is accompanied by transmitter release, measured as a postsynaptic depolarization. The delay between the onset of the sharp tail of calcium current in the presynaptic terminal and the onset of the postsynaptic response has been found to be as little as 200 μsec (Fig. 7-7b). This means that the release process at this synapse is so rapid that the actions of calcium are unlikely to involve complex multistep biochemical reactions.

Domains of calcium entry. Because calcium channels are discrete membrane proteins, calcium concentrations do not rise uniformly throughout the terminal following a depolarization. Instead, calcium levels are initially much higher directly under the membrane where the calcium channels are located (see Fig. 7-6c). It is important to consider this spatial inhomogeneity when thinking about the relation between the amount of calcium that enters a cell and the effect of that calcium on processes such as transmitter release.

Quantitative models have been useful in predicting the way calcium is distributed in the presynaptic terminal after entry through calcium channels. Figure 7-8a shows the pattern of intracellular calcium expected near the mouth of a calcium channel that has been open for a millisecond or so. The computed profile of calcium concentration takes the form of a "volcano" centered on the mouth of the channel. Near the channel itself, cal-

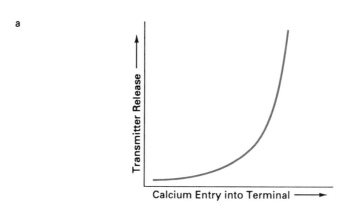

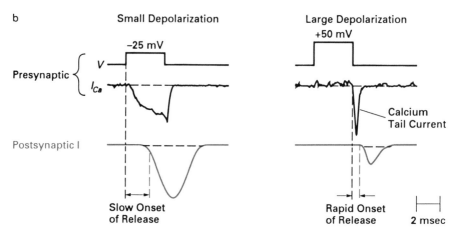

FIGURE 7-7. Calcium dependence of release. a: Steep dependence of release on amount of calcium entry into the squid terminal. b: The postsynaptic response occurs very rapidly after a calcium tail current (after Augustine et al., 1985).

cium concentrations may reach as high as 100 μM (the resting level of calcium is usually about 0.1 μM). As calcium diffuses from the channel, its concentration drops. Proteins located close to the mouth of the channel, including neurotransmitter release sites, may therefore be exposed to higher concentrations of calcium ions than are seen further from the channel.

This idea of calcium domains allows certain predictions about transmitter release. For example, it suggests that the same amount of calcium enter-

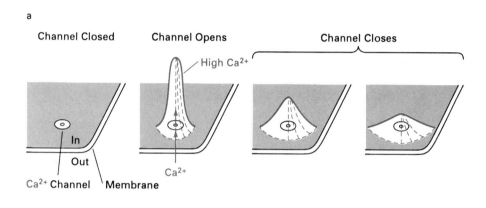

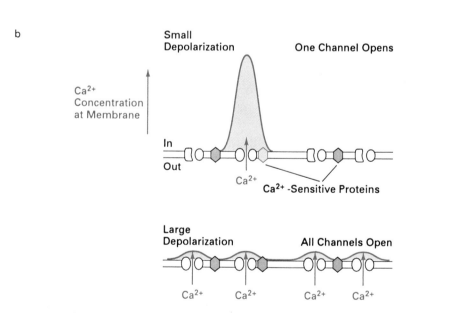

FIGURE 7-8. High levels of calcium occur at the mouth of calcium channels. *a*: "Volcanos" of calcium forming inside a cell near the mouth of an open channel. *b*: Comparison of the spatial profiles of intracellular calcium expected with small and large depolarizations.

ing a cell at two different voltages can, in theory, produce different effects on transmitter release or on other calcium-sensitive reactions. We know that as a cell is progressively depolarized, calcium current first increases as calcium channels open, and then decreases as the voltage approaches the equilibrium potential for calcium ions. Figure 7-8b shows the profile of calcium concentrations expected at two different voltages that produce the same *net* calcium influx into the terminal. In the first case, a small depolarization of the membrane allows only a few calcium channels to open. Because the driving force on calcium ions is high, however, relatively high concentrations of calcium are achieved near the mouths of these few open channels. In the second case, a large depolarization causes the opening of nearly all of the channels. At this very positive potential, however, calcium driving force and hence flux through each open channel is greatly reduced. In the first case, large but spatially inhomogeneous transients of calcium are attained. These may act on calcium-sensitive proteins near the open channels. In the second case, although a more uniform change in calcium levels is attained, levels at any one site may not reach the critical level needed for triggering key calcium-activated processes. We can see, therefore, that the relative localization of calcium channels and release sites may be a very important factor in the efficient organization of neurotransmitter release.

We have now described some of the dynamic features of transmitter release at synaptic terminals. It should be emphasized that the above conclusions about the mechanisms of release were obtained using two very specialized synapses, the squid giant synapse and the neuromuscular junction. We know much less about the release process at the smaller and relatively inaccessible synapses in the vertebrate central nervous system. Furthermore the mechanisms of peptide release at synaptic junctions have not been studied in as much detail, and it is quite possible that slower and more elaborate biochemical processes regulate release at some synapses.

Homosynaptic Plasticity:
Facilitation, Potentiation, and Depression of Transmitter Release

As a series of action potentials invades a nerve terminal, the amount of neurotransmitter released with each action potential does not always remain constant. Depending on the synapse that is studied, and the frequency at which it is stimulated, a train of action potentials in the presynaptic terminal may produce either a progressive increase or a progressive decrease in the amount of release. This property, which allows the amount of transmitter release to change as a result of previous activity in the terminal, has been termed *homosynaptic plasticity*. Three major forms of homosynaptic plasticity are *facilitation, potentiation,* and *depression.* These may be contrasted with a change in release induced by the action of other cells. The latter phenomenon has been termed *heterosynaptic plasticity,* and examples will be provided in Chapter 17. Facilitation, potentiation, and depression may all occur at a single type of synapse, resulting in a complex time course of changes in neurotransmitter release following the onset of a train of action potentials. Figure 7-9 shows the relative time course of these different homosynaptic processes. We will now describe each of them in turn.

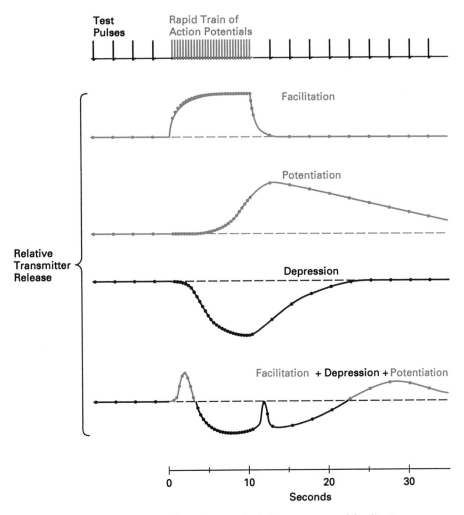

FIGURE 7-9. Homosynaptic plasticity. Typical time courses of facilitation, potentiation, and depression.

Facilitation describes a progressive increase in the amount of transmitter released by successive action potentials *during* a brief stimulus train lasting up to a few seconds (Fig. 7-9). A leading hypothesis for the cause of facilitation is that the calcium concentration at the release sites does not have time to return to basal levels between the action potentials. Thus, at the occurrence of each action potential after the first, a small amount of *residual calcium* remains near the release site. The level of residual calcium is, of itself, insufficient to cause a significant amount of transmitter release during the interval between action potentials. Because release is very nonlinearly dependent on calcium concentration (Fig. 7-7a), the residual calcium adds to the calcium that enters during the next action potential to produce a progressive increase in release. Recovery from facilitation is rapid, occurring within a few hundred milliseconds after the end of stimulation.

Potentiation describes an increase in transmitter release by an action potential *following* repetitive stimulation of a synapse (Fig. 7-9). Unlike facilitation, potentiation is long-lasting and is slow in onset, usually requiring seconds to develop. For example, following a rapid train of action potentials, termed a *tetanus,* the amount of release in response to a single action potential may be enhanced over that prior to the tetanus for up to several minutes. This phenomenon can be observed at a wide variety of synapses including the neuromuscular junction, and has been termed *post-tetanic potentiation* (PTP). Although its mechanism is not known, there is evidence to suggest that a rise in the intracellular sodium concentration plays a role in PTP. At many central synapses, potentiation may endure for periods ranging from tens of minutes to hours or days, depending on the duration and intensity of the stimulus train. This has been termed *long-term potentiation* (LTP). The mechanisms of LTP appear to be distinct from those of PTP, and differ from synapse to synapse. The phenomenon of LTP has been studied as a model for the establishment of memories by the nervous system, and we will discuss LTP further in this context in Chapter 17.

Synaptic depression describes a progressive decrease in the amount of transmitter released during a train of action potentials (Fig. 7-9). This phenomenon is often encountered when a synapse is stimulated at a high rate and, at least in some cases, results from the depletion of neurotransmitter from the terminal.

Calcium/Calmodulin-Dependent Protein Kinases and Neurotransmitter Release

Although the chemical events underlying neurotransmitter release are still poorly understood, there is evidence to suggest that a calcium-dependent enzyme termed the *multifunctional calcium/calmodulin-dependent protein kinase* plays a role in the regulation of release. We have already seen that the rapid release of neurotransmitter at synaptic junctions occurs too rapidly to be mediated by a complex enzyme reaction. This enzyme is, therefore, not likely to be involved in triggering release directly, but, rather, may determine the amount of release that occurs with each action potential. We shall first give an account of the properties of this enzyme, and then describe the evidence that implicates it in the release process.

Protein kinases. There exist enzymes within cells that transfer the terminal phosphate group from ATP to a variety of proteins. These enzymes are termed *protein kinases.* The addition of phosphate to a protein alters the electrical charge on the protein (phosphate groups are negatively charged), and may also change the three-dimensional shape, or conformation, of the protein. This, in turn, may induce a change in the biological activity of the phosphorylated protein (for a more detailed discussion, see Chapter 10). The activity of protein kinases can be detected in a homogenate of the nervous system by incubating the homogenate with ATP that has been radiolabeled in the terminal (γ) phosphate. The transfer of the radioactive phosphate group to specific proteins can then readily be measured. If calcium ions are also added to the homogenate, it is found that the incorporation of phosphate into certain proteins is very much enhanced. Further enhancement is also usually observed on addition of *calmodulin,* a small (16.7-kDa) calcium-binding protein that is found in all eukaryotic cells. The calcium-dependent phosphorylation of proteins is carried out by several different protein kinases, one of which is the multifunctional calcium/calmodulin-dependent protein kinase. This enzyme is one of the myriad of cellular activities that are regulated by calcium and calmodulin. It has also been called the calcium/calmodulin-dependent protein kinase type II, and its name may be abbreviated to Ca^{2+}/Cam kinase II.

Ca^{2+}/Cam kinase II. Ca^{2+}/Cam kinase II is an enzyme found in most cells. It is, however, particularly abundant in the nervous system where it accounts for 0.5–1.0% of total protein, an extraordinarily high concentration for an enzyme. It is a large, multisubunit complex consisting of two different subunits, α and β, with molecular weights of 50,000 and 60,000, respectively. The active enzyme appears to contain 12 subunits (Fig. 7-10), and the relative ratio of α to β-subunits in the active enzyme varies in different types of cells. For example, in some cells, the enzyme is composed entirely of α-subunits, whereas in other cells the larger β-subunit predominates.

Following an elevation of the calcium concentration inside a cell, the calcium ions bind to calmodulin in the cytoplasm. The calcium/calmodulin complex, in turn, binds to the individual subunits of the Ca^{2+}/Cam kinase. When the enzyme is activated in this way, it is able to transfer phosphate groups from ATP to a variety of different proteins. One of the best substrate proteins for this enzyme is the synaptic vesicle protein *synapsin I* (see Table 6-2), which we mentioned briefly in the last chapter. in addition to binding to synaptic vesicles, synapsin I also binds actin molecules of the cytoskeleton, and may serve to link vesicles to the cytoskeleton. It has been found that when synapsin I is phosphorylated by Ca^{2+}/Cam kinase II, it binds to synaptic vesicles less tightly than when it is unphosphorylated (Fig. 7-10b).

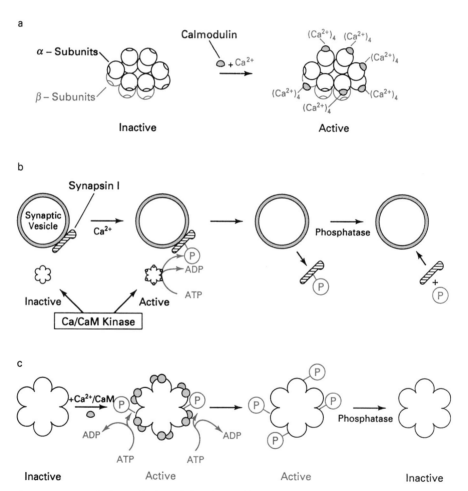

FIGURE 7-10. Calcium/calmodulin-dependent protein phosphorylation. *a*: Activation of Ca^{2+}/Cam kinase II. *b*: Phosphorylation of synapsin I. *c*: Autophosphorylation of Ca^{2+}/Cam kinase II.

One interesting feature of Ca^{2+}/Cam kinase II is that it is able to phosphorylate itself. This is termed *autophosphorylation*. When this happens, the enzyme remains active but becomes independent of calcium and calmodulin until the phosphate groups on the enzyme are removed by other enzymes termed *phosphoprotein phosphatases*. The autophosphorylation may therefore allow Ca^{2+}/Cam kinase II to remain active for a considerable time even after the calcium concentration in the cell has returned to its basal level (Figure 7-10c).

Ca^{2+}/Cam kinase II and transmitter release at the squid giant synapse. Evidence that Ca^{2+}/Cam kinase II plays a role in neurotransmitter release has been provided by injecting this enzyme into the giant presynaptic terminal of the squid (Fig. 7-11). This causes an increase in neurotransmitter release. However there is *no* change in the amplitude of the presynaptic calcium current, suggesting that the injection of the enzyme increases the efficiency of transmitter exocytosis (Fig. 7-11). It is possible that this action of Ca^{2+}/

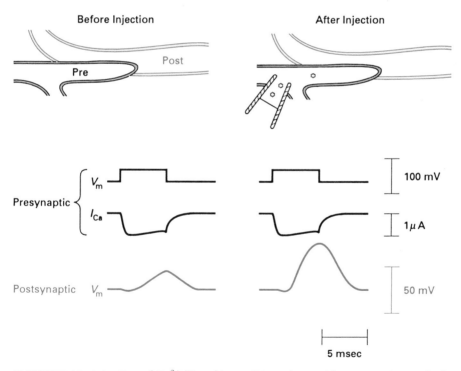

FIGURE 7-11. Injection of Ca^{2+}/Cam kinase II into the squid presynaptic terminal. In a collaborative experiment between the laboratories of Paul Greengard and Rodolfo Llinas, injection of this enzyme caused increased transmitter release, measured as an increase in the postsynaptic depolarization (Llinas et al., 1985).

Cam kinase II at the squid synapse occurs through the phosphorylation of synapsin I or a related molecule. Injection of unphosphorylated synapsin I into the presynaptic terminal decreases or blocks release, whereas injection of synapsin I that has previously been phosphorylated by Ca^{2+}/Cam kinase II has no apparent effect on release. A hypothesis that has been advanced to account for these and related observations is that, under resting conditions (i.e., at low levels of intracellular calcium), dephosphorylated synapsin I is associated tightly with many small synaptic vesicles. This tight association hinders these vesicles from fusing with the plasma membrane. On phosphorylation by Ca^{2+}/Cam kinase II, the synapsin I molecule dissociates from the vesicles, making them available to undergo exocytosis.

Because Ca^{2+}/Cam kinase II is so abundant and so widespread, it is likely that it also plays a number of different roles unrelated to secretion. It is, for example, the major protein to be found in the region of a neuron immediately under the neurotransmitter receptors at the *postsynaptic* junction, where its role is not yet known (but see Chapter 17 for one possible role). In addition, transmitter release may be altered at many synapses by the activation of another calcium-dependent protein kinase, termed *protein kinase C*. This latter enzyme does not phosphorylate synapsin I, and may therefore act through a mechanism that differs from that of Ca^{2+}/Cam kinase II.

Summary

Two specialized synapses, one in the squid stellate ganglion and the other at the frog neuromuscular junction, have been instrumental in advancing our understanding of the release of neurotransmitters from presynaptic terminals. Studies of rapid synaptic transmission have shown that neurotransmitters are released in packets or quanta, which may correspond to the exocytosis of individual synaptic vesicles. Following an action potential, release is very closely linked in space and in time to the entry of calcium though voltage-dependent channels. With repetitive stimulation, changes in the amount of release occur. We are still far from a complete understanding of the mechanisms that induce neurotransmitter release, and of the factors that modulate release. Furthermore, the characteristics of the release of transmitter from small synaptic vesicles at active zones may differ substantially from those of neuropeptides in large dense core vesicles. In the years to come, biochemical experiments that characterize the components of synaptic vesicles and release sites, coupled with physiological experiments that test the effects of these components at specific synapses, will provide further insights into the release process.

Neurotransmitters
and neurohormones

We have seen in the previous two chapters that a neuron is in large part an elaborate and intricately regulated machine for the secretion of a variety of chemicals. Why do cells go to all this trouble? Although direct electrical connections between nerve cells also play an essential role (see Chapter 6), it is chemical signaling that mediates much of the intercellular communication among nerve cells within the central nervous system. In addition, the transfer of information into the nervous system from sensory organs and the output from the nervous system in the form of muscular contraction are mediated by extracellular chemical messengers. In this chapter we will discuss in a systematic way the different classes of *neurotransmitters* and *neurohormones*, some of which we have already met in other contexts. Subsequent chapters will deal with *receptors* on the target cell that recognize and bind these substances and with the *transduction mechanisms* that are involved in converting the extracellular chemical signal into an appropriate response (usually electrical) in the target cell.

What Is a Neurotransmitter and What Is a Neurohormone?

Which neuroactive substances are transmitters and which are hormones? Classically, synaptic neurotransmitters have been thought of as substances that are released locally into an anatomically well-defined *synaptic cleft*, and influence the activity of only one or a few adjacent cells (Fig. 8-1). The prototype neurotransmitter is acetylcholine, which was first demonstrated to be the chemical mediating nerve-to-muscle synaptic transmission in cardiac and skeletal muscle (see Chapter 1), and was subsequently found to be an important neuron-to-neuron transmitter as well. In contrast, hormones have been defined as substances that are released from the tissue in which they are synthesized, and travel via the blood to other (often remote) organs whose activities they influence (Fig. 8-1). Another criterion that has often been used to distinguish between transmitters and hormones is a temporal one: transmitters have been thought to produce rapid-onset and rapidly reversible responses in the target cell, whereas the actions of hormones can be slower and much longer lasting.

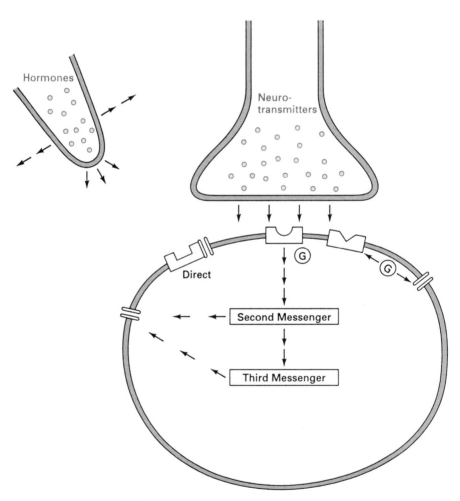

FIGURE 8-1. Intercellular communication. This chapter will emphasize the neurotransmitters and hormones (red) that mediate much cell-to-cell communication in the nervous system.

Because hormone synthesis and secretion had been studied most thoroughly in organs such as the adrenal gland and the gonads, it came as something of a surprise when it was demonstrated more than 30 years ago that the nervous system also synthesizes and releases hormones—*neurohormones*—that act on remote organs. The classical neurohormones are the closely related nine amino acid peptides *oxytocin* and *vasopressin,* which are (1) synthesized in the so-called *magnocellular neurons* in the hypothalamus, (2) transported down the axons of these neurons to the posterior pituitary where they are released, and (3) dispersed via the bloodstream to regulate smooth muscle contraction and water balance, respectively.

More recently it has become evident that oxtyocin, vasopressin, and other neuroactive peptides can have local transmitter-like actions as well. In addition to being secreted into the blood, they may be released at synapses instead of—or even together with—more classical transmitter substances. Furthermore we know now that the actions of some classical neurotransmitters may be slow in onset and long in duration, and indeed they may be released into the bloodstream to act as true hormones, whereas many peptide "hormones" may produce rapid "transmitter-like" effects on target cells. In other words the classical distinctions between neurotransmitters and neurohormones are becoming obsolete. One person's transmitter is another person's hormone, and it may be more useful for many purposes to classify neuroactive substances on the basis of other criteria, for example, the nature and mechanism of the response they evoke in the target cell (see Chapters 9 and 10).

Acetylcholine

The first chemical to be implicated as a neurotransmitter was *acetylcholine,* the structure of which is shown in Figure 8-2. It had been known since the early part of this century that acetylcholine could influence the physiological properties of nerve and muscle cells. In the 1920s it was demonstrated that acetylcholine is the transmitter at *neuromuscular synapses,* synapses between neurons and cardiac, smooth and skeletal muscle, as well as at a

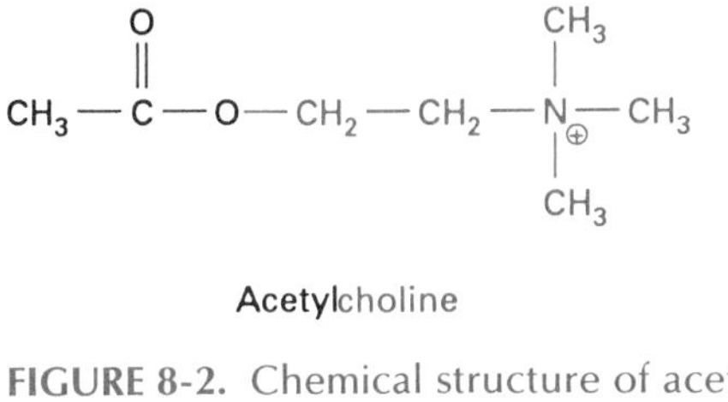

$$CH_3-\overset{\overset{\displaystyle O}{\|}}{C}-O-CH_2-CH_2-\overset{\overset{\displaystyle CH_3}{|}}{\underset{\underset{\displaystyle CH_3}{|}}{N}}\overset{}{\underset{\oplus}{}}-CH_3$$

Acetylcholine

FIGURE 8-2. Chemical structure of acetylcholine.

variety of neuron–neuron synapses in the central and peripheral nervous systems. It is instructive to recall at this time the first experiment to demonstrate unequivocally the chemical mediation of synaptic transmission by acetylcholine, the classic double heart experiment published by Otto Loewi in 1921 (See Fig. 1-11), because it is so beautiful an example of a *bioassay,* the use of a physiological response to assay for a biologically active compound. The bioassay remains an essential tool by which neuropharmacologists identify and quantitate neuroactive substances for which no sufficiently sensitive or specific chemical assay is available.

Synthesis and release. Unlike the other neurotransmitters and neurohormones that we shall discuss below, acetylcholine is not simply one member of a class of closely related compounds. Rather it has some unique properties that place it in a class by itself. It is synthesized in nerve terminals from acetyl-CoA and choline, in a reaction catalyzed by the enzyme *choline acetyltransferase* (Fig. 8-3). Although acetyl-CoA and choline are common metabolites present in all cells, choline acetyltransferase (and hence acetylcholine) is not. In fact, the presence of choline acetyltransferase in a neuron is sufficient to define it as a cholinergic neuron. The acetylcholine is packaged in synaptic vesicles (see Chapter 6) and is released following the arrival of an action potential at the nerve terminal via vesicle exocytosis. However, not all the acetylcholine in nerve terminals is in vesicles. Some is in the cytoplasm, and evidence exists for the direct release of acetylcholine from these cytoplasmic stores at some synapses.

Degradation and resynthesis. Following its release into the synaptic cleft, acetylcholine can bind to at least two distinct classes of receptor molecule and produce different responses in different target cells (see Chapter 9). However, the neurotransmitter does not remain at a high concentration in the synaptic cleft for very long. The fate of the released acetylcholine is to be destroyed by a powerful hydrolytic enzyme, acetylcholinesterase, which produces acetate and choline (Fig. 8-3). This enzyme is clustered at high concentration in the synaptic cleft. In addition its catalytic rate (of the order of 10^4 to 10^5 substrate molecules hydrolyzed per second) ranks it among the most rapid enzymes known, ensuring that the concentration of acetylcholine in the cleft drops very quickly following its release. Much of the choline is then taken up again into the nerve terminal and utilized for the replenishment of the terminal's acetylcholine. Cholinergic nerve terminals contain a high affinity sodium-dependent choline uptake system (Fig. 8-3) that provides a large proportion of the choline in the terminal, and the activity of which is probably rate limiting for acetylcholine synthesis. As we shall see below, sodium-dependent uptake systems specific for particular neurotransmitters, or for their precursors or metabolites, are ubiquitous in nerve terminals, and play an essential role in regulating the amount of neurotransmitter available for release.

The Amine Neurotransmitters

Epinephrine and norepinephrine. During the period in which it was established by Loewi and others that acetylcholine is the chemical transmitter in peripheral *parasympathetic* nerves such as the vagus nerve, other neuropharmacologists were investigating the nature of synaptic transmission in the *sympathetic* nervous system, another component of the peripheral

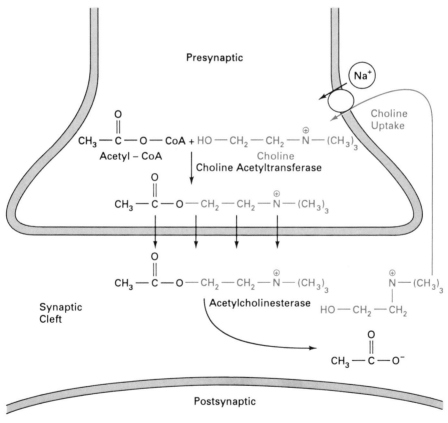

FIGURE 8-3. Synthesis, release, and degradation of acetylcholine. The transmitter is synthesized from choline and acetyl-CoA by the enzyme choline acetyltransferase. Following its release it is broken down rapidly by the enzyme acetylcholinesterase to choline and acetate. The choline can be taken back up into the terminal via a sodium-coupled choline transport system.

autonomic nervous system. There is a classic set of criteria that must be fulfilled to conclude that a particular compound is indeed a physiological transmitter substance. It was demonstrated early that the compound *norepinephrine* (Fig. 8-4) fulfills all the criteria for the sympathetic neurotransmitter, in that it

1. is synthesized and stored in high concentration in sympathetic nerve terminals;
2. is released on stimulation of sympathetic nerves; and
3. mimics the action of the endogenous transmitter when it is applied exogenously to target organs.

In addition, pharmacological agents that block the response to sympathetic nerve stimulation also antagonize the actions of exogenously applied norepinephrine. Norepinephrine can bind to several different classes of *adrenergic receptor* on the postsynaptic cell; these receptors can be distinguished on the basis of their pharmacological properties and the mechanisms they use to transduce the signal from the neurotransmitter into a response in the target cell (see Chapter 10).

For technical reasons it was much more difficult to establish a neurotransmitter role for norepinephrine in the central nervous system, but this is also now widely accepted. It is now known that most of the brain synapses that use norepinephrine arise from neurons whose cell bodies lie in the *locus ceruleus,* a cluster of only several thousand neurons located in the midbrain.

Adrenergic synapses use either norepinephrine (noradrenaline) or its N-methylated derivative epinephrine (adrenaline) as their neurotransmitter (Fig. 8-4). The fuzziness in the distinction between neurotransmitters and hormones, to which we have referred above, arises even in this most classic case of adrenergic transmission. Although most adrenergic synapses use *norepinephrine* as their neurotransmitter, *epinephrine* is synthesized and stored at high concentration in the *chromaffin cells* of the adrenal medulla, and is released into the bloodstream to act as a hormone on many of the same target organs that receive sympathetic innervation. During embryonic development the adrenal medulla arises from the same precursor cells as do sympathetic neurons. They share many properties with sympathetic neurons, and thus are widely used as a readily accessible model system for investigating adrenergic transmission (see Chapter 13).

Dopamine. Both norepinephrine and epinephrine belong to the general family of compounds known as *catecholamines,* organic molecules that contain an amine group as well as a catechol nucleus (a benzene ring with two adjacent hydroxyl substitutions—see Fig. 8-4). Another important

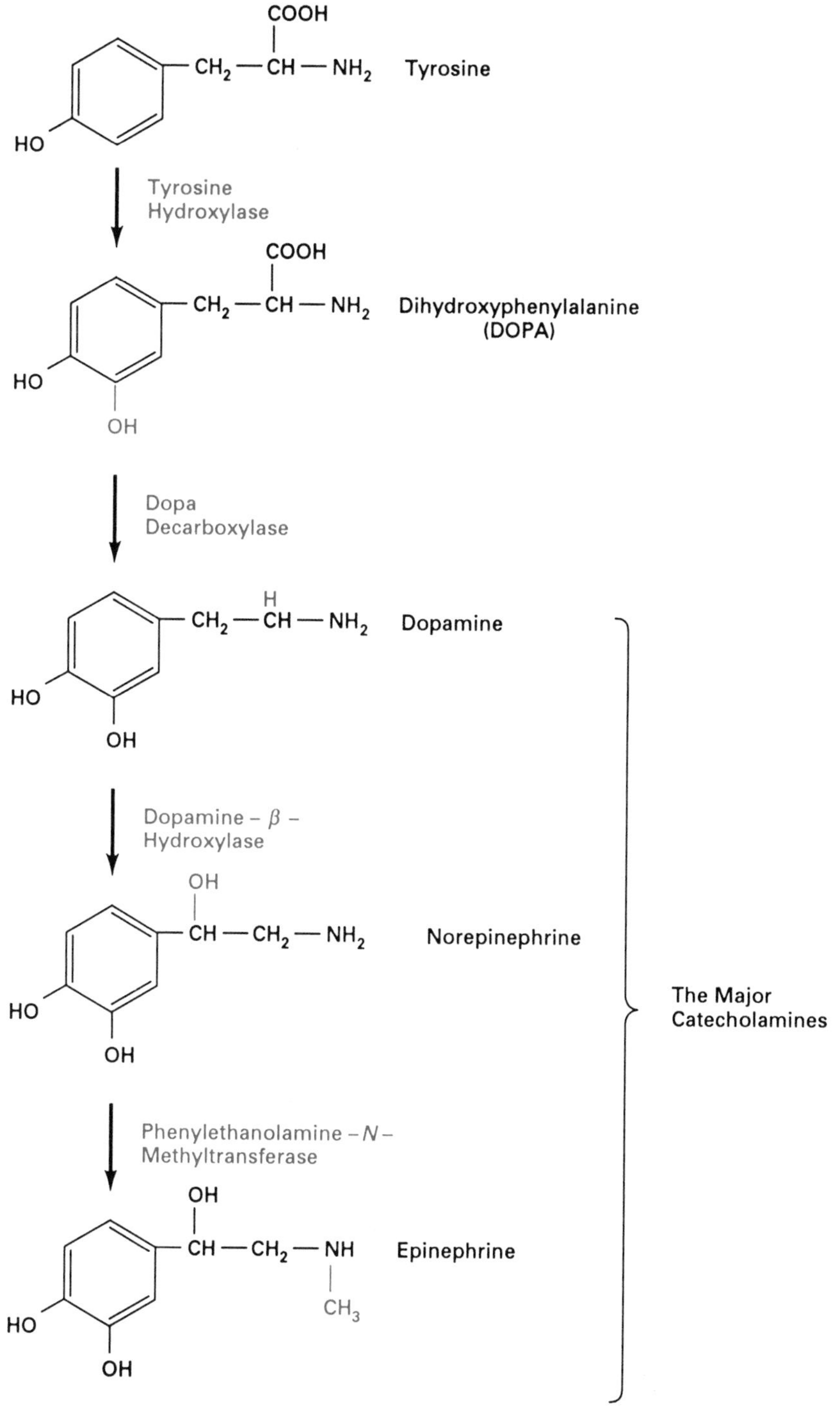

FIGURE 8-4. Biosynthesis of the catecholamines. The starting point for the synthesis of the catecholamines is the amino acid tyrosine. Hydroxylation of tyrosine to dihydroxyphenylalanine by the enzyme tyrosine hydroxylase is the rate-limiting step in the biosynthetic pathway.

member of this family is *dopamine,* which is an intermediate in the biosynthesis of norepinephrine and epinephrine, and is a major central nervous system neurotransmitter in its own right. In contrast to the relatively few adrenergic neurons in mammalian brain, there are several distinct nuclei—large collections of neurons—that contain dopamine, and dopaminergic axons are distributed in complex patterns to many parts of the brain. Dopamine systems play an essential role in certain motor functions as well as in behavior, mood, and perception. As we shall discuss briefly below, a number of debilitating diseases, including Parkinson's disease, schizophrenia, and bipolar depressive illness, can be attributed at least in part to a dysfunction in dopaminergic pathways.

Synthesis, storage, and release of the catecholamines. The biosynthetic pathway for the catecholamines is diagrammed in Figure 8-4. The first step is the hydroxylation of the common amino acid tyrosine to dihydroxyphenylalanine (DOPA) via the enzyme *tyrosine hydroxylase.* Tyrosine hydroxylase is found only in adrenal chromaffin cells and sympathetic neurons, and thus its presence is diagnostic of an adrenergic cell (analogous to choline acetyltransferase for cholinergic cell types). It is the rate-limiting enzyme for catecholamine biosynthesis and is subject to complex regulatory control; for example, its activity can be modulated by products in the biosynthetic pathway, and its synthesis is controlled by factors (such as nerve growth factor, see Chapter 13) that affect the growth and differentiation of sympathetic neurons.

The subsequent step in the pathway involves the removal of the carboxyl group from DOPA, via the enzyme DOPA decarboxylase, to produce dopamine, the first of the major catecholamine neurotransmitters. Dopamine can then be hydroxylated on the β carbon by dopamine β-hydroxylase (dopamine β-oxidase) to produce norepinephrine, which can in turn be methylated to epinephrine by a phenylethanolamine-*N*-methyltransferase (Fig. 8-4).

The catecholamines, like acetylcholine, are packaged into vesicles. The properties of catecholamine-containing vesicles have been most thoroughly studied in adrenal chromaffin cells (where they are called *chromaffin granules*) and sympathetic noradrenergic neurons, and it is assumed that the properties of the storage vesicles in central catecholaminergic neurons are similar. In addition to the neurotransmitter, catecholamine storage vesicles contain high concentrations of ATP (perhaps as high as 100 mM, about one-quarter the intravesicular concentration of the neurotransmitter), as well as a protein called *chromogranin,* whose function is only poorly understood but which may be involved in packaging and storage. Dopamine β-hydroxylase is also present in the granules, suggesting that at least one of the steps in the biosynthesis of norepinephrine occurs within the vesicles.

The release of catecholamines from chromaffin cells occurs by exocytosis following the entry of calcium into the cell, presumably via voltage-dependent calcium channels as is the case for cholinergic neurons. Because of the large size of chromaffin granules (in the range of 0.1 μm), exocytosis can be observed directly in the microscope, and can be studied with techniques such as capacitance measurements as described in Chapter 6. Exocytosis has, however, not yet been established for other catecholamine-releasing neurons. In fact it seems somewhat puzzling that a neuron would dump the energetically expensive contents of its adrenergic granules into the extracellular space. There is evidence that the ATP may play some role in communicating with adjacent cells, but this is probably not the case for the chromogranin and dopamine β-hydroxylase, which would have to be replaced via new protein synthesis and axonal transport, processes that are slow and might place an apparently unnecessary demand on the cell's energy resources. It is possible that the proteins remain associated with the vesicle membrane in some way that allows their reuptake, perhaps via active transport or endocytosis. Alternatively the suggestion has arisen, as in the case of acetylcholine, that nonvesicular release occurs.

Uptake and metabolism of the catecholamines. The actions of catecholamines on the target cells are terminated much more slowly than those of acetylcholine. There is no rapidly acting extracellular enzyme analogous to acetylcholinesterase. Instead catecholamines are removed from the synaptic cleft by reuptake into the presynaptic cell. The uptake is the sodium-dependent process we have come across in our discussion of choline uptake and will meet again often before this chapter ends. The high affinity active uptake mechanism moves catecholamines against a concentration gradient and causes them to accumulate inside the presynaptic cell at a much higher concentration than that in the extracellular space.

The two major enzymes involved in the catabolism of catecholamines are *monoamine oxidase* (MAO) and *catechol O-methyltransferase* (COMT). The former catalyzes the metabolism of catecholamines to their corresponding aldehydes (Fig. 8-5), which can then be further broken down to products that leave the brain and are excreted. MAO is localized in the mitochondrial membrane of catecholaminergic terminals, and helps to regulate the levels of catecholamines in these terminals. It is also present in other cell types in which its function is not understood. COMT catalyzes the methylation of one of the hydroxyl groups of the catechol nucleus (Fig. 8-5), again producing a product that is metabolized further and then excreted. Although COMT functions to metabolize catecholamines throughout the body, its precise localization and function in catecholaminergic transmission have yet to be determined. Inhibitors of MAO and COMT are important psychoactive drugs.

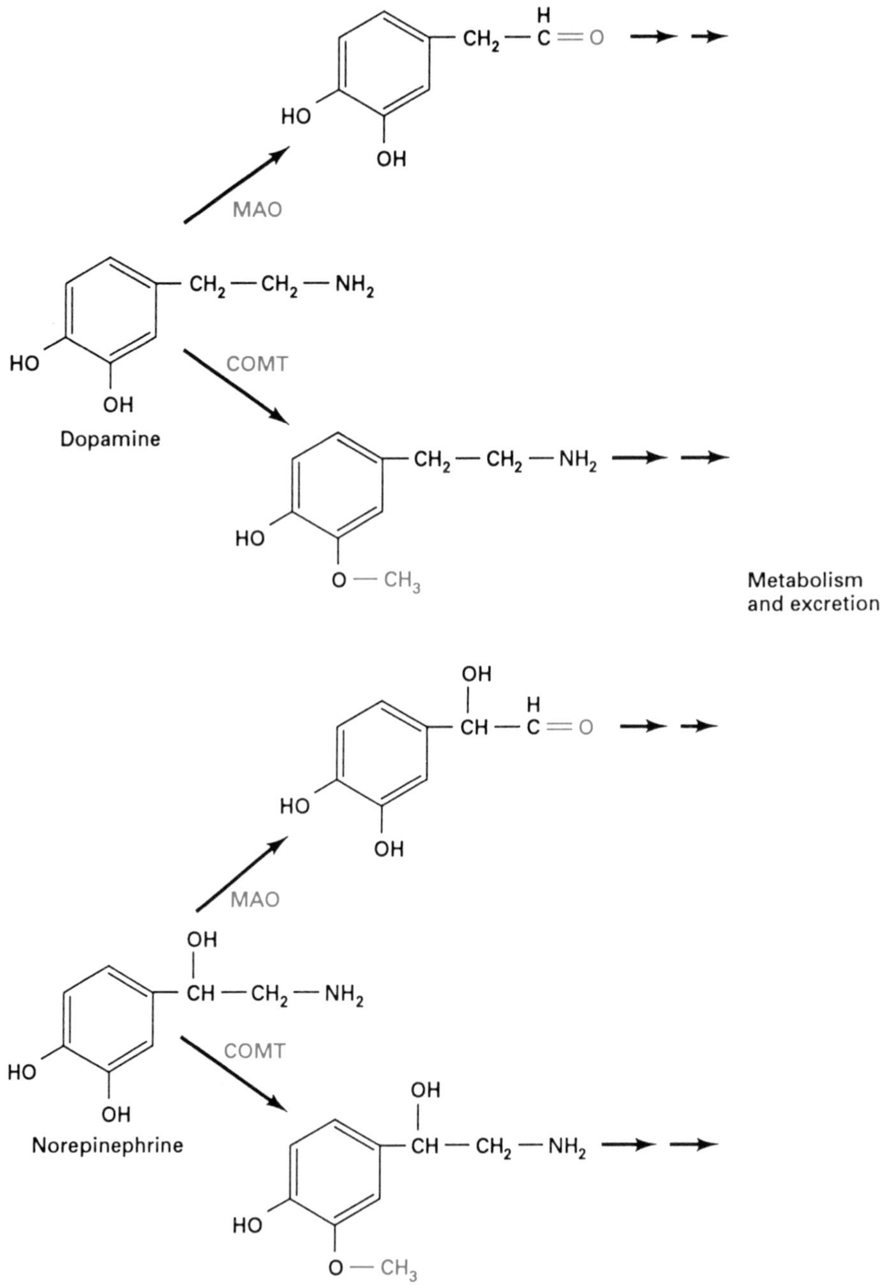

FIGURE 8-5. Metabolism of the catecholamines. Two enzymes, monoamine oxidase (MAO) and catechol-O-methyltransferase (COMT), catalyze the first steps in the degradation of the catecholamines.

Catecholamine pharmacology and nervous system dysfunction. There is evidence that dysfunction in brain catecholamine pathways contributes to *bipolar disorder* and to *schizophrenia.* The evidence is, however, indirect and is based largely on the fact that drugs that ameliorate the symptoms of these diseases interact with catecholamine systems. The classic finding (which is now more than 30 years old) is that a variety of MAO inhibitors such as pargyline, which cause a rise in brain catecholamine levels, are clinically effective antidepressants. A separate class of clinically effective compounds, the so-called *tricyclic antidepressants* such as imipramine, appear to prolong catecholamine action (predominantly at noradrenergic synapses) by inhibiting the high affinity reuptake system. Findings such as these have given rise to the *catecholamine theory of affective disorder,* which in essence states that decreased activity at certain central noradrenergic synapses causes behavioral depression. In addition, according to this theory mania results from excess activity at these synapses. The latter hypothesis is supported by the fact that the stimulant drug *amphetamine* increases activity in noradrenergic pathways, either by increasing transmitter release or (more likely) by inhibiting its reuptake. All of these data point to the sodium-dependent high-affinity neurotransmitter uptake systems as essential for the proper functioning of central nervous system synapses. More recently serotonin has also been implicated in bipolar disorders.

The *catecholamine theory of psychotic illness* focuses on dysfunction at dopaminergic synapses. A variety of antipsychotic drugs, the classic example being chlorpromazine, are effective blockers of postsynaptic dopamine receptors. It is particularly compelling that several chemically diverse groups of compounds (such as the butyrophenones, thioxanthenes, and phenothiazenes) all block dopamine receptors and ameliorate the symptoms of schizophrenia, giving rise to the hypothesis that excessive activity in dopaminergic pathways is responsible for schizophrenia. Again the evidence is indirect, but this has not deterred pharmaceutical companies from devoting enormous resources to the search for novel dopamine receptor blockers that might serve as antipsychotic drugs.

The role of dopaminergic pathways in certain motor functions is more firmly established in that there are anatomical findings to correlate with the pharmacology. The *basal ganglia,* several *nuclei* or collections of nerve cells deep in the brain, play an essential role in the control of body movements. The basal ganglia themselves receive inputs from the *substantia nigra,* a major dopamine-containing center in the brain, and influence the activity of the motor cortex via the thalamus (Fig. 8-6a).

Dopamine released by this pathway is essential for normal motor activity. The dopamine neurons that project from the substantia nigra to the *caudate nucleus,* one of the basal ganglia, can undergo pathological degeneration of unknown cause. As a result the influence of the substantia nigra is removed (Fig. 8-6b), and this leads to characteristic motor dysfunction, including low-frequency tremor of the extremities at rest and impairment of postural reflexes. This degenerative disorder is called *Parkinson's disease*, after the English physician Edward Parkinson who first described the symptoms more than a century ago. With the realization that Parkinson's disease involves a degeneration of dopaminergic neurons, it was reasoned that it might be possible to alleviate the symptoms by introducing dopamine into the brain. Because dopamine cannot readily enter the brain from the blood, the treatment of choice is to administer DOPA, which does enter the brain and is converted to dopamine via the action of DOPA decarboxylase

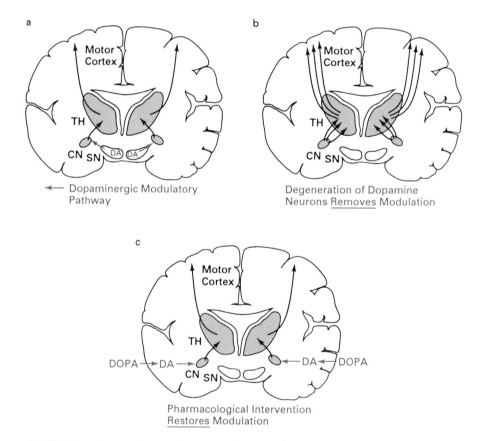

FIGURE 8-6. Dopamine pathways and motor functions (a–c). Dopamine (DA) neurons in the substantia nigra (SN) provide an important influence on motor functions. The DA neurons project to the caudate nucleus (CN), which in turn communicates with neurons in the thalamus (TH) that influence output from the motor cortex.

(Fig. 8-4). When Parkinson's patients are administered DOPA, together with an MAO inhibitor (Fig. 8-6c), there is a dramatic, albeit brief, alleviation of symptoms. This is a striking example of a disease of the nervous system that is attributable to the loss of a particular neurotransmitter and that responds to a rational treatment. Unfortunately DOPA therapy does not reverse the course of the disease, which involves a progressive degeneration and deterioration over a period of several years, but simply relieves the symptoms for some brief interval.

Serotonin. Serotonin, or 5-hydroxytryptamine (5-HT), is another neuroactive compound that is neither exclusively a hormone nor exclusively a classical neurotransmitter. Large quantities of serotonin are found in the circulation, for example in platelets, and also in the central nervous systems of vertebrates and invertebrates. Like the catecholamines, serotonin is synthesized from one of the common amino acids, in this case tryptophan. Tryptophan is taken up into neurons via a specific sodium-dependent uptake system and is hydroxylated in a reaction catalyzed by *tryptophan hydroxylase* (Fig. 8-7). As in the case of catecholamine biosynthesis, this initial reaction is the rate-limiting step in the formation of serotonin. The resulting 5-hydroxytryptophan is then immediately decarboxylated to produce serotonin. Serotonin is packaged into secretory granules and is presumed to be released by a calcium-dependent exocytotic mechanism, then taken back up into the presynaptic terminal where it can be degraded by MAO (Fig. 8-7).

Serotonin is localized in discrete groups of neurons, largely in the brainstem, that send projections all over the brain. However, the precise role of serotonergic pathways remains obscure. Serotonin has been implicated in the regulation of sleep and "vigilance," and it is widely believed that hallucinogenic drugs such as lysergic acid diethylamide (LSD) may produce their effects by interacting with serotonin pathways. The experimental evidence is complex and does not fall into a readily interpreted pattern, but it seems possible that LSD may antagonize some and mimic other actions of serotonin. In some invertebrates, for example lobsters and the marine snail *Aplysia,* serotonin acts as a neurotransmitter/neurohormone to influence certain behaviors via mechanisms that are becoming reasonably well understood at the cellular and even molecular level (see Chapters 16 and 17).

More amines. We have not yet exhausted the category of amine neurotransmitters. For example, evidence is rapidly accumulating that histamine can affect neuronal activity in the mammalian (as well as in the invertebrate) central nervous system. Another important compound in many species is octopamine, which is closely related to dopamine and norepinephrine. The nucleotide ATP may be released at certain synapses together with

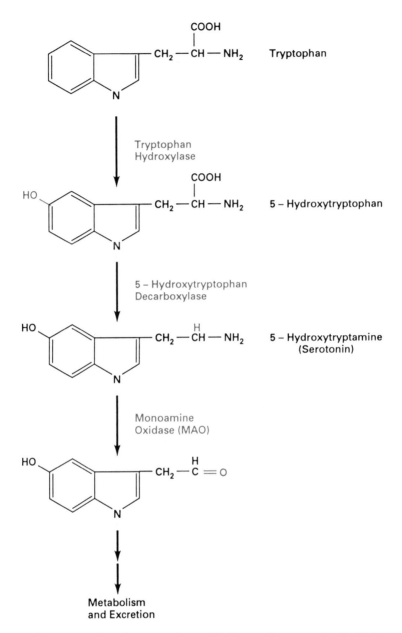

FIGURE 8-7. Synthesis and metabolism of serotonin (5-hydroxytryptamine or 5HT).

catecholamines and influence neuronal properties, and the nucleoside adenosine can also regulate neuronal activity. However, the evidence is not yet
as clear for other putative amine transmitter candidates as it is for the catecholamines and serotonin, and so we will not discuss these other candidates further. We simply remind the reader that the list of amine neurotransmitters is almost certain to be extended in the future.

Amino Acid Neurotransmitters

It should come as no surprise that it has been difficult to demonstrate a
neurotransmitter role for amino acids. Although such a role had been suspected for many years, how does one test the relevant criteria, for example
the *presence* of the compound in synaptic terminals, for compounds that
are present in high concentration not only in neurons but in all cell types?
Of the three major amino acid neurotransmitters in the mammalian nervous system, γ-aminobutyric acid (GABA), glycine, and glutamic acid (Fig.
8-8), this problem has been particularly acute for the latter two, which are
ubiquitous constituents of proteins; GABA, in contrast, is present almost
exclusively in brain and was recognized early on as a probable neurotransmitter. In any event accumulating evidence has more or less put the doubts

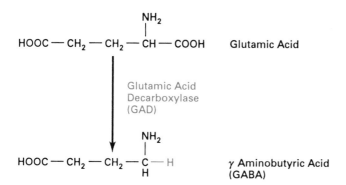

FIGURE 8-8. The major amino acid neurotransmitters.

to rest, and it is now widely accepted that the large majority of central nervous system synapses use amino acids as their neurotransmitters.

Glutamate and other excitatory amino acid transmitters. Glutamic and aspartic acid, and various synthetic analogs of these amino acids, produce excitatory (depolarizing) responses on neurons in virtually every part of the mammalian brain. Although glutamate has long been established as an excitatory neurotransmitter at insect and crustacean neuromuscular junctions, only more recently has it become evident that its pharmacological actions in mammalian brain reflect the fact that it is the major excitatory brain neurotransmitter. Much of the evidence has been based on the demonstration of several different classes of receptor for glutamate in brain (see Chapter 9), and there is a great deal of current excitement about the possibility that one of these receptor classes may play an essential role in long-term plastic changes at synapses, both during development and in the adult (see Chapters 15 and 17). It is not clear whether aspartate can be classified as a physiological neurotransmitter in its own right, or whether it simply interacts with one or more of the glutamate receptor classes.

The details of glutamate synthesis and catabolism may be found in any standard textbook of biochemistry, and we will not dwell on them here. The way in which the synaptic actions of glutamate are terminated is not understood, but, as is the case for all the neurotransmitter candidates we have discussed, there is a sodium-dependent high-affinity glutamate uptake system in neurons (and probably in glia) that is likely to be involved.

GABA and glycine: the inhibitory amino acid neurotransmitters. GABA has been much easier to deal with. Again it was first established as an inhibitory neurotransmitter at an invertebrate synapse, in crustacean muscle. Attention was focused on its role in mammalian brain when it was found that its concentration there is much higher than in any other tissue. GABA is synthesized from glutamic acid via a reaction catalyzed by the enzyme glutamic acid decarboxylase (GAD) (Fig. 8-8), the presence of which is considered to be positive identification of a GABAergic neuron.

The location of proteins within the brain (or other tissues) can be found using a technique called *immunohistochemistry*. This is an important technique that we shall encounter again, particularly in the chapters dealing with neuronal development and plasticity. An antibody that binds specifically to a protein is incubated with fixed or frozen sections of brain tissue. The sections are then incubated again with a fluorescent or colored reagent that binds to and reveals the location of the antibody, and the pattern of distribution of the protein to which the antibody binds can be observed in the microscope. Such immunohistochemical investigations, using specific anti-GAD antibodies, have revealed that GABAergic terminals are present throughout the brain. Thus GABA appears to be the major inhibitory neurotransmitter in the mammalian central nervous system.

Synaptic terminals contain a sodium-dependent high-affinity GABA uptake system, and the major route of catabolism is via a mitochondrial GABA-α-oxoglutarate transaminase that regenerates glutamic acid. GABA is a particularly interesting neurotransmitter because its actions are subject to modulation by a variety of pharmacological agents that have profound effects on brain function; we shall discuss in the next chapter the ways some of these agents interact with GABA receptors to modulate GABAergic transmission.

Glycine has not been easy to deal with because, like glutamate, there is too much of it. It appears, however, to be the most important inhibitory neurotransmitter in the spinal cord and lower brainstem (and probably in the retina as well). As in the case of glutamate, much recent progress has come from the identification and isolation of specific receptors that mediate the postsynaptic actions of glycine. Glycine can also interact with at least one class of brain glutamate receptor and modulate glutamatergic transmission. The actions of glycine in the spinal cord appear to be terminated by a sodium-dependent uptake system in presynaptic endings.

Sodium-Dependent Uptake Systems

Perhaps the most striking characteristic common to all of the small neurotransmitters is the existence of uptake systems that are responsible for halting neurotransmitter action and replenishing their supply in their presynaptic terminal. In the case of acetylcholine at the neuromuscular junction its action is terminated by enzymatic hydrolysis, but the choline that is produced is then taken up in the same way. Although the uptake systems are specific, in that a neuron that synthesizes and releases a particular transmitter will take up only that transmitter, they do have common features. In particular all require sodium to be present in the external medium. They use the energy provided by the sodium gradient across the membrane to concentrate transmitters within the terminal. That is, the movement of the transmitter molecule into the cell is coupled obligatorily to the movement of sodium ions across the membrane. In the case of the GABA transporter, which has been purified and is probably the best understood, chloride is also an obligatory cosubstrate (Fig. 8-9). The *stoichiometry* of the transport, the ratios of the three transported species, is 2 Na^+:1 Cl^-:1 GABA, and as a result the transport is electrogenic (i.e., it will tend to change the membrane voltage). The combination of the chloride and sodium gradients allows GABA to be concentrated inside the synaptic terminal to concentrations as much as 10,000-fold higher than those in the extracellular space. Although the other neurotransmitter uptake systems have not been studied as thoroughly as that for GABA, these properties seem to be common to all of them.

The Peptides

By the early 1970s a consistent pattern was emerging of neurotransmitter candidates in the central nervous system. Acetylcholine, serotonin, and the catecholamines were well established as major brain neurotransmitters, and although the transmitters at many central synapses remained unidentified, suspicions were beginning to arise that amino acids might fill this gap. The recognition that *neuroactive peptides* are more than simply hormones that are released to carry out actions on targets outside the brain, that they play a crucial role *within* the central nervous system, has made it clear just how naive this simple picture was.

Neuropeptides share some of the characteristics of the small molecule

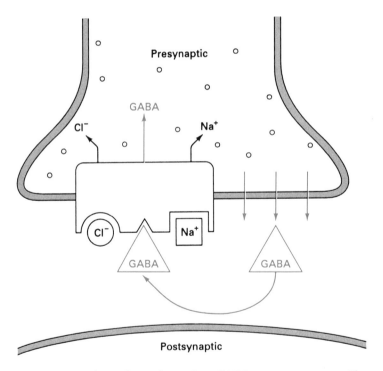

FIGURE 8-9. The sodium-dependent GABA transport system. The coupled transport of GABA, Na⁺, and Cl⁻ is depicted (see Kanner and Shuldiner, 1987).

neurotransmitters. For example, like many classical transmitters, some neuropeptides are released to the circulation and act at a distance, while others are confined to a discrete synaptic cleft. As mentioned at the beginning of this chapter, it will not be profitable to focus on the largely semantic distinction between *neurotransmitters* and *neurohormones*. Nevertheless, the neuropeptides do differ in many respects from their smaller counterparts. Table 8-1 lists some of the properties of peptide neurotransmitters that tend to distinguish them from the classical neurotransmitters. Because these are not all true at every location at which peptidergic transmission occurs, they are more appropriately termed trends rather than properties.

The roster of putative peptide neurotransmitters continues to grow at an enormous pace (Table 8-2), and we cannot do justice to this rapidly developing field in a short space. We will limit ourselves here to describing just a few of the neuroactive peptides, with emphasis on the historic development of this field and some examples of the trends listed in Table 8-1.

Substance P. The story of the peptides actually began almost 60 years ago, with the accidental discovery of *substance P*. While screening various tissues for acetylcholine, Ulf von Euler and John Gaddum found a compound in brain and intestine that caused a lowering of blood pressure and contraction of intestinal smooth muscle. Because its actions were not blocked by the cholinergic antagonist atropine they concluded that this compound could not be acetylcholine, and named it substance P (for *p*owder). Although von Euler suggested as early as 1936 that substance P might be a protein, it was to be more than 30 years before its structure was determined. Substance P is an 11 amino acid peptide that is present, as determined by direct bioassay and immunohistochemical studies, throughout the mammalian brain. It is thought to be a synaptic transmitter in sensory pathways concerned with pain and touch. It is, in fact, found in neurons that innervate tooth pulp, where the only known sensory modality is pain. Substance P was the first, and remains the best understood, of the so-called

TABLE 8-1 Properties of Peptide Neurotransmitters

Synthesized as a larger precursor protein at the soma and transported
   to release sites; must be replenished by synthesis at soma

Slow postsynaptic effects

Actions terminated by extracellular proteases or by diffusion

Co-released with classical neurotransmitters

Can trigger complex coordinated behaviors

Actions do not require point-to-point synaptic connections

brain–gut peptides, neuroactive substances found in both the central nervous system and the gastrointestinal tract. There are now more than a half dozen examples of this localization pattern, which reflects the fact that the gut is densely innervated.

Hypothalamic peptides. As previously mentioned, the hypothalamus is the site of synthesis of the two classical neurohormones, oxytocin and vasopressin. In addition to affecting peripheral tissues, oxytocin and vasopressin can alter the firing patterns of central neurons, and vasopressin in particular has been reported to have behavioral effects in rats and humans. The magnocellular neurons in which these peptides are synthesized project not only to the pituitary circulation, but also to a variety of sites within the brain, and the immunohistochemical localizations of oxytocin and vasopressin are consistent with their involvement in synaptic transmission (Fig. 8-10).

Also synthesized in the hypothalamus are other biologically active peptides, the so-called *releasing factors* or *releasing hormones*. The releasing hormones enter the portal circulation, which brings them to the anterior pituitary where they regulate the release of another group of pituitary peptide hormones. These in turn act on peripheral tissues. For example the release of *thyroid-stimulating hormone* (thyrotropin or TSH), which enters the general circulation from the pituitary to act on the thyroid gland, is promoted by *thyrotropin-releasing hormone* (TRH) from the hypothalamus

TABLE 8-2 Examples of Some Thoroughly Studied Neuroactive Peptides

Peptide	Sequence[a]
Substance P	RPKPQQFFGLM-NH$_2$
Neurotensin	pELYENKPRRPYIL
Vasoactive intestinal peptide	HSDAVFTDNYTRLRKQMAVKKYLNSILN
TRH	pEHB-NH$_2$
LHRH	pEHWSYGLRPG-NH$_2$
Oxytocin	CYIQNCPLG-NH$_2$
Vasopressin	CYFQNCP$_K^R$G-NH$_2$
α-Endorphin	YGGFMTSEKSQTPLVT
Met-enkephalin	YGGFM
FMRFamide	FMRF-NH$_2$
Proctolin	RYLPT
Egg-laying hormone	ISINQDLKAITDMLLTEQIRERQRYLADLRQRLLEKG-NH$_2$

[a]Single letter amino acid code: A,Ala; R,Arg; N,Asn; D,Asp; C,Cys; Q,Gln; E,Glu; G,Gly; H,His; I,Ile; L,Leu; K,Lys; M,Met; F,Phe; P,Pro; S,Ser; T,Thr; W,Trp; Y,Tyr; V,Val. A small p at the N-terminal indicates a pyroglutamate; some C-terminals are amidated (NH$_2$).

(Fig. 8-10). Much TRH is also found in other brain regions where it may act as a local hormone or neurotransmitter. This is true for other releasing factors as well.

Luteinizing hormone-releasing hormone (LHRH) is another of the releasing factors. It was in fact the first peptide to be shown convincingly to be the neurotransmitter at a particular synapse, an excitatory synapse in the frog sympathetic ganglion. This synapse illustrates many of the trends listed in Table 8-1. Stimulation of the presynaptic fibers produces a slow

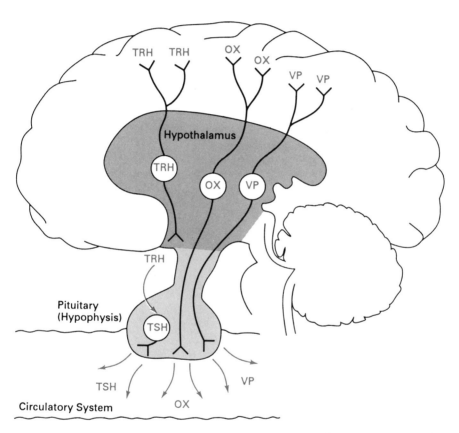

FIGURE 8-10. The hypothalamus is an important source of neuroactive peptides. Oxytocin (OX) and vasopressin (VP) neurons project to the posterior pituitary (also called the neurohypophysis) where the peptides are released to the general circulation. Releasing hormones, here exemplified by thyrotropin-releasing hormone (TRH), reach the anterior pituitary (adenohypophysis) via the portal circulation and stimulate release of hormones such as thyrotropin (TSH). The hypothalamic peptides can also act as brain neurotransmitters (see Silverman and Zimmerman, 1983).

and long-lasting depolarization of the postsynaptic cell, which is known as the *late, slow excitatory postsynaptic potential.* This response is mimicked by application of LHRH, and blocked by drugs that are known to block other (nonneuronal) actions of LHRH. The LHRH-mediated postsynaptic potential is observed in two types of sympathetic ganglion neurons, the large *B* cells and the smaller *C* cells. Interestingly the LHRH-containing presynaptic fibers do not synapse on the B cells directly. The postsynaptic response in the B cells occurs because the released LHRH can *diffuse* to them from the release sites micrometers away. This illustrates that point-to-point wiring of synapses is not required for this synaptic response.

The slow time course of the postsynaptic response can also be shown to be due to the *lingering presence* of LHRH at the postsynaptic cells. If a drug which blocks the actions of LHRH is applied *during* a postsynaptic potential, the response is terminated immediately, indicating that the continued action of LHRH is required to maintain the response. Finally, LHRH is one of the peptides that can produce prolonged and complex behavior in an animal. There is no "acetylcholine behavior" or "GABA behavior," but there is an "LHRH behavior." If LHRH is administered into the brains of female rats, it evokes a complex and stereotyped pattern of sexual behavior.

Opioid peptides. The age of the peptides was ushered in with a vengeance in the mid-1970s with the description of the *enkephalins* and other opioid peptides. The story of the opioid peptides is particularly interesting because their discovery was not simply serendipitous, but resulted from a rational search for endogenous compounds that might mimic the actions of morphine and other *opiates.*

Throughout recorded history it has been known that the juice of the poppy produces feelings of euphoria and is a highly effective pain killer. The active ingredient, morphine (named after *Morpheus,* the Greek god of dreams), was first isolated almost 200 years ago. By the early 1970s it was reasoned that the actions of morphine and other opiates must result from their binding to specific receptor sites in the brain, and it became possible to demonstrate the binding of radioactive opiates to brain membranes. The next assumption was that if the brain contains specific receptor sites, there must be endogenous ligands that bind to and activate these receptors.

Two compounds with such properties were indeed found. They were first isolated by their ability to produce *analgesia,* relief from pain, in bioassays for pain relievers. These compounds are pentapeptides of similar structure, which were named *Met-enkephalin* and *Leu-enkephalin* because they differ only in their carboxyl terminal amino acids:

Tyr-Gly-Gly-Phe-*Met*   *Met*-enkephalin
Tyr-Gly-Gly-Phe-*Leu*   *Leu*-enkephalin

Synthetic enkephalins were also soon shown to be potent opiates in standard bioassays for analgesics.

It was noted that the entire structure of Met-enkephalin is contained as amino acid residues 61–65 of a much larger pituitary hormone, the 91 amino acid β-lipotropin, which had been isolated and sequenced several years earlier but whose function was obscure. Within a year after the description of the enkephalins it had become evident that at least three additional longer peptides with opioid-like activity, the so-called *endorphins*, are contained within the sequence of β-lipotropin. Immunohistochemical experiments have demonstrated that the enkephalins and endorphins are distributed widely in the brain, but do not appear to be co-localized within the same neurons.

The enkephalin/endorphin story emphasizes another important way the peptides differ from the more classical neurotransmitter candidates (see Table 8-1). The latter tend to be small molecules that can be synthesized and metabolized by enzymes present in nerve terminals, whereas the peptides must be synthesized by the protein biosynthetic machinery, presumably in the cell bodies, and undergo complex processing and packaging (see Chapter 6) before they can exert their biological effects. Questions as to how peptide stores can be replenished rapidly to sustain release for long periods of time have not yet been resolved, whereas this does not seem to be a problem for the classical neurotransmitters.

The finding that the endorphins and enkephalins are contained within a larger precursor was not entirely surprising, since it was known that many peptide hormones, for example, insulin and the original neurohormones oxytocin and vasopressin are synthesized as larger *prohormones* and then processed proteolytically to produce the active hormone (see Chapter 6). In fact β-lipotropin is contained within a much larger and more complex precursor. Molecular cloning approaches have allowed the demonstration that the actual precursor is a polypeptide that contains within its sequence the structure of the pituitary hormone corticotropin (ACTH), as well as of the opioid peptides and several other naturally occurring peptides of unknown function. This rather astonishing discovery has many ramifications. For example, in Chapter 16 we shall discuss the physiology of neurons that use several peptide neurotransmitters. Each of these is cleaved from a different part of a single precursor protein (see Fig. 6-6b).

Some invertebrate peptides. Invertebrates have been a rich source of neuroactive peptides, some of which are also present in vertebrate nervous systems (Table 8-2). Among the best studied of the invertebrate neuropeptides are the pentapeptide *proctolin,* first isolated from the cockroach hindgut, the snail tetrapeptide *FMRFamide,* and the 36 amino acid *Aplysia egg-laying hormone.* The genes encoding the precursors for some of these peptides show remarkable homologies to certain mammalian neuropeptides. Furthermore the large size and ready identifiability of many invertebrate neurons provide experimental advantages that have allowed the physiological role of some of these peptides to be investigated in considerable detail. For example, the synthesis, processing, and release of egg-laying hormone by the neurosecretory *bag cell neurons* of *Aplysia,* as well as its mechanism of action on target neurons, are becoming well understood. Similarly the actions of proctolin in modulating behavioral patterns in the lobster have been thoroughly analyzed at the cellular level. Some of these systems will be discussed in more detail in subsequent chapters.

Co-localization of peptides and classical neurotransmitters. The pharmacologist Henry Dale suggested many years ago that a given neuron synthesizes only a single neurotransmitter and releases only that transmitter at all its terminals. However, there is now convincing histochemical evidence that some neurons contain one or more neuropeptides *and* a classical neurotransmitter, packaged in different vesicles but often present in the same synaptic terminal.

It is not known what flexibility this offers a neuron. It has become clear, however, that the different transmitters need not be released at the same time. In several cases it has been found that only the "classical" transmitter is released by low-frequency stimulation, and co-release of the peptide requires short bursts of high frequency stimulation. It is not evident how the selective release of one transmitter, and not another, comes about, but it may have to do with the very different patterns of calcium distribution expected in a nerve terminal invaded by low-frequency and high-frequency action potentials. At low firing frequency there will be brief and highly localized changes in calcium immediately under the membrane and adjacent to the voltage-dependent calcium channels, whereas calcium levels elsewhere, for example, away from the membrane, will be little affected. At higher rates of stimulation, however, calcium levels will increase progressively with each action potential during a burst, and a substantial elevation will also occur away from the immediate vicinity of the calcium channels. Thus if the peptide-containing vesicles undergo exocytosis only at high levels of calcium, or if they are preferentially located at a distance from the membrane, peptide release will occur only with higher frequencies of presynaptic stimulation (Fig. 8-11). We can see, then, that the coexistence of different neurotransmitters in a single neuron allows it to produce different effects on a postsynaptic target, depending on the pattern of stimulation.

Questions about the functional significance of such co-localization of neurotransmitters can be broadened to ask why, in fact, there are so many neuroactive substances. In principle a nervous system could function with just two neurotransmitters, one to mediate excitation and the other for inhibition. Indeed, as we shall see in the next two chapters, since there are several different receptors for most (if not all) neurotransmitters, even a *single* neurotransmitter might be sufficient and the excitatory or inhibitory nature of the response could be determined by the type of receptor that

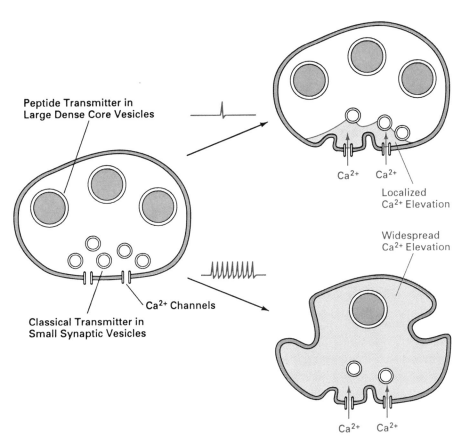

FIGURE 8-11. The release of classical and peptide neurotransmitters at synaptic endings. A hypothesis for the selective release of classical neurotransmitters at low frequencies of stimulation and for co-release of both types of transmitters during rapid bursts of action potentials.

happens to be present on the target cell. We do not have a definitive answer to this dilemma, but again a reasonable explanation is that multiple neuroactive substances provide both a wide range of *times* over which the response endures and a wide range in the *character* of the response. It is a mistake to think of neurons as simply on or off, active or inactive. Rather, as we have seen in Chapter 4, neurons contain a variety of different ion channels that allow their activity to be modulated in subtle ways, and it may be that a multiplicity of transmitters (and receptors) has evolved to participate in this fine tuning of neuronal function.

Summary

A multitude of chemicals mediate intercellular communication in the nervous system. Some of them have been well understood for more than half a century, yet new ones are still appearing regularly. Although they exhibit great diversity in many of their properties, all are stored in vesicles in nerve terminals and are released to the extracellular space via a process requiring calcium ions. Their role is to alter the properties, chemical and/or electrical, of some target cell. With the arrival on the scene of the peptides over the last 15 years it has become evident that signaling in the nervous system occurs through the use of rich and varied forms of chemical currency, and that some neurons use more than one type of currency simultaneously.

Receptors and transduction mechanisms I. Receptors coupled directly to ion channels

All of the neuroactive substances we discussed in the previous chapter alter some property of a target neuron. This effect on the target cell is highly specific for the particular neurotransmitter or neurohormone. How does a neuron know just which of the many possible chemical signals it is being tickled by, and how does it decide on a response appropriate for that signal? Obviously the answers to these questions are crucial for understanding intercellular communication in the nervous system. In this and the following chapter we will discuss the *neurotransmitter* and *hormone receptors* that recognize the signaling molecules and the *transduction mechanisms* that convert the extracellular signal into a response, usually (but not always) electrical, in the target neuron. The focus of this chapter is on neurotransmitter receptors that are coupled *directly* to the ion channels whose activity they regulate (Fig. 9-1). We emphasize here, as we have previously, that these mechanisms are not restricted to nerve cells. Other cell types have receptors for intercellular signaling molecules and convert the signal into some biological response. Neurons are unusual in that their biological response is normally a change in voltage that ultimately alters transmitter release, allowing the signal to be passed along from one neuron to another in a multineuronal pathway.

Specificity of Responses

The pharmacological concept of specific receptor structures has been with us for almost a century. Pharmacologists recognized many years ago that drugs that produce specific effects must interact with specific sites on or in the cell. It became possible to build up a pharmacological profile of a receptor by identifying drugs that produce a particular response (receptor *agonists*), and other drugs that inhibit the response to agonists (receptor *antagonists*). Furthermore the concentrations at which these various agonists and antagonists are effective can be measured, to provide a quantitative pharmacological profile. The earliest examples of such pharmacological distinc-

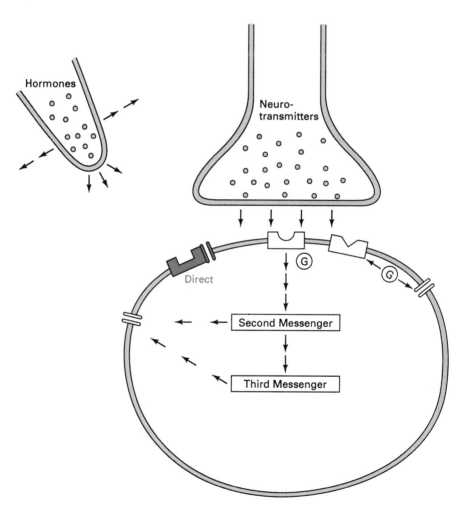

FIGURE 9-1. Intercellular communication. This chapter will discuss receptors for neurotransmitters that are coupled directly to the ion channels that they regulate (red).

tions arose from studies with the first identified neurotransmitter, acetylcholine.

Although acetylcholine was found to be the transmitter substance at neuromuscular junctions, in sympathetic and parasympathetic ganglia of the peripheral autonomic nervous system, and in postganglionic parasympathetic neurons, the pharmacological profile was not identical for all responses to acetylcholine. Instead it was found that nicotine and related compounds can act as agonists at neuromuscular junctions in skeletal muscle, but not in cardiac muscle, and only at *some* cholinergic synapses in the autonomic nervous system (Fig. 9-2). Furthermore responses at these so-called *nicotinic* synapses, but not at the others, could be blocked by compounds such as hexamethonium and the plant alkaloid curare (Table 9-1). It is worth noting that this pharmacology has been understood for a very

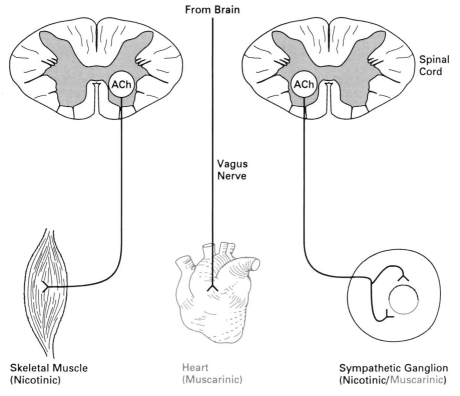

FIGURE 9-2. Acetylcholine activates two different classes of receptors. Some actions of acetylcholine (ACh) released from cranial or spinal neurons can be mimicked by nicotine, whereas others can be mimicked by muscarine. The receptors that mediate these different classes of response are termed nicotinic and muscarinic, respectively.

long time by South American tribes that use curare at the tips of their arrows to paralyze prey.

In contrast the action of acetylcholine at other cholinergic synapses can be mimicked by muscarine but *not* by nicotine (Fig. 9-2) and blocked by atropine and quinuclidinylbenzylate but *not* by curare (Table 9-1). It now appears that such *muscarinic* synapses are far more numerous than nicotinic synapses, and probably account for most cholinergic synapses in the mammalian central nervous system.

Although this breakdown of acetylcholine responses into two distinct groups seems to make life more complicated for the harassed neuroscientist trying to understand transmitter actions, this is only the tip of the iceberg. Virtually every neurotransmitter is now known to interact with more than a single class of receptor, and as shown in Table 9-2 acetylcholine not only has two major classes of receptor, but *within* the muscarinic class there are several different subtypes. About the time the pharmacology of acetylcholine responses was being clarified, norepinephrine was also shown to interact with more than a single class of receptor, and the number of adrenergic receptor subtypes has increased over the years as new, more specific drugs have become available. Similarly the other amines, the excitatory and inhibitory amino acids, and the peptides all can mediate multiple responses in target cells (Table 9-2). As we shall see below, the structural basis for these multiple responses is becoming understood, as methods for measuring receptor properties have become available, and as the techniques of molecular cloning have been brought to bear on the receptor molecules.

Dual-Action Neurons

The finding that a single neurotransmitter can produce more than a single response led to the concept of *multiaction neurons,* neurons that synthesize

TABLE 9-1 Two Pharmacologically Distinct Receptors for Acetylcholine

| Receptor Type | Some Examples of | | Responding Cell Type |
	Agonists	Antagonists	
Nicotinic	Acetylcholine Nicotine	Hexamethonium Curare	Skeletal muscle Sympathetic neurons Parasympathetic neurons Some brain neurons
Muscarinic	Acetylcholine Muscarine Oxotremorine	Atropine Scopolamine Quinuclidinylbenzylate	Smooth muscle Sympathetic neurons Gland cells Heart cells Many brain neurons

and release only a *single* neurotransmitter but still produce *different* responses in different target cells that contain distinct classes of receptor for this transmitter. Such a neuron is present in the central nervous system of the sea snail *Aplysia*. The large identifiable nerve cells in the central ganglia of this creature provide an ideal system in which to test the concept of a multiaction neuron. When an identified neuron, called L10, in the *Aplysia* abdominal ganglion is stimulated, different synaptic responses can be seen in two identified follower neurons, L3 and R15 (Fig. 9-3a; see also Fig. 11-5). A single transmitter, acetylcholine, mediates both of these responses. Furthermore several kinetically and pharmacologically distinct responses to a single neurotransmitter (again acetylcholine) can sometimes be observed in the *same* postsynaptic neuron (Fig. 9-3b). Such findings demonstrate that a single neuron may express on its surface several distinct receptor subtypes for a given transmitter. These findings are important, because the ability of a single presynaptic neuron to elicit more than one response has profound implications for the way we must think about integration and computation in networks of interacting neurons.

Receptor-Binding Assays: From Concept to Physical Entity

We have mentioned that the pharmacological concept of specific receptor sites for transmitters has been around for some time. It is only in the last 20 years, however, that techniques have become available for the direct measurement of membrane receptors. The development of *ligand-binding assays* for this purpose has confirmed the diversity of receptors inferred from the pharmacology of neurotransmitter responses. Furthermore bind-

TABLE 9-2 Neurotransmitters That Interact with More Than One Class of Receptor[a]

Neurotransmitter	Receptor Class
Acetylcholine	Nicotinic
	Muscarinic
Norepinephrine	α-Adrenergic
	β-Adrenergic
Dopamine	DA_1
	DA_2
GABA	$GABA_A$
	$GABA_B$
Glutamate	NMDA
	Kainate/quisqualate

[a]There are also subtypes within some of the receptor classes shown here.

ing assays have enormously expanded our understanding of receptor structure and function, and so it is important to understand this approach and the kinds of information it can provide.

The fundamentals of the ligand-binding assay are shown in Figure 9-4a. The term *ligand* (from the Latin *ligare,* to tie or bind) is used to denote a molecule that will bind to the receptor under study. The ligand is labeled in some way, usually with radioactivity but occasionally with a fluorescent probe, and is incubated together with plasma membrane fragments from cells that contain the receptor. After an appropriate period of time to allow the ligand to bind to receptors on the plasma membrane fragments, the bound ligand is separated from the remaining free ligand (the ligand is added in large excess, so that even when all receptor sites are occupied there will still be some free ligand). The amount of radioactivity bound can be measured, and under appropriate conditions it provides a measure of the number of specific binding sites. Binding studies can also provide an estimate of *affinity,* a measure of how tightly the ligand binds to the receptor (Fig. 9-4b).

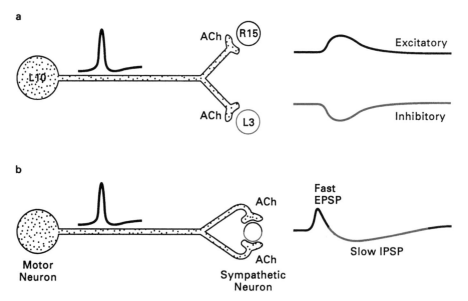

FIGURE 9-3. Dual action neurons. *a*: Eric Kandel and his colleagues (see Kandel, 1976) showed that the identified neuron L10 in the *Aplysia* abdominal ganglion can produce different actions on two different synaptic targets, neuron R15 and neuron L3 (see Fig. 11-5). *b*: In the vertebrate sympathetic ganglion, a presynaptic neuron may have more than one effect on a *single* postsynaptic target cell.

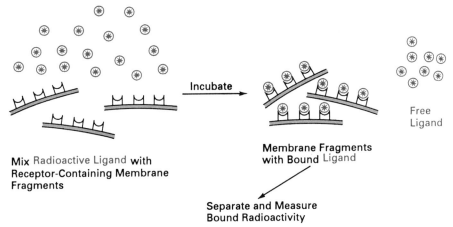

a

Mix Radioactive Ligand with
Receptor-Containing Membrane
Fragments

Incubate

Membrane Fragments
with Bound Ligand

Free
Ligand

Separate and Measure
Bound Radioactivity

b

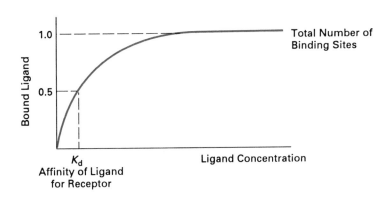

$$R + L \underset{k_{on}}{\overset{k_{off}}{\rightleftarrows}} RL \qquad K_d = \frac{k_{off}}{k_{on}}$$

FIGURE 9-4. Ligand-binding assays for receptors. *a*: Radioactive ligands (L, red) that bind to the membrane receptor under investigation (R, black) can be used to characterize some properties of the receptor. *b*: Under appropriate conditions the ligand-binding assay can provide a measure of the number of receptors, and of their affinity for the ligand.

Such binding assays are easy to perform, usually far easier than assays of the biological response mediated by the receptor. Moreover it is possible to test rapidly whether a wide variety of compounds are possible receptor agonists and antagonists, by determining whether they can compete with the radioactive ligand for binding to the receptor (Fig. 9-5a). When the amount of bound ligand is plotted as a function of the concentration of the different competitors (A_1–A_4 in Fig. 9-5b), a family of curves is generated. The concentration ($K_{1/2}$) at which these compounds displace half the radioactive ligand is characteristic for a given receptor. In other words, a detailed pharmacological profile of the binding site can be generated readily, without the need for radioactive labeling of each compound to be tested. Such studies have allowed the precise definition of various receptor subtypes, some of which are summarized in Table 9-2. It will be evident that the ligand for these experiments must be chosen with care. Using the neurotransmitter itself is generally inappropriate, since most transmitters will bind simultaneously to several different classes of binding site, and this will hopelessly confound attempts to define the pharmacology. However it usually is possible to find a selective ligand that binds only to one subset of the receptors for a particular neurotransmitter.

It is essential to remember that *a binding site is not necessarily a physiological receptor.* Recall that neurotransmitters must bind to a variety of proteins including uptake systems and metabolic enzymes. To determine whether a binding site is a real receptor, it is necessary to compare its pharmacology with that of the biological response mediated by the receptor. Only if the pharmacological profiles coincide can one be confident that the binding site and the receptor are the same. This can be illustrated by one of the earliest examples of a ligand-binding assay, the use of [^{3}H]naloxone to quantify opiate receptors. Naloxone is well known as an opiate antagonist, and in the early 1970s several groups demonstrated that a specific binding site for [^{3}H]naloxone is present in the brain. When the ability of various opiate agonists and antagonists to compete for naloxone binding was compared with their potencies in producing or antagonizing analgesia, it was found that their effective concentrations were very similar in the two assays (Fig. 9-5c). These findings suggest strongly that the [^{3}H]naloxone binding sites in brain are indeed opiate receptors. Subsequently, competition for the binding of radioactive opiates was one of the assays used to identify and purify endogenous opiates, the enkephalins, as described in Chapter 8.

Why Are There So Many Receptors?

In Chapter 8 we pointed out that the enormous number of chemical neurotransmitters comes as quite a surprise. Now we find that most and per-

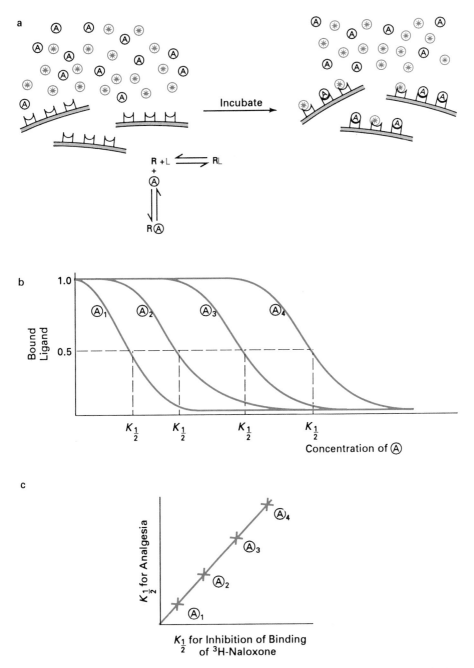

FIGURE 9-5. Ligand-binding assays can be used to determine the pharmacological profile of the receptor. *a*: The equilibrium between the ligand (L) and the receptor (R) can be disturbed by including in the assay a receptor agonist or antagonist (A). *b*: A plot of the amount of radioactive ligand bound at different concentrations of various agonists or antagonists ($A_1–A_4$). *c*: The pharmacological profile for inhibition of binding can be compared with the pharmacological profile of the same series of agonists and antagonists in some biological response mediated by the receptor.

haps all of these transmitters bind to several different classes of receptor. Again we do not understand the reasons for this diversity, but it seems likely that it provides a nerve cell with greater flexibility in responding to extracellular signals. Let us consider, as an example, the different receptor subtypes for glutamate, the major excitatory neurotransmitter in the mammalian brain. Receptors for glutamate are particularly interesting, in part because glutamate has only recently been established unequivocally as a neurotransmitter, but also because glutamate receptors play an important role in one model of learning and memory (see Chapter 17).

At least two distinct classes of glutamate receptor are known from ligand-binding studies and from physiological responses. One class binds and responds to the agonists kainic acid and quisqualic acid (Fig. 9-6a), and hence is known as the *kainate/quisqualate* class of receptor (there may even be further subtypes within this class). The other class is activated by N-methyl-D-aspartate (NMDA) and related compounds (Fig. 9-6b) and is known as the *NMDA receptor*. Both of these receptor subtypes mediate depolarizing responses to glutamate in the target cell, and in some parts of the brain both appear to be present on the same neurons. Why are both required?

The answer appears to lie in the properties of the different ion channels that are activated by the two receptor classes. Kainate/quisqualate receptors activate cation channels that allow sodium and potassium ions to flow. Near a neuron's resting potential the driving force for potassium is low and that for sodium is high, so activation of this channel leads to depolarization as a result of an inward sodium current (Fig. 9-7a). In contrast, NMDA receptors activate cation channels that allow not only sodium and potassium but also calcium ions to flow (Fig. 9-7b). It might be thought that activation of these channels would also cause a depolarization that would simply add to that produced by the kainate/quisqualate receptor channels. Near the resting potential, however, this does not occur because of an interesting property of the NMDA receptor channels—they are blocked by extracellular magnesium ions in a voltage-dependent manner.

The way the NMDA receptor/channel works is shown in Figure 9-7. When a neuron is near its resting potential, magnesium ions bind to the outside of the channel and effectively prevent the movement of other ions through the pore (Fig. 9-7a). When the cell is depolarized, however, the magnesium ions are driven out of the channel, allowing the other ions free access (Fig. 9-7b). The amount of current passing through NMDA receptor channels is therefore much greater at depolarized than at hyperpolarized membrane potentials. Thus when the cell is depolarized in the presence of glutamate, calcium (as well as sodium) flows into the cell through the NMDA receptor channels (Fig. 9-7b).

The presence of two distinct receptor types coupled to different ion channels allows the target cell to respond differently to different intensities of synaptic stimulation. A moderate stimulus produces only a small depolarization, and no calcium entry, as a result of activation only of the kainate/quisqualate receptor channels. Although glutamate binds to the NMDA

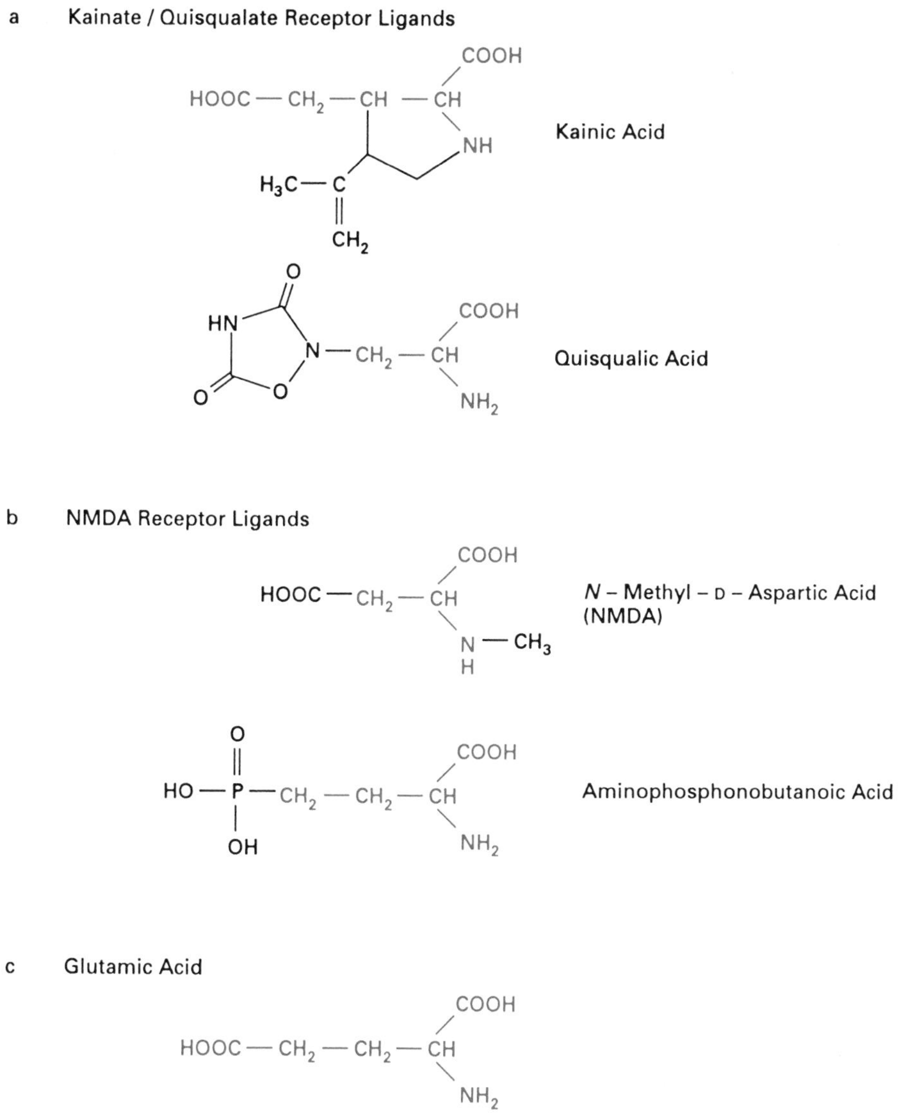

a   Kainate / Quisqualate Receptor Ligands

Kainic Acid

Quisqualic Acid

b   NMDA Receptor Ligands

N – Methyl – D – Aspartic Acid
(NMDA)

Aminophosphonobutanoic Acid

c   Glutamic Acid

FIGURE 9-6. Ligands for different classes of glutamate receptors. The portion of the ligand molecule that resembles glutamate is shown in red. a: Kainic acid and quisqualic acid are both agonists at the K/Q subclass of glutamate receptors. b: N-Methyl-D-aspartate (NMDA) is the agonist that defines the NMDA receptor subtype, and aminophosphonobutanoic acid is an antagonist at this receptor subtype. c: The structure of glutamic acid, for comparison.

a   Weak Stimulus Activates Only K/Q Receptor Channels

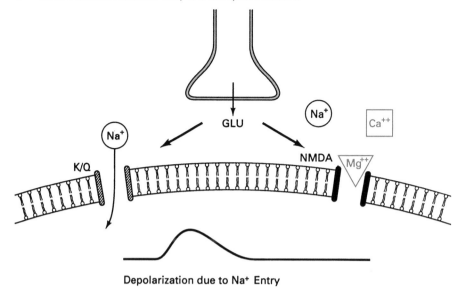

Depolarization due to Na⁺ Entry

b   Strong Stimulus Depolarizes Sufficiently to Relieve
Voltage-Dependent Magnesium Block of NMDA Receptor Channels

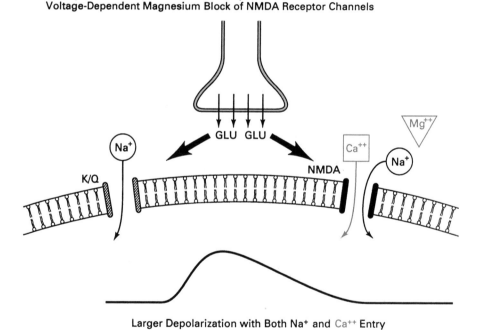

Larger Depolarization with Both Na⁺ and Ca⁺⁺ Entry

FIGURE 9-7. Different kinds of responses mediated by two glutamate receptor
subtypes. *a:* The response to a weak presynaptic stimulus is activation of only the
K/Q receptor subtype. *b:* In contrast, a stronger stimulus causes ions to flow
through both the K/Q and NMDA receptor channels. Voltage-dependent Mg^{2+}
block of the NMDA receptor channel was first demonstrated by Philippe Ascher
and his colleagues (see Nowak et al., 1984).

receptors, no current flows through the NMDA channels under these conditions. With more intense stimulation the depolarization becomes sufficient to relieve the magnesium block of the NMDA receptor channels, resulting in further depolarization and calcium entry.

The presence of the two glutamate receptor subtypes also allows a target cell to respond differently depending on whether or not it is actively firing action potentials. A moderate stimulus to a silent cell will not produce sufficient depolarization to activate the NMDA channels. In contrast the same stimulus to an active neuron, whose membrane is of course relatively depolarized by the action potentials, causes calcium to enter through these channels. As we shall see in Chapter 17, it is believed that the calcium that enters through the NMDA receptor channels may trigger long-lasting changes in the properties of the postsynaptic neurons.

Pharmacological Diversity of Receptors Reflects Structural Diversity

What is the molecular underpinning for these pharmacologically distinct classes of receptor for a single neurotransmitter? Are there distinct receptor molecules, or can a single receptor molecule take on different pharmacological guises? These questions could be answered definitively only by isolating receptor molecules. It did not escape the attention of receptorologists that, in addition to providing an assay for receptors in their normal membrane-bound state, ligand binding might be used to measure numbers of receptor molecules during purification. Furthermore appropriate ligands might be exploited not only for assay, but also as reagents to be used in purification—so-called *affinity reagents*. Let us take as an example the first neurotransmitter receptor to be purified by methods based on these concepts, the nicotinic acetylcholine receptor and its associated ion channel. It was the first because it was the easiest; a high-affinity and highly selective ligand was available, and there was a rich source of receptor in the electric organs of certain eels and fish.

A toxin that binds to nicotinic acetylcholine receptors. Certain creatures appear to have been designed by evolution as gifts to neurobiologists. The story of the nicotinic acetylcholine receptor, like that of the ionic mechanism of the action potential, belongs to two groups of such animals, each with its unique specializations for survival. The first group is a family of snakes whose venoms contain toxins that bind with extremely high affinity and selectivity to the nicotinic acetylcholine receptor. These snakes kill their prey by blocking neuromuscular transmission with the toxins. The most thoroughly characterized of these toxins is *α-bungarotoxin*, from the venom of the Taiwanese cobra *Bungarus multicinctus*. It is a polypeptide of 74 amino acids, which binds to the nicotinic receptor with a dissociation constant in the range of $10^{-15}M$ (Fig. 9-8). Once it binds to the receptor the half-time for its dissociation is many days! This makes it an ideal reagent for receptor assay and affinity purification.

A rich source of nicotinic acetylcholine receptors. The second set of creatures that has contributed so greatly to our understanding of the nicotinic receptor are various species of the genus *Torpedo,* the electric rays, and the genus *Electrophorus,* the electric eels. The electric organs in these animals consist of modified muscle cells, which are flattened and organized in parallel arrays known as *electroplaques.* These cells have lost their contractile apparatus and have neuromuscular junctions that cover virtually the entire surface of one of the flattened faces of the cell, but not the other. The geometry of the electroplaque is such that voltage differences across the many cells add in series, producing a very large voltage that can stun prey.

The advantage of using the electric organ as a source of nicotinic receptor becomes obvious when we examine the density of the receptor in the membrane of the *Torpedo* electroplaque cells. Most receptors and ion channels are relatively rare membrane proteins. Typically, one or a few receptor molecules are found per square micron of membrane surface area. In contrast the *Torpedo* electroplaque membrane is packed tightly with nicotinic receptor, to a density as high as 10,000 to 30,000 per square micron, and indeed there appears to be little room for any additional protein in the membrane! Thus the rays have done a significant part of the purification for us, by inserting the protein and very little else into a membrane that can be isolated with relative ease.

Structure of the nicotinic acetylcholine receptor. With this rich source of material and the rapid, sensitive, and specific α-bungarotoxin assay, large amounts of homogeneous nicotinic receptor can be obtained. Biochemical analysis demonstrates that the receptor is a large complex consisting of five subunits (Fig. 9-9). There are two α-subunits of molecular weight approximately 40,000, and one each of a β (MW 48,000), γ (MW 58,000), and δ (MW 64,000) subunit. Each α-subunit contains a site for binding acetylcholine (and α-bungarotoxin) (Fig. 9-9b). This is consistent with physiolog-

$$NAChR + \alpha\,BT \underset{k_{on}}{\overset{k_{off}}{\rightleftharpoons}} NAChR - \alpha BT$$

$$K_d = \frac{k_{off}}{k_{on}} \approx 10^{-15}M$$

FIGURE 9-8. α-Bungarotoxin (α-BT) binds very tightly to the nicotinic acetylcholine receptor (NACLR). The affinity of the toxin for the receptor is extremely high.

ical data, which indicate that the binding of two acetylcholine molecules is necessary for the opening of the nicotinic receptor's associated ion channel.

The amino acid sequences of the various subunits have been found by molecular cloning techniques. Mutagenesis and heterologous expression experiments, such as those described in Chapter 5, have begun to provide a detailed picture of structure/function relationships in the nicotinic receptor. The different subunits have substantial amino acid sequence homology with one another, which suggests that they may have evolved from a single ancestral protein. Each subunit has four hydrophobic domains that are predicted to be membrane-spanning sequences (Fig. 9-10). As we shall discuss

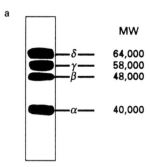

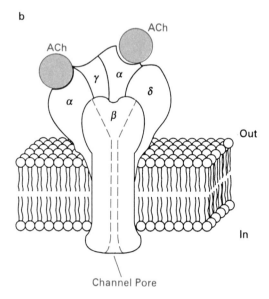

FIGURE 9-9. The nicotinic acetylcholine receptor from *Torpedo.* a: Polyacrylamide gel separation of the individual subunits of the purified nicotinic receptor. b: Cross-linking studies have demonstrated that the functional receptor molecule contains two α-subunits, each with an acetylcholine binding site, and one each of the β-, γ-, and δ-subunits.

below, clues are beginning to emerge about regions of the molecule that are involved in ion conduction.

The important point we wish to make here is that when, considerably later, the muscarinic acetylcholine receptor from various sources was purified and cloned, its structure was found to be totally different from that of the nicotinic receptor. The muscarinic receptor has only a single kind of subunit. Although the functional membrane receptor is probably an oligomer, this has not yet been established with certainty. The subunit molecular weight is 51,000, similar to that of the nicotinic receptor subunits, but a comparison of the amino acid sequences shows no relationship. From hydrophobicity measurements it is inferred that the muscarinic receptor has seven membrane-spanning domains, in contrast to the nicotinic receptor subunits that are each thought to cross the membrane four times (Fig. 9-10).

Common Structural Motifs in Receptors: Receptor "Superfamilies"

It would appear that the nicotinic and muscarinic acetylcholine receptors have virtually nothing in common other than the fact that they bind and are activated by acetylcholine. They are present in different kinds of nerve, muscle, and gland cells (Fig. 9-2), mediate very different physiological

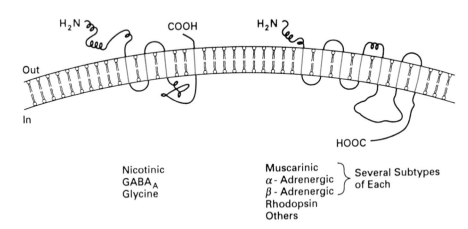

FIGURE 9-10. Topology of acetylcholine and other receptors. Hydrophobicity profiles predict that nicotinic acetylcholine receptor subunits, as well as those of the GABA$_A$ and glycine receptors, contain four membrane spanning domains (red). In contrast, muscarinic and related receptors are predicted to have seven membrane spanning domains.

responses in terms of kinetics and mechanisms, and their amino acid sequences and inferred structures show no similarities (Fig. 9-10). Because this has been a disappointment to those biologists who seek unifying concepts in nature, it is all the more satisfying that unifying concepts have indeed emerged. Just as there are commonalities among the voltage-dependent ion channels (Chapter 5), there are indeed families of receptor molecules with common structural features. However, the family groupings are not based on the neurotransmitter that binds to the receptor; rather they reflect the *transduction mechanism,* the molecular mechanism that the receptor uses to translate the extracellular signal into a response in the target cell. We shall now discuss one family, that of the *ligand-gated ion channels,* to which the nicotinic acetylcholine receptor belongs. We will describe the structural features that the family members share, as well as the transduction mechanism that they use to excite or inhibit a neuron. In the next chapter we shall describe other families in which the transduction mechanisms are more complex. These include the family to which the muscarinic acetylcholine receptor belongs, the *G protein-coupled receptors.*

The Family of Directly Coupled Receptor/Ion Channel Complexes

The simplest way of transducing an extracellular signal into a change in excitability of the target neuron is by *direct coupling* of the neurotransmitter receptor to the ion channel whose activity it regulates. This is the mechanism used by the various ligand-gated ion channels. The ligand-binding site and the ion channel are part of the same molecule or macromolecular complex. Occupation of the receptor by the neurotransmitter leads to a conformational change that is passed along to the closely associated ion channel, and as a result channel properties are altered (Fig. 9-11). Note that in Figure 9-11 we show a channel that is closed in the resting state, and that *opens* as a result of transmitter action, allowing the ion (X^+) to flow across the membrane. Although this is the way most of the known ligand-gated ion channel systems appear to function, in principle it is possible that a transmitter might *close* some channel that is open in the resting state.

An important feature of this mechanism is that the modulation of ion channel properties is dependent on continued occupation of the receptor by the transmitter. The conformational change induced by receptor occupancy is readily reversible, and as soon as the receptor is no longer occupied by transmitter, the channel returns to its normal resting state (Fig. 9-11). Accordingly it might be expected that the ligand-gated ion channel systems would mediate rapid onset and rapidly reversible synaptic transmission, and this is indeed the case.

The nicotinic receptor. It is now clear that the nicotinic acetylcholine receptor complex is one of these ligand-gated ion channels, in that it contains not only the acetylcholine-binding site, but also the ion channel that is activated by acetylcholine binding. Most of the data that support this conclusion, including the amino acid sequence information from which the transmembrane topology has been inferred, come from work on the *Torpedo* and muscle nicotinic receptors. However, studies of several neuronal nicotinic receptors place them in this same category of ligand-gated ion channels as well.

Biochemical evidence for direct receptor/channel coupling. How do we know that the receptor and ion channel of the ligand-gated systems are indeed part of the same macromolecular complex? The evidence is most complete for the nicotinic acetylcholine receptor, and we shall consider some of this evidence here. First, recall from the discussion above that affinity chromatography, using a ligand that binds to the acetylcholine-binding site, purifies a macromolecular complex consisting of four polypeptide subunits (Fig. 9-9). Early experiments to test the direct coupling hypothesis involved *reconstitution* of the purified complex into liposomes (phospho-

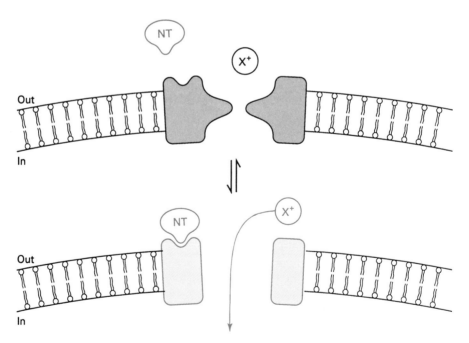

FIGURE 9-11. Direct receptor/channel coupling. In directly coupled receptor/ channel systems the neurotransmitter (NT) binding site and the ion channel are intimately associated in a single macromolecular complex. Contrast with Figures 10-2 and 10-6.

lipid vesicles), and measurement of the transport of radioactive cations across the liposome membrane (Fig. 9-12a). The phospholipid bilayer membrane of the liposome is impermeant to ions, and hence ion flux will occur only if there is an open ion channel in the membrane. Experiments such as this demonstrate that the complex consisting of the α-, β-, γ-, and δ-protein

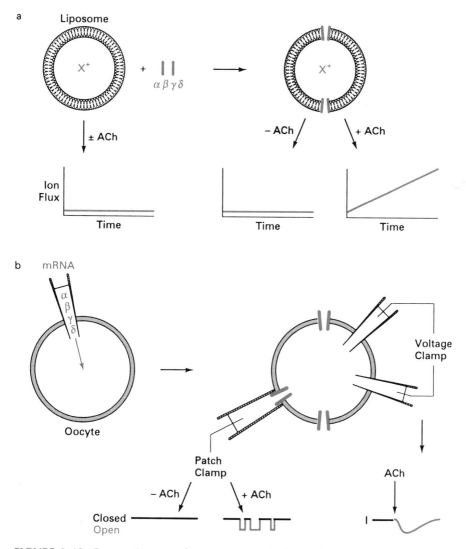

FIGURE 9-12. Reconstitution of receptor-coupled ion channel activity. a: Liposomes, spherical membrane vesicles made from artificial phospholipid, can be loaded with an ion (X^+). The flow of X^+ out of the liposome can be measured under various experimental conditions. b: Eric Barnard and Ricardo Miledi first demonstrated heterologous expression, in Xenopus oocytes, of messenger RNA for the nicotinic acetylcholine receptor subunits (see Barnard et al., 1982).

subunits is sufficient to reconstitute acetylcholine-dependent ion flux (Fig. 9-12a). In other words, the acetylcholine binding site and its associated ion channel copurify on the affinity column, indicating that they must be intimately associated.

Heterologous expression and site-directed mutagenesis. These data were confirmed and extended once the nucleotide sequences coding for the different subunits became available. A series of striking experiments have begun to elucidate the role of each subunit and of particular amino acids in channel gating and conduction. These experiments involve injection of messenger RNA for the α-, β-, γ-, and δ-subunits into *Xenopus* oocytes, a heterologous expression system that we have met previously in our discussion of the molecular analysis of voltage-dependent ion channels in Chapter 5. The oocyte has no endogenous nicotinic acetylcholine receptor, and so the functional expression of exogenous messenger RNA can be detected readily by voltage clamp or single channel recording techniques.

Functional acetylcholine receptor/channels with properties identical to those in muscle can indeed be expressed following injection of messenger RNA for the α-, β-, γ-, and δ-subunits into oocytes (Fig. 9-12b). This confirms the conclusion from the biochemical reconstitution data that these subunits are sufficient to produce a complete functional receptor/channel complex. A similar approach, using messenger RNA for two subunits (α and β) of a receptor/channel complex that recognizes GABA, has demonstrated that expression of the α- and β-subunits is sufficient to produce a GABA-activated chloride current. This kind of experiment in itself provides little new information—indeed oocyte expression is no more than a sophisticated reconstitution system—but the ease with which nucleic acid sequences may be manipulated provides the opportunity to ask more detailed questions about structure/function relationships in the receptor/ channel complex.

One way this has been done for the nicotinic receptor/channel is by constructing *chimeric* receptors, in which a particular amino acid sequence from one receptor is inserted into the corresponding region of another. Early experiments took advantage of the fact that the nicotinic receptor/ channels from mammalian muscle and *Torpedo* electric organ have similar but not identical gating properties (i.e., the channel mean open times differ significantly). By injecting messenger RNAs for the mammalian and *Torpedo* subunits in various combinations, it was found that substitution of the *Torpedo* δ-subunit with that from muscle is sufficient to change the gating properties of a *Torpedo* receptor to those of a muscle receptor (Fig. 9-13). A similar approach demonstrated that differences in both gating and conduction between fetal and adult forms of the mammalian muscle receptor/channel can be accounted for by a developmentally regulated switch from the γ-subunit, which is expressed in the fetus, to the closely related ε-subunit that is expressed later in development. The major conclusions from these experiments were that the γ/ε-subunits are important for determining the single channel conductance, and both these and the δ-subunit are involved in channel gating.

More recent experiments have focused on identifying particular amino acid sequences *within* each subunit that are important for the various functions of the receptor/channel complex. Chimeric *subunits,* consisting of various *Torpedo* sequences substituted in the mammalian subunits and vice

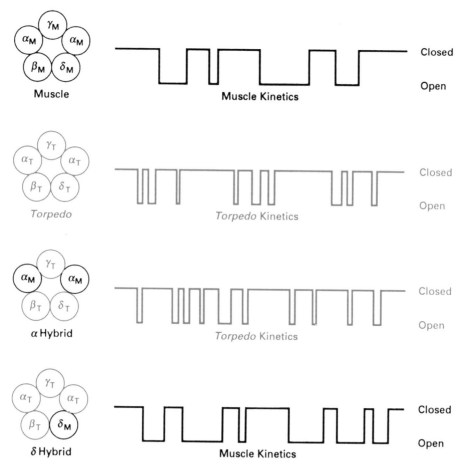

FIGURE 9-13. Chimeric nicotinic receptors in oocytes. The nicotinic receptors from muscle (black) and *Torpedo* (red) display different opening and closing kinetics. Shosaku Numa and Bert Sakmann have constructed chimeric receptors, which contain some muscle and some *Torpedo* subunits, by injecting the appropriate messenger RNAs into *Xenopus* oocytes (see Sakmann et al., 1985).

versa, have demonstrated that the second membrane-spanning domain (known as the M2 segment) determines the single channel conductance and thus probably forms part of the conducting pathway. This has been confirmed by an elegant experiment, in which the conducting pathway was probed with a local anesthetic blocker of the nicotinic receptor/channel that is known to penetrate and plug the channel pore. This experiment showed that particular amino acids within the M2 sequences of all the subunits interact with the blocker. Thus the M2 regions of all the subunits must line the channel pore, as illustrated in Figure 9-14. In other site-directed mutagenesis experiments it has been found that negatively charged amino acids in the intracellular and extracellular loops of all the subunits form a ring of charge (red minus signs in Fig. 9-14) that helps to concentrate permeant cations in the mouth of the channel, and hence influences the single channel conductance.

Experiments of this type have provided an increasingly detailed picture of the relationship between structure and function in the nicotinic acetylcholine receptor/channel, and by analogy in the other ligand-gated ion channels. Most important for our present discussion, they have provided very strong evidence that the transmitter binding site and the ion channel that the transmitter regulates are indeed part of the same macromolecular complex in this prototype member of the ligand-gated channel family. Two of the five subunits in the nicotinic receptor/channel contain ligand-binding sites, and all five contribute at least to some extent to the channel-gating and conduction processes.

The GABA_A receptor. Among the other ligand-gated ion channel systems that have been identified to date is the mammalian brain $GABA_A$ receptor.

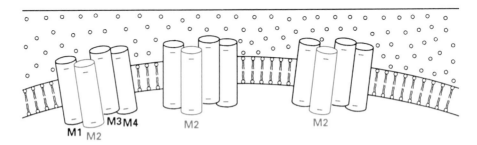

FIGURE 9-14. Inferences about structure of the nicotinic acetylcholine receptor from site-directed mutagenesis. Three of the five subunits of the nicotinic acetylcholine receptor in a lipid bilayer membrane are depicted. Each subunit spans the membrane four times (each cylinder represents one membrane spanning domain). See the summary by Dani (1989).

In fact there are at least two receptors for GABA, the GABA$_B$ receptor, which is not directly coupled to its ion channel, and the GABA$_A$ receptor, on which we will focus. This receptor has now been purified using specific affinity reagents. It exhibits a subunit structure that resembles that of the nicotinic acetylcholine receptor. The inhibitory action of GABA results from activation of a chloride channel, which is directly coupled to the GABA$_A$ receptor (Fig. 9-15a—contrast this with the nicotinic receptors, which are coupled to cation channels that allow sodium and potassium to flow). We have referred in the previous chapter to the fact that the receptor for GABA has a rich pharmacology, in that it is the site of action of a number of clinically important drugs. These include the benzodiazepines (such as Valium and Librium), which are widely used as antianxiety and relaxant drugs, and the barbiturates, which are important anticonvulsants and sedatives. These agents produce their clinical consequences by enhancing the effect of GABA on its directly coupled chloride channel (Fig. 9-15b).

The GABA$_A$ receptor was purified to homogeneity by isolating the proteins that bind to benzodiazepines, through the technique of affinity chromatography. This purifies the GABA$_A$ receptor complex with all its pharmacological binding sites intact. The complex contains α- (MW 53,000) and β- (MW 57,000) subunits with the stoichiometry α_2-β_2. A γ-subunit, which actually contains the benzodiazepine binding site, also exists. The overall molecular weight of the complex approaches that of the nicotinic acetylcholine receptor. Molecular cloning and sequencing of the α-, β-, and γ-subunits of the GABA$_A$ receptor reveal substantial homologies between them in primary amino acid sequence and deduced structure. What is most striking, however, is the similarity to the nicotinic acetylcholine receptor. The number and distribution of predicted transmembrane domains are identical for all the subunits of the nicotinic and GABA$_A$ receptors, and there is significant sequence homology in certain regions of the subunits.

The glycine receptor. In some parts of the central nervous system, in particular in the brainstem and spinal cord, glycine rather than GABA is the major inhibitory neurotransmitter. At these sites glycine functions just like GABA, by activating a receptor linked to a chloride channel (Fig. 9-15a). The convulsant drug strychnine blocks inhibition by selectively antagonizing glycine-mediated (but not GABA-mediated) inhibition, and strychnine has been used for assay and affinity purification of the glycine receptor. The receptor complex consists of α- and β-subunits, with molecular weights of about 48,000 and 58,000, respectively, and cross-linking studies suggest an overall pentameric structure with a molecular weight of about 260,000. The smaller subunit has been cloned and sequenced, and it exhibits convincing sequence and structural homology with the GABA$_A$ and nicotinic receptor subunits.

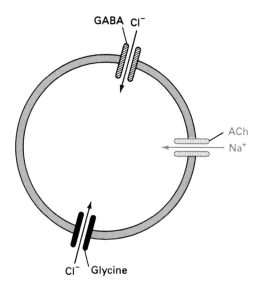

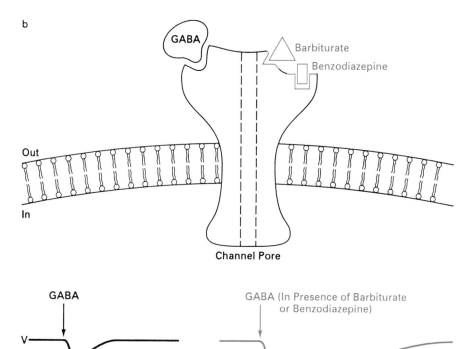

FIGURE 9-15. The GABA$_A$ and glycine receptors. *a*: The GABA$_A$ and glycine receptors are intimately coupled to channels selective for chloride ions. In contrast the nicotinic acetylcholine receptor is coupled to a cation channel. *b*: The GABA$_A$ receptor/channel complex has binding sites for barbiturates and benzodiazepines, which can markedly enhance the hyperpolarizing response to GABA.

Summary

The question of how cells respond to signals from their environment is relevant to all aspects of cell biology. Thus it will come as no surprise that many of the signaling molecules discussed in the previous chapter are not restricted to the nervous system, but can act on many cell types. The common feature that allows neurons and other cells to respond to these extracellular signals is the presence of specific receptors in the plasma membrane that may be very similar in their structure and functional properties in different kinds of cells. It has been known for a long time from pharmacological studies that there may be several different kinds of receptors for individual neurotransmitters. This has important implications for the way information is processed in multineuronal networks. This heterogeneity has been confirmed as the structures of many receptors have been elucidated. A gratifying picture has emerged from these structural studies—many receptors can be grouped into families based on structural, functional, and regulatory homologies that are far more extensive than had been appreciated previously.

Perhaps surprisingly, the family groupings reflect common receptor transduction mechanisms rather than common ligand-binding sites. One such family is that of the ligand-gated ion channels. The members of this family were first linked on the basis of a functional criterion, direct coupling between the receptor and the ion channel whose activity it regulates. Biochemical and molecular studies provide a structural basis for grouping these receptors into a single family. The sequence homologies and remarkably similar predicted arrangement of transmembrane segments suggest that the various subunits of the different receptors have evolved from a single ancestral subunit. Evolution has allowed the ligand specificity of the receptor site, and the ion selectivity of the channel pore, to diverge. However the essential overall structural design of the ligand-gated receptor/channel complex has been preserved. We shall now consider the molecular details of other, more intricate, signal transduction pathways.

Receptors and transduction mechanisms II. Indirectly coupled receptor/ion channel systems

The recognition of extracellular signals by specific receptors on the target cell is not the final step in intercellular communication. The cells must also possess mechanisms to transduce the extracellular signal, to convert it into some biological response that is characteristic of the particular target cell. In neurons, the biological response is often the modulation of the properties of one or more membrane ion channels. We have seen that in the ligand-gated ion channel family, the tasks of recognition and transduction reside in a single protein complex. Other families of receptors alter neuronal excitability through more complicated changes in the biochemistry of the cell (Fig. 10-1).

G Protein-Coupled Receptor/Ion Channel Systems

Most neurotransmitter receptors are not coupled directly to the ion channel whose activity they regulate. Among the indirectly coupled receptor/channel systems is a large (and ever-growing) family coupled via *guanyl nucleotide-binding proteins,* or G proteins. It can be seen from an examination of Table 10-1 that at first glance the various G protein-coupled receptors have little in common. This receptor family includes several peptide receptors, the muscarinic acetylcholine receptor subtypes, and receptors for most of the major classes of biogenic amines. It even contains the visual pigment rhodopsin, which does not respond to a signaling *molecule* but, as we shall see in Chapter 12, can be thought of as a "receptor" for light. Note that cell surface receptors in yeast and in the slime mold *Dic-*

tyostelium, cell types far removed from neurons, are included in the list, emphasizing the commonality of mechanisms in neurons and other kinds of cells. Why would we possibly want to gather these diverse receptors in the same list? As in the example of the directly coupled receptor systems described in the last chapter, many of these were first linked on the basis of

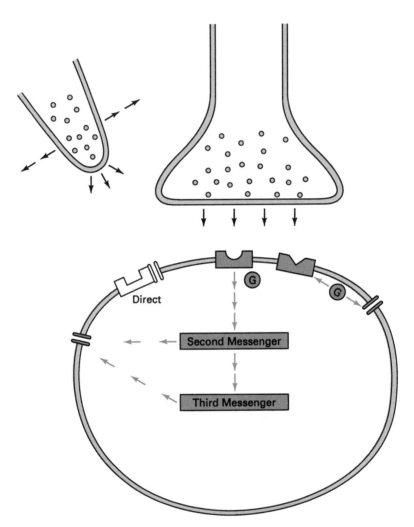

FIGURE 10-1. Intercellular communication. This chapter will deal with transduction mechanisms that are not mediated via direct receptor–channel coupling, but rather involve either protein–protein interactions in the plain of the plasma membrane, or intracellular messengers (red). Guanyl nucleotide binding proteins (G) play an essential role in these transduction mechanisms.

a functional criterion, in this case transduction of the extracellular signal via a G protein.

As we will see later in this chapter, many receptors use G proteins to activate intracellular enzymes that produce *second messengers*. These are diffusible molecules that may influence a variety of cellular constituents, including ion channels. It is now realized that G proteins also interact *directly* with ion channels to regulate their activities (Fig. 10-2), without the mediation of a second messenger. Such modulation of ion channel properties, via protein–protein interaction in the plane of the membrane, may be slower in onset and somewhat longer lasting than that mediated by the intramolecular conformational changes in the directly coupled systems (see Fig. 9-11).

Receptors coupled to G proteins belong to a family. Molecular cloning has made it clear that this grouping of receptors in Table 10-1 is indeed appropriate. The first members of the family to be sequenced were the opsin visual pigments. The most striking feature in the sequence was the presence of seven stretches, each of 20–28 hydrophobic amino acids, which presumably represent membrane-spanning domains. This of course is very different from the four membrane-spanning regions characteristic of the directly coupled receptors (see Fig. 9-10). When one of the mammalian β-adrenergic receptors was subsequently cloned and sequenced, it was also found to have seven hydrophobic membrane-spanning domains, and substantial amino acid homology with bovine opsin. This is also the case for the α-adrenergic receptors, although they differ markedly from the β-receptors in their binding site pharmacology. We have already pointed out in Chapter 9 that the various muscarinic acetylcholine receptor subtypes also exhibit these structural features (see Fig. 9-10).

This structure has now been seen in all other members of the G protein-coupled receptor family for which sequence information is available. What

TABLE 10-1 Some of the Receptors That Use G Proteins
to Transduce Extracellular Signals

Substance K receptor
The *mas* oncogene—a form of angiotensin receptor
Receptors for yeast mating factors
Slime mold cyclic AMP receptor
Luteinizing hormone receptor
Muscarinic acetylcholine receptors
Adrenergic receptors
Rhodopsin

is most interesting is that the pattern is so consistent. New receptors whose transduction mechanism is not understood can be assigned with confidence to this family solely on the basis of their amino acid sequence. A good example of this is the receptor for substance K, one of several related peptides called the *tachykinins*. Among the biological actions of these peptides are important effects on sensory processing. Before its amino acid sequence was determined, the transduction mechanism for the actions of substance K was not known. However, the sequence showed the presence of seven membrane-spanning domains. This allowed the prediction that the receptor is coupled to a G protein.

Novel G protein-coupled receptors. This approach can be taken even one step further. DNA can be isolated that encodes new receptors in the G protein-coupled family, solely on the basis that this DNA encodes proteins with structures similar to those of known members of the family. This is accomplished by allowing DNA for a known receptor to hybridize with cDNA (see Chapter 5) prepared from tissues that may express unknown receptors. This must be done under conditions that allow hybridization between related, but not necessarily identical, DNA strands. For example,

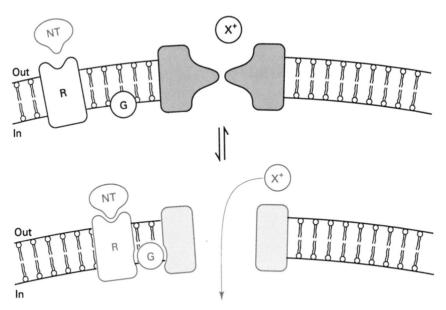

FIGURE 10-2. G protein-mediated receptor-channel coupling. Not all neurotransmitter receptors are intimately associated with an ion channel as in Figure 9-11. In the case shown here binding of neurotransmitter (NT) to its receptor (R) activates a G protein (G), which then interacts with the ion channel, causing it to open.

under such *low stringency* hybridization conditions, with a DNA probe derived from a G protein-coupled β-adrenergic receptor, it has been possible to isolate, clone, and sequence a related cDNA that codes for a protein with seven hydrophobic membrane-spanning sequences. The function of this protein was not known at the time its DNA was isolated and sequenced, although its structure was! It could be predicted with confidence that it would turn out to be a receptor that is coupled to a G protein, and it has now been demonstrated that this is indeed the case: this cDNA codes for one of the subtypes of serotonin receptors, the G protein-coupled $5HT1_a$ receptor.

How do G proteins work? To discuss G protein modulation of ion channels we must first consider the structure of G proteins and some aspects of the molecular mechanism by which they act. G proteins are heterotrimers consisting of one each of an α-, β-, and γ-subunit (Fig. 10-3), and a large num-

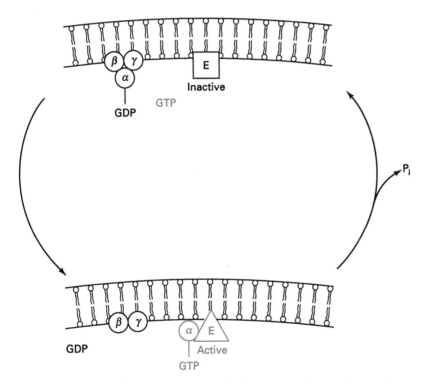

FIGURE 10-3. Mechanism of action of G proteins. G proteins are heterotrimers consisting of one each of an α-, β-, and γ-subunit. In the presence of GTP, the α-subunit can activate some effector (E). A GTPase activity intrinsic to the α-subunit completes the cycle. See the review by Gilman (1987).

ber of G proteins appear to exist. Sequences for at least eight distinct α-subunits have been identified by molecular cloning techniques, and polypeptides corresponding to at least four of these sequences have been isolated from various tissues. In contrast the β- and γ-subunits show little variation from one cell type to another. This has led to the assumption that the α-subunit is the business end of the G protein, and is responsible for its coupling to particular effector systems. Although the β- and γ-subunits were originally thought to subserve the important but relatively humdrum function of anchoring the G protein to the membrane, we shall see below that this picture is an oversimplification. Four different G proteins that differ only in their α-subunits have been characterized most thoroughly. Their nomenclature is based on the effector systems with which they were first shown to couple:

1. G_S (α_S-β-γ), which *s*timulates adenylate cyclase;
2. G_I (α_I-β-γ), which *i*nhibits adenylate cyclase;
3. G_T (α_T-β-γ) also known as *transducin,* which is found in vertebrate photoreceptors and is involved in light *t*ransduction (see Chapter 12 for a detailed discussion); and
4. G_O (α_O-β-γ), for *o*ther, a G protein that was isolated and characterized biochemically before its function was known.

All of the known G proteins operate via the same general mechanism illustrated in Figure 10-3. In its inactive state the G protein has GDP bound to a specific guanyl nucleotide-binding site on the α-subunit. As it moves in the plane of the membrane, the G protein occasionally bumps into an appropriate agonist-occupied membrane receptor (or, in the case of transducin, a light-activated rhodopsin molecule). This receptor–G protein interaction allows GTP to displace GDP from the guanyl nucleotide-binding site. The displacement is accompanied by the dissociation of the α-subunit from the $\beta\gamma$ complex, and the now active α-subunit can interact with and activate its effector system (but see the discussion below about possible direct actions of $\beta\gamma$ on effector systems). The effector system might be an ion channel as in Figure 10-2, or some enzyme (denoted by E in Fig. 10-3). The α-subunit carries an intrinsic GTPase activity, and so after some time the bound GTP is hydrolyzed to GDP. Following this hydrolysis the α-subunit is no longer active and recombines with $\beta\gamma$, allowing the cycle to begin again (Fig. 10-3).

A variety of biochemical probes have been instrumental in proving the above sequence of events and in identifying and analyzing biological responses mediated by G proteins. These include several GTP analogs such as GTPγS and Gpp(NH)p (guanylylimidodiphosphate), which displace

GDP from the guanyl nucleotide-binding site but are not hydrolyzed by the GTPase. These compounds act as essentially irreversible *activators* of the G proteins. In contrast the GDP analog GDPβS, which binds very strongly to the binding site, maintains the G protein in the inactive GDP state, thus *inhibiting* G protein-mediated responses. Two bacterial toxins have also been extremely useful. *Cholera toxin* irreversibly activates G_S by causing ADP-ribose to be covalently attached to an arginine residue on α_S. *Pertussis toxin* irreversibly inhibits G_I and G_O by catalyzing the same reaction with α_I and α_O, respectively (both toxins appear to act on transducin, although the functional consequences are not completely understood).

Some ion channels are modulated directly by G proteins. As we shall see later in this chapter, second messengers whose enzymatic synthesis is mediated by G proteins influence the activity of many ion channels. Such indirect G protein involvement in ion channel modulation appears to be a rather ubiquitous phenomenon. Some ion channels, however, themselves may interact directly with G proteins. That is, the effector system (E in Fig. 10-3) is not an enzyme, but is the ion channel protein itself (Fig. 10-2). The list of ion channels that are regulated directly by G proteins is growing steadily. We shall illustrate this mechanism by considering the inwardly rectifying potassium channel in the heart, whose activity is increased by acetylcholine acting at muscarinic receptors. This channel is responsible for the slowing of the heart on stimulation of the vagus nerve, and thus this story is of historical interest (see the Otto Loewi experiment described in Chapter 1), as well as being the first and most thoroughly characterized example of this kind of channel modulation.

A variety of evidence indicates that there is no second messenger involvement in the muscarinic activation of this cardiac potassium channel. However a role for a G protein was suggested by the finding that potassium current can be activated only when GTP and GTP analogs are added to the electrode in the whole cell recording configuration (Fig. 10-4a). Furthermore the channel can be activated in detached membrane patches by GTP applied to the cytoplasmic membrane surface (Fig. 10-4b). In addition, pretreatment with pertussis toxin prevents the activation of the potassium current by muscarinic agonists, suggesting that the relevant G protein is either G_I or G_O.

Purified G proteins that have been activated by prior treatment with GTP analogs can be applied directly to the cytoplasmic membrane surface of detached patches containing the inwardly rectifying potassium channel (Fig. 10-5). This treatment activates the channel, confirming a direct interaction of the G protein with the channel or with some unknown regulatory protein that is very closely associated with the channel. In an extension of these experiments isolated α-subunits or βγ complexes were applied to the inside of the patch, and an interesting controversy arose. One research group found that α but not $\beta\gamma$ could activate the channel, whereas another found just the opposite, a rather surprising result in view of the prevailing dogma that the α-subunit is responsible for coupling to effector systems. It now appears that both positive results are correct; α does activate the channel directly and, under some circumstances, $\beta\gamma$ can modulate the channel indirectly by activating an enzyme called phospholipase A_2. In addition $\beta\gamma$ can bind to free α_S and thereby inhibit the enzyme adenylate cyclase (see

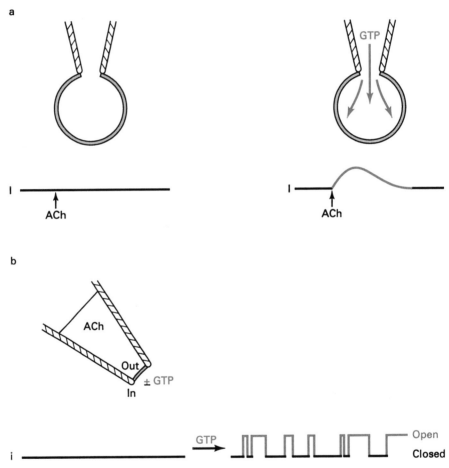

FIGURE 10-4. A G protein is involved in activation of a cardiac potassium channel by acetylcholine. Acetylcholine binds to a muscarinic receptor and activates a potassium channel in cardiac muscle cells. This response, whether measured by whole cell recording (a) or in a detached membrane patch (b), requires GTP (see Pfaffinger et al., 1985; Breitwieser and Szabo, 1985).

below), which requires α_S for its activation. These findings, that $\beta\gamma$ can play more than just a passive role, may necessitate a revision of our relatively simple picture (Fig. 10-3) of the molecular mechanism of G protein action.

Second Messenger-Coupled Receptor/Ion Channel Systems

The third category of receptor/channel coupling is channel modulation via intracellular second messengers. In this case the receptor and channel are not part of a single macromolecular complex (Fig. 9-11), nor do they interact directly in the plane of the membrane (Fig. 10-2). Rather the receptor is coupled via a G protein to an enzyme, the activity of which is regulated by the occupation of the receptor by a neurotransmitter (Fig. 10-6). When the receptor is occupied by transmitter (the so-called first messenger), the enzyme is activated and can catalyze the formation of an intracellular second messenger (Fig. 10-6). The second messenger can then set in motion a sequence of events that ultimately influences the properties of, or *modulates,* an ion channel. Note that the channel properties do not immediately revert to the normal resting state when the transmitter dissociates from the receptor, as is the case for the directly coupled systems. Instead the channel modulation persists (in the example in Fig. 10-6, the channel remains open) until the actions of the second messenger are reversed. This may involve a

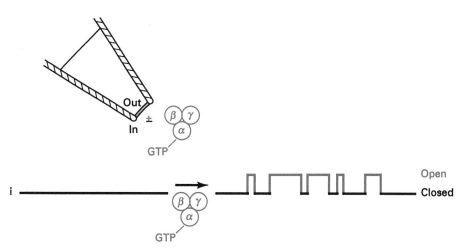

FIGURE 10-5. Purified G proteins can activate the cardiac muscarinic potassium channel. In detached membrane patches in which the endogenous G proteins have been previously inactivated by treatment with pertussis toxin, addition of an activated G protein can produce gating of the potassium channel (see Logothetis et al., 1987; Codina et al., 1987).

sequence of events, requiring seconds, minutes, or even hours. Thus second messenger coupling provides a mechanism for ion channel modulation that may be delayed in onset. In addition it can long outlast the initial stimulus, the occupation of receptor by neurotransmitter.

Several other features of this general scheme deserve mention. First, we have already mentioned that most second messenger pathways involve G protein-activated enzymes. Accordingly second messenger-mediated receptor/channel coupling might be thought of as just a special case of G protein coupling, but in fact the temporal and molecular complexities introduced by having a second messenger mediate the response justify a separate category for this coupling mechanism. A second important feature is that amplification can occur at each step of what may be a multistep second messenger cascade (Fig. 10-7), so that the activity of many target ion channels may be modulated by the occupation of a relatively small number of receptors. A corollary of this is that activation of a second messenger sys-

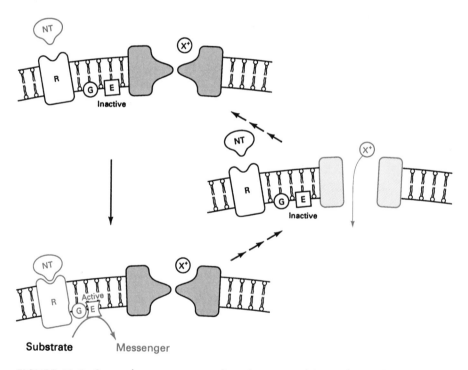

FIGURE 10-6. Second messenger-mediated receptor/channel coupling. In some cases neither the neurotransmitter receptor nor the G protein interacts directly with the ion channel. In these cases an intracellular second messenger causes activation of the ion channel. Contrast with Figures 9-11 and 10-2.

tem may produce coordinated changes in the activity of more than one *kind* of ion channel in the cell membrane (Fig. 10-7; see also Chapter 11). Furthermore, while tuning the ion channels, a second messenger can also influence other cellular processes that have nothing to do with ion fluxes. Clearly these features are not characteristic of directly coupled systems.

Different Kinds of Second Messengers

A comprehensive treatment of the vast field comprising second messenger systems is far beyond the scope of this book. Indeed whole books have been written about individual second messengers. We will confine ourselves in this and the next chapter to a summary of those pathways that are believed to play important roles in the modulation of neuronal excitability. In this chapter we will present an overview of the biochemistry of second messenger production and mechanism of action. Specific experimental examples that illustrate the role of each pathway in neuromodulatory phenomena will be given in the next chapter.

The specific pathways that we will discuss include

1. the adenylate cyclase/cyclic AMP-dependent protein kinase system;
2. the guanylate cyclase/cyclic GMP-dependent protein kinase system;

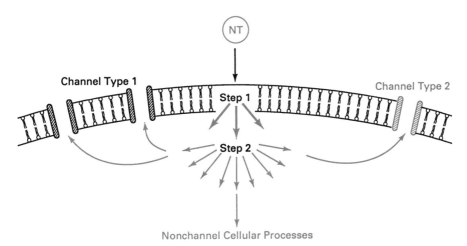

FIGURE 10-7. Amplification in second messenger cascades. With second messenger coupling the binding of a single neurotransmitter can activate many channels of a given class, activate several classes of channels, and affect other cellular processes not associated with ion channels.

3. the phospholipase-mediated turnover of membrane phospholipids, which can lead to the production of the three distinct second messengers inositol trisphosphate (IP$_3$), diacylglycerol (DAG), and arachidonic acid; and

4. the systems that are activated by calcium ion, including the calcium/calmodulin-dependent protein kinases.

Before we discuss these specific pathways, however, let us consider the general question of how changes in ion channel properties mediated by second messengers might be identified. Changes that last for a long time are of course strong candidates, but this criterion is not an unequivocal one. However, the cell-attached mode of the patch clamp technique does provide a convincing test for second messenger-mediated alteration of ion channel properties, independent of what the second messenger might be (Fig. 10-8).

The gigaohm seal between a patch electrode and the plasma membrane prevents the movement of neurotransmitter (and other) molecules between the extracellular bathing medium and the inside of the electrode. Accordingly, transmitters placed in the bathing medium cannot have access to receptors within the patch electrode (Fig. 10-8, top). Therefore directly coupled ion channels within the patch (black in Fig. 10-8) cannot be activated by a transmitter outside the patch electrode. If the transmitter activates ion channels within the patch, it must do so by binding to receptors outside the electrode. The only way these receptors can communicate with the channels in the patch is via some diffusible intracellular messenger (Fig. 10-8, bottom). An important feature of this test is that it requires no assumptions about the identity of the particular second messenger involved.

The adenylate cyclase/cyclic AMP-dependent protein kinase system. Cyclic AMP was first described by Earl Sutherland and his colleagues as the second messenger mediating hormonally stimulated glycogen breakdown in the liver. A surprisingly long time elapsed before it was accepted as the intermediary in some neurotransmitter responses in nerve cells, but many such cyclic AMP-mediated responses have now been described. The major components of this system are illustrated in Figure 10-9a. Cyclic AMP is synthesized from ATP by the enzyme *adenylate cyclase,* which is coupled to receptors via the G protein G$_S$. The second messenger then activates the *cyclic AMP-dependent protein kinase,* which is a tetrameric complex of two each of two kinds of subunit. The holoenzyme complex is completely inactive. Binding of cyclic AMP to the *regulatory subunits* causes them to dissociate from the *catalytic subunits.* The free catalytic subunits are active (Fig. 10-9b) and can catalyze the transfer of the terminal phosphate from ATP to the hydroxyl groups of serine or threonine residues in the target protein, to produce a phosphoprotein. This system is turned off by two kinds of enzymes, *phosphodiesterases,* which break down the cyclic AMP, and *phosphoprotein phosphatases,* which dephosphorylate the substrate proteins (Fig. 10-9a).

As we shall see below, phosphorylation of enzymes, ion channels, or

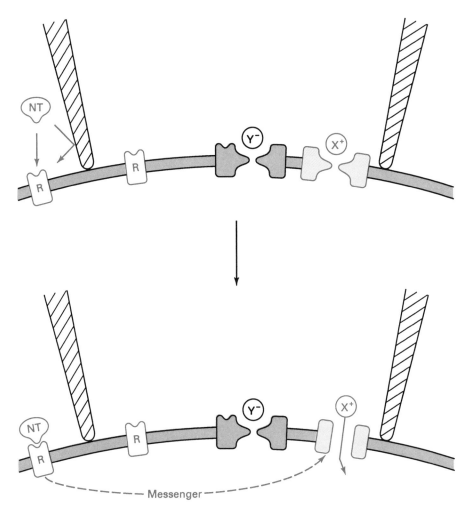

FIGURE 10-8. A test for second messenger mediation of neurotransmitter effects on ion channels. A neurotransmitter applied outside a patch electrode does not have access to receptors within the patch. Accordingly it can affect ion channels within the patch only by interacting with some receptor outside the patch. These receptors can communicate with channels inside the patch only by means of some diffusible intracellular messenger.

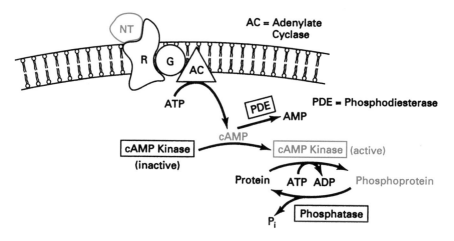

FIGURE 10-9. The adenylate cyclase/cyclic AMP-dependent protein kinase system. *a*: Cyclic AMP (cAMP) is synthesized from ATP by the enzyme adenylate cyclase (AC), and broken down by a phosphodiesterase (PDE). It activates a cyclic AMP-dependent protein kinase (cAMP kinase). *b*: The cyclic AMP-dependent protein kinase is an inactive holoenzyme consisting of two regulatory (R) and two catalytic (C) subunits. Cyclic AMP binds to the regulatory subunits, releasing the catalytic subunits that are now active.

other proteins can lead to large changes in their functional properties. The way in which a particular cell responds to an elevation of cyclic AMP will thus depend on its particular spectrum of cell-specific substrate proteins that can be phosphorylated by the cyclic AMP-dependent protein kinase. Some specific examples of different cyclic AMP-mediated modulations in nerve and muscle cells are presented in Figures 11-3, 11-4, 11-6, 11-8, and 11-12–11-14. It was thought for a long time that all actions of cyclic AMP in eukaryotic cells are mediated by the cyclic AMP-dependent protein kinase as described above. However, it is now believed that cyclic AMP may interact directly with some ion channels, and activate them independently of kinase activation. This somewhat heretical notion will be covered in detail in Chapter 12, when we discuss sensory neurons in olfactory epithelia.

The guanylate cyclase/cyclic GMP-dependent protein kinase system. The components of the cyclic GMP system (Fig. 10-10a) are to a considerable extent analogous to those for cyclic AMP. Some extracellular signaling molecules are known to elevate intracellular cyclic GMP levels, but in most cases the coupling mechanism between the membrane receptor and the guanylate cyclase (which is often a soluble enzyme) is not understood. However, in the case of one receptor, for the peptide hormone atrial natriuretic factor (ANF—Fig. 10-10a), the amino acid sequence contains a guanylate cyclase catalytic domain on the intracellular side.

Like cyclic AMP, cyclic GMP can interact directly with ion channels, for example, the light-dependent ion channel in vertebrate photoreceptors (Chapter 12). However, most of the actions of cyclic GMP are mediated by the cyclic GMP-dependent protein kinase, which is a dimeric protein with two identical subunits linked by disulfide bridges (Fig. 10-10b). Each subunit contains a cyclic GMP-binding domain, whose amino acid sequence is homologous to the cyclic AMP-binding region of the cyclic AMP-dependent protein kinase regulatory subunit. Similarly the catalytic domain in each subunit is homologous to the cyclic AMP-dependent kinase catalytic subunit. However, there is no subunit dissociation involved in the activation of the cyclic GMP-dependent kinase. Instead cyclic GMP binding induces a conformational change that exposes and activates the catalytic domain (Fig. 10-10b). The role of the cyclic GMP system in neuromodulatory phenomena is less well understood than that of cyclic AMP, but there are now several examples of ion channel modulation in nerve cells that clearly are mediated by cyclic GMP-dependent protein phosphorylation.

Turnover of membrane phospholipids. It was first demonstrated in 1952 by Lowell and Mabel Hokin that acetylcholine and other neurotransmitters and hormones can stimulate the breakdown and resynthesis *(turnover)* of the minor membrane phospholipid *phosphatidylinositol* (PI). However it was to be almost 25 years before it was widely recognized that PI (and other phospholipids) might participate in signal transduction, and only the last 7 years or so have seen rapid—indeed explosive—growth in this field. This may be contrasted with the cyclic nucleotides, which were discovered later but very quickly became established as important second messenger molecules. Why was the PI story so long in developing? It is interesting to speculate that it might have been because lipid biochemistry is tough to do.

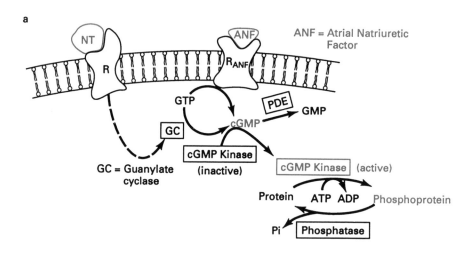

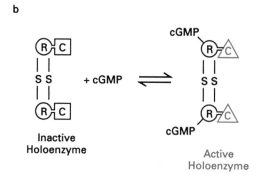

FIGURE 10-10. The guanylate cyclase/cyclic GMP-dependent protein kinase system. *a*: The enzyme guanylate cyclase (GC) produces cyclic GMP (cGMP) from GTP. Cyclic GMP can activate a cyclic GMP-dependent protein kinase. *b*: The cyclic GMP-dependent protein kinase is an inactive holoenzyme, and the regulatory and catalytic domains reside on a single subunit.

Only a small brave band of biochemists was prepared to struggle with these complex, water-insoluble phospholipids, whereas the majority extracted and discarded the lipid from their preparations as quickly as possible so it would not interfere with their study of proteins. In any event the latent period is now behind us, and the 1990s provide us daily with new findings that speak to the fundamental importance of phospholipid metabolism in signal transduction.

A critical breakthrough came when it was demonstrated that the phospholipid species whose metabolism is stimulated by agonists is not PI itself,

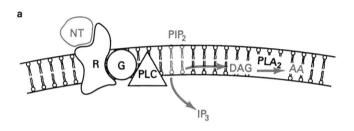

FIGURE 10-11. Three second messengers from polyphosphoinositides. a: Some neurotransmitter receptors are coupled (via a G protein) to the enzyme phospholipase C (PLC). b: The chemical structures of the second messenger molecules derived from the breakdown of the polyphosphoinositides (see the review by Berridge, 1987).

but rather is its doubly phosphorylated derivative *phosphatidylinositol-4,5-bisphosphate* (PIP$_2$). As shown in Fig. 10-11a, many receptors are coupled (via a G protein) to the membrane enzyme *phospholipase C* (PLC). This acts as a phosphodiesterase to split PIP$_2$ into two products, the water-soluble *inositol 1,4,5-trisphosphate* (IP$_3$), and *diacylglycerol* (DAG), which remains in the membrane. *Arachidonic acid* (AA), which is usually the fatty acid present in the 2 position of the gylcerol backbone of PIP$_2$ (see Fig. 10-11b), can be released from DAG through the action of another membrane phospholipase, *phospholipase A$_2$*. Phospholipase A$_2$ may also release AA directly from PIP$_2$. All three of these products—IP$_3$, DAG, and AA—are important second messengers.

The mechanisms by which these three second messengers work are summarized in Figure 10-12. IP$_3$ diffuses in the cytoplasm and binds to specific receptors on the endoplasmic reticulum. This binding opens a calcium channel in the endoplasmic reticulum membrane, and as a result calcium ions are released into the cytoplasm from storage sites in the lumen of the endoplasmic reticulum (Fig. 10-12a). As we shall discuss below, the elevated cytoplasmic calcium can influence ion channel activity and a myriad of other cellular functions.

The DAG, in contrast, remains associated with the membrane. An increase in DAG results in a translocation from cytoplasm to membrane of *protein kinase C* (often called C kinase), another in the family of protein kinases that catalyze the phosphorylation of serine and threonine residues in substrate proteins (some of which are ion channels). Protein kinase C becomes active only when it associates with the membrane and binds DAG (Fig. 10-12a). A specific example of neuromodulation mediated by protein kinase C can be found in Figure 11-7.

In the test tube, protein kinase C can be activated by the addition of DAG together with phosphatidylserine, the latter apparently serving as a substitute for the cell membrane. In addition there exists a series of *phorbol esters,* first described as tumor-promoting agents, that can substitute for DAG in activating protein kinase C both in the test tube and when applied to intact cells. PIP$_2$ may not be the most important source of DAG. PIP$_2$ is a minor phospholipid, and the amounts of DAG produced in response to an agonist are often too large to be accounted for solely by PIP$_2$ breakdown. However, some agonists activate a *phospholipase D* (again via a G protein), which catalyzes the release of phosphatidic acid from *phosphatidylcholine,* the major membrane phospholipid. The phosphatidic acid can then be metabolized further to DAG. This pathway provides a mechanism for producing DAG and activating protein kinase C without concomitant activation of the IP$_3$ calcium release system.

The actions of AA, the third second messenger product of phospholipid

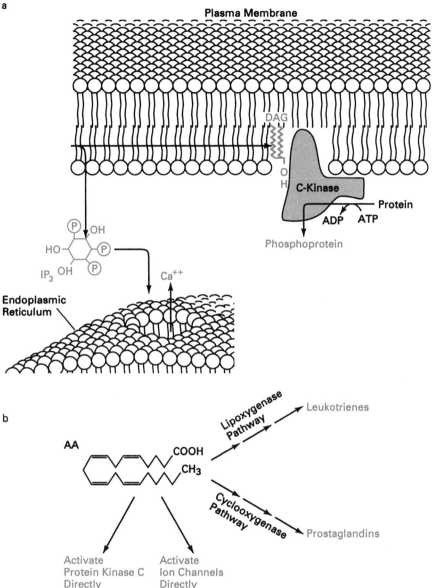

FIGURE 10-12. Mechanisms of action of the phospholipid-derived second messengers. *a*: DAG activates a protein kinase called protein kinase C (C kinase). IP_3 opens calcium channels in the endoplasmic reticulum. See the reviews by Berridge (1987) and Nishizuka (1986). [Drawing modified from Berridge, 1985.] *b*: Arachidonic acid (AA) may directly activate protein kinase C or some ion channels. It can also be metabolized to the leukotrienes and the prostaglandins.

turnover, are less well understood. AA can be metabolized via several different pathways, each of which gives rise to biologically active products including the prostaglandins and the leukotrienes (Fig. 10-12b). In addition AA can act as an activator of one form of protein kinase C in the brain. AA or its metabolites can have direct effects on some kinds of ion channels and an example is given in Figure 11-4.

Calcium as a second messenger. Calcium plays a central role in the activity of all cells. In nerve cells and other cells that possess calcium channels in their plasma membranes, calcium acts as a charge carrier to modulate beating or bursting activity, action potential shape, and other aspects of electrical activity (see Chapter 4). But charge transfer may be the most humdrum of calcium's many functions (Fig. 10-13). Its entry across the plasma membrane or release from intracellular stores produces such diverse effects as

1. triggering of secretion;

2. muscle contraction; and

3. activation of protein kinases, other enzymes, and calcium-dependent ion channels.

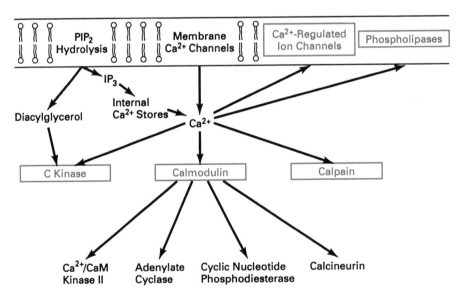

FIGURE 10-13. Some actions of intracellular calcium ions. Calcium can interact directly with several membrane and cytoplasmic proteins and regulate their properties. Among its cytoplasmic binding proteins is calmodulin, which in its calcium-bound form regulates the activities of many enzymes. Modified from Kennedy (1989).

Changes in intracellular calcium can be induced directly by certain neurotransmitters, such as those linked to formation of IP_3, or others that activate calcium-permeable ion channels such as the NMDA receptor channel (see Chapter 9). Let us not forget, however, that changes in this second messenger also occur as a *direct result of neuronal activity*. Calcium entry through voltage-dependent calcium channels results in transduction of an *electrical* signal—a change in voltage—into a *chemical* signal.

The intracellular calcium signal is received and decoded by calcium-binding proteins such as the ubiquitous, small, protein *calmodulin*, which we discussed in Chapter 7. Calcium binding induces a large conformational change in calmodulin, exposing a hydrophobic domain that can interact with a variety of effector proteins (Fig. 10-13). Among these are at least four different types of calcium/calmodulin-dependent protein kinase. One of them, the *multifunctional calcium/calmodulin-dependent protein kinase* (Ca^{2+}/Cam kinase II, see Chapter 7), is present at high concentration in both presynaptic terminals and postsynaptic densities, and in fact comprises as much as 1% of total brain protein. Like the cyclic AMP-dependent protein kinase, this enzyme acts on a wide range of substrates, among them proteins that may be involved in synaptic transmission. We described in Chapter 7 the modulation of neurotransmitter release at the squid giant synapse by Ca^{2+}/Cam kinase II in the *presynaptic* terminal, and its high concentration in *postsynaptic* densities has led to speculation about its role in receptor and/or ion channel modulation. One feature of this enzyme is that, like many other kinases, it can undergo autophosphorylation. That is, it phosphorylates itself in a calcium/calmodulin-dependent manner. Particularly intriguing is the fact that once several of the subunits are phosphorylated, the enzyme becomes active *independent* of calcium/calmodulin (Fig. 7-10). Thus a transient calcium signal can produce long-lasting activation of this kinase, a finding that has led to hypotheses that the enzyme acts as a "calcium switch" to produce long-lasting changes in neuronal properties (see Chapter 17).

Interactions and commonalities among second messenger systems. The diversity in second messenger systems does not (yet) approach that of the first messengers—the extracellular signaling molecules—or their membrane receptors. However, there is considerable complexity due to the fact that the different second messenger systems can interact at various levels. For example G proteins may participate in the formation of several different second messengers. In addition, one second messenger may influence the activities of enzymes that are involved in the turning on or off of another messenger pathway. To give just one example, there are calcium/calmodulin-dependent cyclic AMP phosphodiesterases and phosphoprotein phosphatases. In the face of this complexity it is gratifying to find common features among these pathways. Most second messengers are generated via G protein-mediated enzymatic reactions, and most use protein phosphorylation as their final common mechanism for producing a biological response. Nevertheless one must always take into account the *interactions* among second messenger systems, in drawing conclusions about their roles in physiological responses. We shall elaborate on this theme in our discussion of transduction in specialized nerve cells, and mechanisms of modulation of neuronal electrical behavior, in subsequent chapters (e.g., see Figs. 11-4 and 11-9).

Structure and Function in the G Protein-Coupled Receptors

All of the second messenger pathways that we have considered can be set in motion by different members of the G protein-coupled family of receptors. Is there anything about the amino acid sequence of a particular receptor that predicts either the exact nature of the G protein coupling mechanism, or the type of ligand to which the receptor binds? For example, it is particularly interesting that the nicotinic and muscarinic acetylcholine receptors, which are so very different in their amino acid sequences, predicted structures and coupling mechanisms, are activated by the same ligand. Are the ligand-binding sites constructed in the same way, or did evolution arrive at a similar three-dimensional binding pocket via different routes?

Definitive answers to these fascinating questions are still in the future, but an important step in this direction has been taken recently with the construction of *chimeric* receptors as described in Chapter 9. The adrenergic receptors are particularly well suited for this kind of analysis, because all of them bind and are activated by epinephrine, but they can be distinguished by their affinities for various other adrenergic agonists and antagonists, as well as by the G proteins to which they couple. By constructing and expressing a series of adrenergic receptor genes that are chimeras of

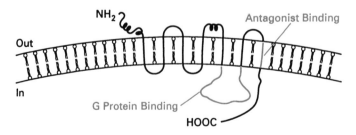

FIGURE 10-14. Structure and function in the G protein-coupled receptors. The construction of chimeras has been used to probe the relationship between structure and function in the G protein-coupled receptor family.

the α_2- and β_2-adrenergic receptor subtypes, one can ask which structural domains are responsible for particular functions. Using this approach the seventh transmembrane domain has been identified as being of importance for determining antagonist specificity, whereas coupling to a specific G protein is determined to a considerable extent by the amino acid sequence in the large cytoplasmic loop between the fifth and sixth transmembrane domains (Fig. 10-14). Approaches such as these are beginning to be applied to other receptor families.

Down-regulation of G protein-coupled receptors. We mentioned briefly in Chapter 3 that many neurotransmitters are subject to *down-regulation,* also often called *desensitization.* These terms refer to a progressive decrease in the response to a neurotransmitter during maintained exposure or multiple exposures of the cell to the same transmitter. In some cases desensitization results from receptor *internalization,* the removal of receptors from the plasma membrane by the pinching off and internalization of membrane vesicles that contain receptor molecules (Fig. 10-15a). This agonist-dependent process is best understood for certain growth factor receptors, and may be important for regulating β-adrenergic receptors as well. Internalization may result in degradation of the receptors within the cell, although in at least some cases the internalized receptors can be recycled back to the plasma membrane (Fig. 10-15a).

Several members of the G protein-coupled receptor family can also be down-regulated by phosphorylation. Photoreceptors contain a rhodopsin kinase that specifically phosphorylates light-activated rhodopsin and decreases its sensitivity to light. The agonist-occupied β-adrenergic receptor can be phosphorylated and down-regulated by the cyclic AMP-dependent protein kinase as well as by a more specific β-adrenergic receptor kinase (βARK; see Fig. 10-15b). There are a number of serine and threonine residues that are potential sites for phosphorylation near the carboxyl terminal of the G protein-coupled receptors (on the cytoplasmic side of the membrane). These residues and adjacent amino acid sequences are well conserved in the various members of this receptor family, and so it seems likely that other members (e.g., the muscarinic receptors) are also down-regulated by phosphorylation. As we shall see in Chapter 11, phosphorylation by several different protein kinases is also involved in desensitization of at least one member of the ligand-gated receptor/channel family, the nicotinic acetylcholine receptor, and thus this mechanism may be widespread.

Receptors Linked to Tyrosine Kinases

Some receptor families transduce their signals through pathways other than those we have described thus far. The *tyrosine kinases* are a specialized class

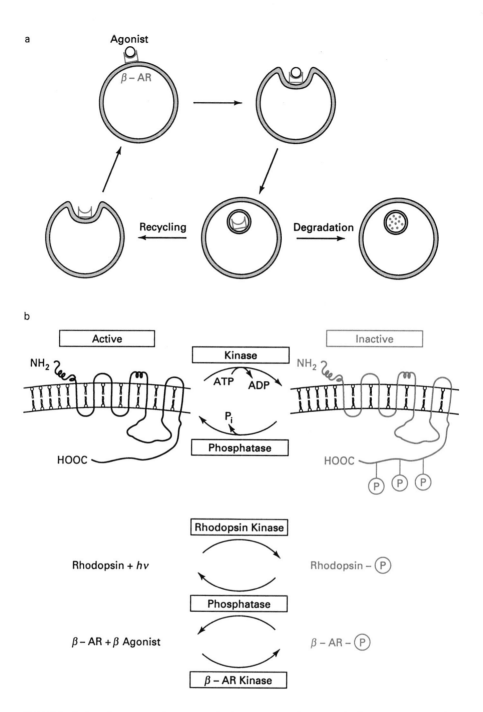

FIGURE 10-15. Receptor down-regulation. *a*: Agonist occupied receptors may be internalized by endocytosis, and either recycled to the cell surface or degraded. *b*: Members of the G protein-coupled receptor family, such as rhodopsin or the β-adrenergic receptor (β-AR), may also be down-regulated by phosphorylation. These down-regulations can be reversed by phosphoprotein phosphatases.

of protein kinases that phosphorylate proteins exclusively on tyrosine residues. In contrast to the other protein kinases we have discussed, the activity of the tyrosine kinases is not regulated by intracellular second messenger molecules. Instead the enzymatic activity resides in an intracellular catalytic domain, which is part of the membrane receptors for certain growth factors (Fig. 10-16). In other words the growth factor receptors themselves are tyrosine kinases, which are activated directly by the binding of the growth factor. Once activated, these kinases phosphorylate themselves as

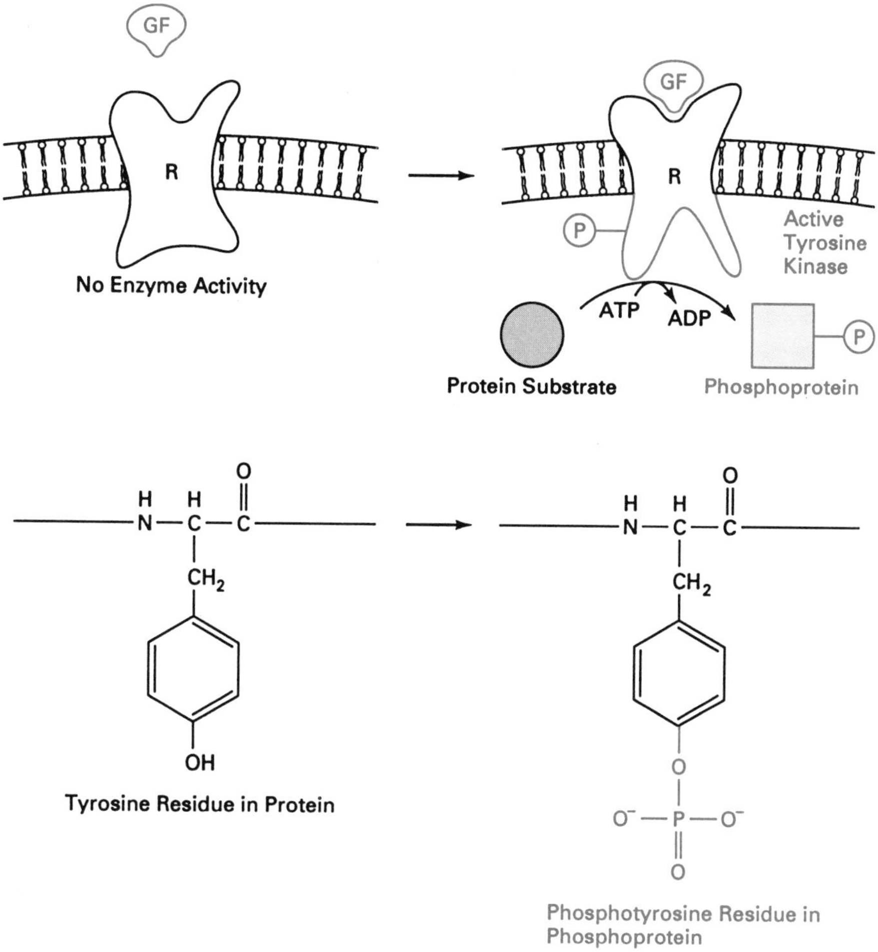

FIGURE 10-16. Receptors linked to tyrosine kinases. The receptors for a number of growth factors (GF) contain an intracellular domain that can catalyze the phosphorylation of proteins on tyrosine residues. This tyrosine kinase activity is observed only when the receptor is occupied by the growth factor.

well as other substrates, which include at least one ion channel (see Fig. 11-14). Among the growth factors that act via such receptors are those related to *epidermal growth factor* (EGF). As we shall see in Chapter 13, these play a crucial role in neuronal differentiation during development of the nervous system. It is known from site-directed mutagenesis experiments that the tyrosine kinase activity is essential for the biological actions of EGF (and other growth factors that act through tyrosine kinase-linked receptors).

Other Receptor Families

Although we have focused our attention on the G protein-coupled receptor family in our discussion of second messengers, it is important to realize that there exist other receptor families whose activity can also influence the same second messenger pathways. For example, some growth factor receptors, structurally unrelated to the G protein-coupled family, also stimulate the IP_3/DAG pathway directly. In addition the receptor for one peptide hormone, atrial natriuretic factor, contains a cytoplasmic domain that is a guanylate cyclase (Fig. 10-10). This may be the forerunner of an entirely new class of membrane receptors.

The fact that we have not dealt with other receptors in as much depth as the directly coupled or G protein-coupled families should not be taken to imply that other receptor systems are any less important. Rather it reflects the breadth of the field and the fact that more is known about how these two families regulate neuronal excitability. We have also chosen the directly coupled and G protein-coupled receptors as examples of receptor families in which the relationship between structure and function is beginning to be understood. Other families are known, and we can be confident that there are even more out there waiting to be discovered.

What Determines the Biological Response?

Where does the specificity lie in the response of a target cell to a particular extracellular signal? This is a dilemma that has been with us since it was realized that cyclic AMP can mediate different biological responses in different kinds of cells. The question has become even more pressing with the recognition that G proteins are involved in the production of a wide variety of responses. We are still without a satisfactory answer, but important clues have emerged from experiments in which receptors have been expressed in cell types in which they are not normally found. These receptors can couple to the cell's endogenous transduction mechanisms, and activation by agonist then produces a biological response that is characteristic of the host cell rather than of the exogenous receptor and its ligand. For example, serotonin receptors, which may be coupled to ion channels in the nerve cells from which they are isolated, can cause malignant transformation when they are expressed in fibroblasts and activated by serotonin. Similarly, changes typical of the response to sperm can be evoked by serotonin when one of its receptors is expressed in eggs. These experiments tell us that the specificity resides in the target cell and in the particular biological response system(s) that the cell makes available to interact with the receptor and transduction mechanism.

Summary

Extracellular signals must be recognized by the target cell and transduced into an appropriate biological response. Signal recognition is accomplished by the specific membrane receptors that are coupled to different kinds of transduction mechanisms, which in nerve cells usually regulate the activity of ion channels. The simplest receptor/ion channel coupling system, discussed in the last chapter, consists of the ligand-binding site and channel within a single protein molecule or macromolecular complex. A coupling mechanism of intermediate complexity involves protein–protein interactions, between a G protein and ion channel, in the plane of the membrane. Finally many ion channels are coupled to receptors via diffusible intracellular second messengers. The purpose of this diversity in the categories of receptor/channel coupling may be to provide a wide temporal range in the responses of neurons to neurotransmitters, hormones, and sensory stimuli.

Diversity also exists *within* the category of second messenger-mediated coupling. There are a variety of second messenger systems that at first glance appear to bear little relationship to one another. However, several of these share a common final mechanism of action on response systems, namely protein phosphorylation on serine or threonine residues via one of several second messenger-dependent protein kinases. Many receptors for growth factors also operate via protein phosphorylation, in this case on tyrosine residues. As we shall see in Chapter 11, modulation of neuronal excitability by protein phosphorylation may in some cases involve direct phosphorylation of the ion channel protein itself or of some closely associated regulatory component.

Neuromodulation:
mechanisms of induced changes
in the electrical behavior
of nerve cells

All neurons are not created equal. Even neighboring nerve cells may be distinct in their electrical properties, and exhibit very different patterns of endogenous electrical activity. As we know from the discussion in Part I of this book, these diverse patterns of activity reflect the complement of ion channels that are active under a given set of conditions. The important issue that we address now is the fact that these patterns of electrical activity are not fixed, but are subject to *modulation* resulting from synaptic or hormonal stimulation. Neurons may undergo long-lasting changes in the shape and amplitude of their action potentials, in the temporal pattern of action potential firing, and in the ways they respond to synaptic stimulation (Fig. 11-1). Such modulation of neuronal electrical properties, mediated by the transduction mechanisms described in Chapters 9 and 10, not only allow the nervous system to adapt its output in the face of a continually changing environment, but also are the basis for many long-lasting changes in behavior. Because neuromodulation underlies the choice of different patterns of behavior at different times, it is of fundamental importance for the functioning of the nervous system.

In this chapter we will discuss several examples of neuromodulation that are understood, at least partly, in terms of their biochemical mechanisms and physiological consequences. The theme here, as it has been in a number of previous chapters, is one of diversity. Different kinds of ion channels may be modulated in different cells, and a variety of molecular mechanisms are employed to effect the modulation. Because there is not yet a rosetta stone for neuromodulation, a single system for which the physiological significance and molecular details of a modulatory process are understood in depth, we have chosen to present several examples of modulatory phenomena to provide a feel for the diversity. We will also consider the possibility that direct phosphorylation of ion channels is a mechanism mediating some of these modulatory events.

It is important to emphasize that this chapter, in contrast with some of the earlier ones, is by necessity more a review of an ongoing and rapidly evolving area of research than a presentation of a complete and neatly packaged story. Although the choice of examples may be somewhat arbitrary, it is no accident that many—although by no means all—of the most thoroughly investigated examples of neuromodulation are in molluscan nervous systems. The large size and ready identifiability of many molluscan neurons permit the combined biochemical and electrophysiological approaches that are essential for a detailed understanding of long-term neuromodulatory phenomena.

Modulation of the Size and Shape of Action Potentials

When we discussed the ionic mechanisms of the action potential in Part I, we emphasized that the action potential is an all-or-none phenomenon, the

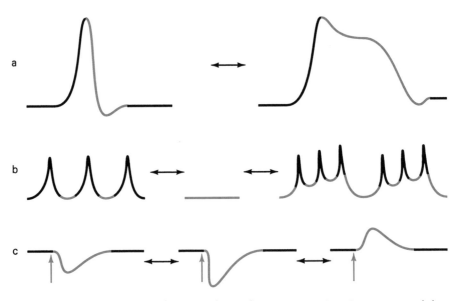

FIGURE 11-1. Modulation of neuronal membrane properties. Common modulatory changes in neuronal membrane properties include (a) changes in the amplitude and/or duration of action potentials, (b) changes in endogenous firing patterns between beating (left), silent (middle), and bursting (right) modes, and (c) changes in the efficacy of synaptic inputs.

amplitude of which is invariant because it depends only on the sodium concentration gradient. Although this is essentially correct for the special case of the squid giant axon, it is an oversimplification for most nerve cells. We know now that in many neuronal somata there is a calcium current that contributes to the depolarizing phase of the action potential, and a series of different potassium currents that participate in repolarization and help to determine action potential shape. We also know that many of these currents can be modulated, leading to changes in the size and shape of action potentials. An important functional consequence of such changes is the modulation of transmitter release triggered by the action potentials when they invade presynaptic terminals.

Dorsal root ganglion neurons. One of the earliest examples of modulation of action potential duration was in neurons of the chick *dorsal root ganglion* (DRG), a way-station in the pathway through which sensory information from the periphery reaches the spinal cord. A variety of neurotransmitters can cause a narrowing of action potentials in DRG neurons (Fig. 11-2a). Among the transmitters that produce this effect are the peptides

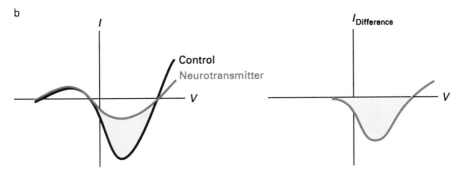

FIGURE 11-2. Modulation of action potential duration in dorsal root ganglion neurons. *a*: As first demonstrated by Dunlap and Fischbach (1978), a variety of neurotransmitters can cause a shortening of the action potential in dorsal root ganglion neurons. *b*: Voltage clamp analysis indicates that this results from a decrease in inward current. At the right the difference current elicited by the neurotransmitter is plotted.

enkephalin and somatostatin, as well as serotonin, GABA, and norepinephrine. There are two possible ways such a shortening could come about. On
the one hand there might be an increase in the potassium currents that are
responsible for spike repolarization; alternatively, the transmitters might
directly decrease the calcium or sodium currents that underlie the depolarizing phase of the spike. Voltage clamp experiments strongly support the
latter explanation. As shown in Figure 11-2b, the various transmitters cause
a decrease in a current with a voltage dependence expected for a calcium
current (compare with Fig. 4-10c).

What is the molecular mechanism mediating this shortening of action
potentials in DRG neurons? Intracellular GDPβS or pretreatment of the
DRG neurons with pertussis toxin blocks the actions of norepinephrine or
GABA. In addition application of a diacylglycerol activator of protein
kinase C can decrease the calcium current in these neurons, and a protein
kinase C inhibitor can block the transmitter-induced spike shortening.
Taken together these results suggest that a G protein-mediated activation
of phospholipase C, with the ensuing release of DAG and activation of protein kinase C, is responsible for the decrease in calcium current and shortening of the action potential. The physiological consequence of this shortening is likely to be a decrease in transmitter release at the DRG neuron
terminals, and an attenuation of the amount of sensory information that is
allowed to reach the spinal cord. This is of particular interest in the case of
enkephalin. An enkephalin-induced decrease in action potential duration,
with a consequent decrement in release of the pain pathway sensory transmitter, substance P, from DRG neurons, might account for some of the
analgesic actions of enkephalin and other opiate agonists.

Cardiac myocytes. The best understood example of modulation of action
potential duration is the spike prolongation evoked by β-adrenergic agonists in cardiac cells. We do recognize that cardiac muscle cells are not,
strictly speaking, neurons. However, we may define them as honorary neurons because there are so many parallels between cardiac and nerve cells
with respect to mechanisms of electrical signaling and its modulation. The
resemblance between neurons and other types of cells, which we have
emphasized throughout, is nowhere more evident than it is here.

The fundamental mechanism of the cardiac action potential and its modulation are well understood. The action potential is predominantly a calcium spike, and it can be prolonged dramatically by treatment with noradrenaline or other β-adrenergic agonists (Fig. 11-3a). This prolongation,
together with a series of other biochemical and electrophysiological
changes, contributes to the multifaceted consequences of β-adrenergic stimulation that include changes in the rate and force of contraction of the
heart.

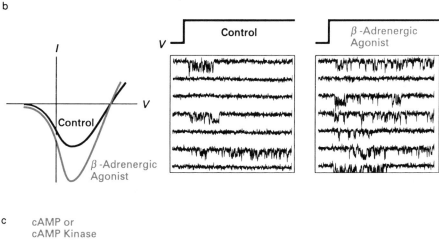

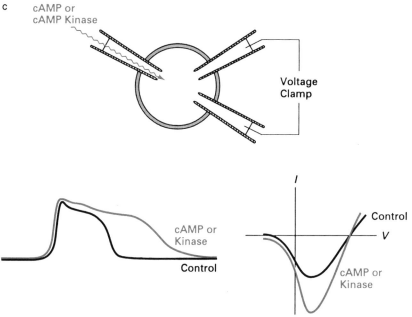

FIGURE 11-3. Prolongation of the action potential in cardiac muscle cells. *a*: Studies from the laboratories of Harald Reuter, Richard Tsien, and Wolfgang Trautwein have shown that β-adrenergic agonists increase the duration of the cardiac action potential. *b*: This is due to an increase in calcium current *(left)*, resulting from an increase in the activity of voltage-dependent calcium channels *(right)*. *c*: Intracellular injection of cyclic AMP or cyclic AMP-dependent protein kinase elicits the same changes (see Tsien, 1987).

As in the case of the DRG neurons discussed above, it seems possible that changes in either calcium or potassium currents could account for action potential prolongation. Voltage clamp experiments demonstrate that the ionic mechanism of the change in spike duration is an increase in the calcium current, which is responsible for the plateau phase of the depolarization (Fig. 11-3b). Single channel analysis confirms that β-adrenergic stimulation produces an increase in the activity of calcium channels, which leads to spike broadening. All of these effects can be mimicked by the intracellular injection of cyclic AMP or the extracellular application of membrane-permeant cyclic AMP analogs, as well as by the injection of the active catalytic subunit of the cyclic AMP-dependent protein kinase (Fig. 13-3c). Clearly cyclic AMP-dependent protein phosphorylation mediates this example of modulation of action potential duration. Later in this chapter we shall consider the possibility that the phosphorylatable regulatory protein is actually part of the ion channel protein itself.

The theme of diversity, to which we referred at the beginning of this chapter, will already be evident in these first two cases of modulation of spike shape. Both involve changes in calcium current, but in one case the current is decreased whereas in the other it is enhanced, and entirely separate transduction mechanisms mediate the modulatory responses. Now we shall introduce another level of complexity as we move on to discuss two examples from the *Aplysia* nervous system, in which action potential amplitude and duration can be regulated by several different ion currents and/or several different transduction mechanisms *within* an individual neuron.

Aplysia *sensory neurons.* Several groups of sensory neurons, the cell bodies of which are located in the abdominal and pleural ganglia of the marine snail *Aplysia,* undergo modulation of action potential duration. The mechanisms of modulation in these neurons have been investigated particularly thoroughly, because of their involvement in neural circuits that mediate a simple form of behavior in *Aplysia* (see Chapter 17). Here we shall focus on changes in action potential duration evoked by two putative neurotransmitters, the amine serotonin and the tetrapeptide FMRFamide.

Serotonin can cause an increase in action potential duration in abdominal and pleural sensory neurons (Fig. 11-4a). Voltage clamp analysis indicates that this increase results not from a direct action on the calcium current, but rather as a consequence of a decrease in the amplitude of a potassium current that normally contributes to spike repolarization (Fig. 11-4a). Thus, in the presence of serotonin, repolarization is delayed, allowing more calcium entry during the action potential. The increase in intracellular calcium contributes to enhanced release of neurotransmitter from the sensory neuron terminals.

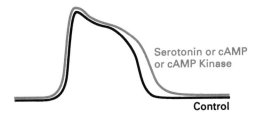

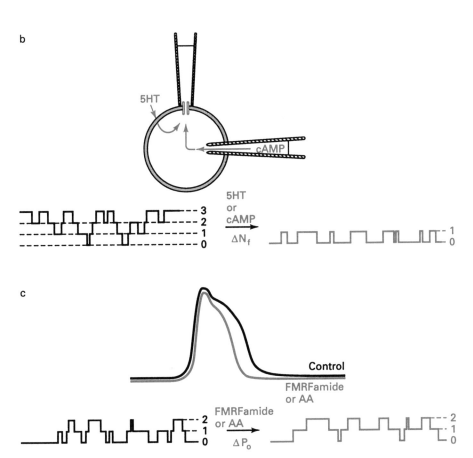

FIGURE 11-4. Modulation of action potential duration in *Aplysia* sensory neurons. *a*: As shown by Steven Siegelbaum, Eric Kandel, and their colleagues (Belardetti and Siegelbaum, 1988), extracellular application of serotonin (5HT), or intracellular injection of cyclic AMP or the cyclic AMP-dependent protein kinase, increases the duration of the action potential in *Aplysia* sensory neurons. This is due to inhibition of a largely voltage-independent potassium current (S current or I_S—*right*). *b*: Serotonin and cyclic AMP decrease the number (N_f) of functional S potassium channels. *c*: FMRFamide decreases the duration of the action potential by increasing the open probability (P_o) of single S channels. This effect can be mimicked by arachidonic acid (AA).

The particular potassium current affected by serotonin (called the S current) is different from many of the potassium currents that we discussed in Chapter 4 in that it exhibits only limited voltage dependence (Fig. 11-4a). Recent evidence suggests that serotonin might in addition decrease a voltage-dependent delayed rectifying potassium current, and that both of these changes contribute to prolongation of the action potential. The actions of serotonin on the S current have been investigated at the single channel level. Recall from Chapter 4 that the membrane current carried by a population of ion channels is dependent on the single channel current (i), the number of functional channels (N_f), and the single channel open probability (P_o). In patch clamp experiments using the cell-attached patch configuration, serotonin applied outside the patch electrode causes a decrease in the number of functional S channels within the patch, without altering the open probability of those channels that remain functional (Fig. 11-4b). We mentioned in Chapter 10 that such an effect of a transmitter applied outside the patch electrode *must* be mediated by an intracellular second messenger. As in the case of modulation of the cardiac action potential described above, several lines of evidence point to cyclic AMP as the mediator of this action of serotonin. Cyclic AMP can mimic the actions of serotonin on action potential duration (Fig. 11-4a) and on the S current and S channels (Fig. 11-4b). Furthermore serotonin can increase cyclic AMP levels in sensory neurons. Injection of the catalytic subunit of cyclic AMP-dependent protein kinase can produce spike prolongation (Fig. 11-4a), and the action of serotonin can be blocked by the injection of a specific inhibitor of this kinase.

It has been found that application of FMRFamide produces the opposite effect, namely a decrease in action potential duration (Fig. 11-4c). This phenomenon involves an *increase* in S current (as well as a decrease in calcium current). Again a second messenger must be involved in the effect on S current, because FMRFamide applied in the bathing medium modulates the activity of S channels in cell-attached patches (Fig. 11-4c). An increase in P_o is observed, in contrast to the change in N_f produced by serotonin. In other words, the cell uses very different molecular mechanisms for the up and down modulations of this single ion channel.

The pattern of diversity persists when one examines the identity of the second messenger that mediates the action of FMRFamide. Cyclic AMP is not involved here, but the effects of FMRFamide on spike duration, S current, and S channels can be mimicked by extracellular application of arachidonic acid (AA) (Fig. 11-4c). Experiments using pharmacological inhibitors of the different pathways of AA metabolism suggest that products of the lipoxygenase metabolic pathway (see Fig. 10-12b) are required, and several hydroperoxy acid intermediates in this pathway can decrease sensory neuron spike duration.

We can see then that it is no simple matter to predict the duration of the action potential in these sensory neurons. The balance between at least two different modulatory transmitters, coupled to at least two distinct second messenger systems, will determine the activity of the several ion channels that regulate spike duration. It is particularly intriguing that one of these transmitter/second messenger systems produces all-or-none closures of individual ion channels, whereas the other modulates the open probability of the same channels in a graded fashion. It would appear that the neuron will use all the tricks in its repertoire to achieve subtle regulation of neuronal activity.

Aplysia *bag cell neurons.* This conclusion is strengthened when one examines modulation of the bag cell neurons, the two clusters of homogeneous neurosecretory neurons associated with the *Aplysia* abdominal ganglion (Fig. 11-5). As we shall discuss in Chapter 16, the bag cell neurons synthesize and release several neuroactive peptides that trigger a series of events necessary for reproduction. Secretion of the peptides is evoked by a long-lasting discharge during which both the amplitude and duration of the

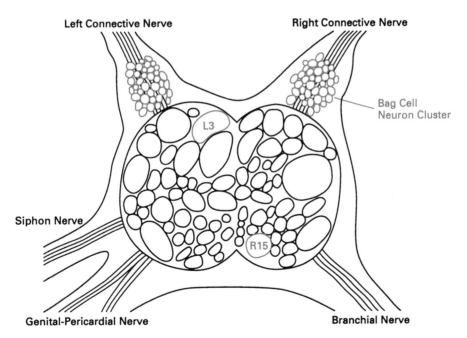

FIGURE 11-5. Identified neurons and groups of neurons in the *Aplysia* abdominal ganglion. Drawing, after one by Eric Kandel and his colleagues (Kandel, 1976), of the dorsal surface of the *Aplysia* abdominal ganglion. The bag cell neuron clusters and several identified neuronal cell bodies described in this and other chapters are shown in red.

action potentials are enhanced (see Fig. 16-9). We shall consider first the mechanisms involved in the modulation of spike size and shape; the after-discharge will be discussed below in the context of modulation of neuronal firing patterns.

Because all the bag cell neurons in an intact cluster are coupled to one another via electrical synapses, it is impossible to voltage clamp the cells to examine their electrical properties. However, the neurons retain many of their morphological and electrical characteristics when they are isolated in primary cell culture, and this preparation has been exploited to investigate neuromodulatory phenomena. Although such isolated neurons generally exhibit no spontaneous electrical activity, they can be induced to fire action potentials in response to depolarizing current pulses. Both the amplitude and duration of these action potentials are enhanced when cyclic AMP levels in the cell are increased (Fig. 11-6a). This can be effected by treatment

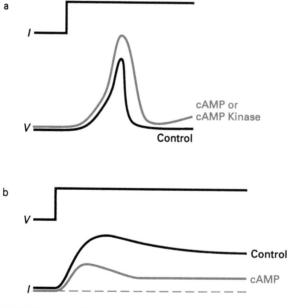

FIGURE 11-6. Changes in action potential amplitude and duration in bag cell neurons. a: The amplitude and duration of action potentials, evoked by a depolarizing current pulse, change when the cell is injected with cyclic AMP analogs or the catalytic subunit of the cyclic AMP-dependent protein kinase. b: Spike modulation is accompanied by changes in the kinetics and amplitude of the delayed rectifying potassium current (see Strong and Kaczmarek, 1987).

with a peptide transmitter synthesized and released by the bag cell neurons themselves (see Chapter 16), or by membrane-permeant cyclic AMP analogs. The modulation of the action potential by cyclic AMP, which presumably acts as an intracellular second messenger for the peptide, is mimicked by the injection of the catalytic subunit of cyclic AMP-dependent protein kinase (Fig. 11-6a) and blocked by injection of a protein kinase inhibitor.

What ion current is modulated by cyclic AMP-dependent protein phosphorylation to produce these changes in the action potential in the bag cell neurons? Again both calcium and potassium currents are a priori candidates, and one could not predict the answer based on the examples we have examined thus far. Voltage clamp experiments (Fig. 11-6b) have demonstrated that the answer in this case is potassium. The bag cell neurons have several components of the delayed rectifier potassium current involved in spike repolarization. Decreases in these contribute to the marked prolongation of the action potential following treatment with cyclic AMP.

Cyclic AMP is not the only second messenger that modulates the action potential in bag cell neurons. Treatment of isolated neurons with phorbol ester activators of protein kinase C causes an increase in action potential amplitude, in this case with little or no change in the duration (Fig. 11-7a). This effect can be mimicked by direct intracellular injection of protein kinase C (Fig. 11-7a) and blocked by protein kinase C inhibitors. In contrast to the actions of cyclic AMP, phorbol esters do *not* alter voltage-dependent potassium currents in the bag cell neurons. Instead the increase in action potential amplitude can be accounted for by an increase in the calcium current, which is a major contributor to the rising phase of the spike (Fig. 11-7b). When the microscopic mechanism of this change is examined by single channel analysis, it can be seen that the enhancement of the whole cell calcium current by activators of protein kinase C involves the recruitment of a novel calcium channel. In control neurons the calcium current is carried by a class of voltage-dependent calcium channels with a single channel conductance of about 12 pS. After exposure of the neurons to phorbol ester or diacylglycerol these channels are still present. In addition, there is a new 24 pS calcium channel that is never seen in control cells (Fig. 11-7b). Although the possibility that the small channels are converted into large ones cannot be ruled out entirely, this seems unlikely because the smaller conductance channels are seen with about the same frequency in control and phorbol ester-treated bag cell neurons. Furthermore the two channels exhibit different spatial distributions over the neuronal membrane surface. Images made using the fura-2 technique to measure the distribution and levels of intracellular calcium (see Chapter 7) reveal new sites of calcium entry following activation of protein kinase C. One interesting possibility is that protein kinase C may trigger the recruitment of the 24 pS channel to the plasma membrane from intracellular vesicles (Fig. 11-7c).

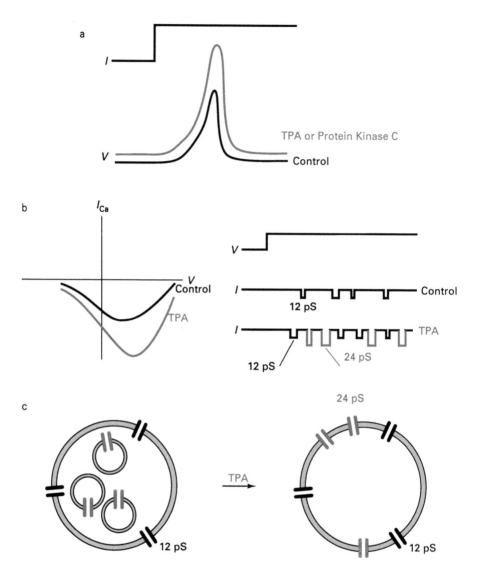

FIGURE 11-7. Modulation of bag cell neuron action potentials by protein kinase C. *a*: Protein kinase C injection or phorbol ester (TPA) treatment increases the amplitude but not the duration of the action potential. *b*: TPA increases voltage-dependent calcium current *(left)*. This is due to the recruitment of a novel calcium channel, which is not seen in the absence of TPA *(right)*. *c*: The possible recruitment of new calcium channels from intracellular vesicles is illustrated (see DeRiemer et al., 1985; Strong et al., 1987).

Modulation of action potential size and shape—are there any rules? We could go on to provide other examples of modulation of the amplitude and duration of action potentials in neurons and other excitable cells, but those above should suffice for the messages we wish to convey. The first is that this kind of modulation is ubiquitous, and can be seen in many different types of nerve and muscle cells. On reflection this is not surprising. Such changes in action potentials will result in changes in the amount of calcium entering the cell, and calcium is of central importance to many cell functions including secretion. The second message is that different cells choose different mechanisms to modulate their action potentials. The cyclic AMP, diacylglycerol, and arachidonic acid second messenger systems may be involved, sometimes cooperating and sometimes opposing each other's actions in a single cell. Furthermore changes in the action potential can result from modulation of either calcium channels or of several distinct potassium channels. Again occasionally more than one kind of channel is modulated in a single cell. The one feature that appears to be common to many of these systems is that the modulations involve protein phosphorylation (although the actions of arachidonic acid are not yet well understood). We shall see now that this diversity of mechanisms is also characteristic of modulation of neuronal firing patterns.

Modulation of Spontaneous Neuronal Discharge

The intrinsic electrical activity of a neuron can be examined by isolating it from hormonal or synaptic inputs from other cells. This can be achieved by physically removing the cell from the nervous system and placing it alone in a tissue culture dish, or by using pharmacological treatments to block intercellular interactions. When this is done it becomes evident that some neurons display no spontaneous electrical activity, whereas others fire action potentials at more or less regular intervals (see Fig. 2-10). We shall see now that the nervous system uses biochemical modulatory mechanisms to turn on, turn off, or otherwise alter these patterns of spontaneous activity.

Aplysia *bag cell neurons.* The bag cell neurons of *Aplysia,* which we described above, not only exhibit modulation of action potential size and shape, they also provide one of the best studied examples of alteration of neuronal firing patterns. In addition to its effects on action potentials, cyclic AMP plays a critical role in the generation of the afterdischarge in these neurons. At the onset of the discharge there is an increase in cyclic AMP levels in the bag cell neurons. Discharges can be both triggered and prolonged by treatments that elevate cyclic AMP. Moreover cyclic AMP can elicit a voltage-dependent inward current, seen as a region of negative slope resistance in the steady-state current–voltage relationship of the cell. Such a negative slope resistance is characteristic of cells that exhibit endogenous repetitive firing activity (see for example Fig. 11-8b). It is not yet clear which of the modulations of bag cell ionic currents discussed above is most important for the generation and maintenance of an afterdischarge. Probably all of them contribute to some extent, including the protein kinase C-mediated recruitment of the novel calcium channel. It is of interest that at the end of an afterdischarge, bag cell neurons enter a prolonged refractory state (see Chapter 16) that may last 18–24 hours, during which further long-lasting afterdischarges cannot be generated either by electrical stimulation or elevation of cyclic AMP levels. There is some evidence that calcium may play a role in the expression of the inhibited state, but the mech-

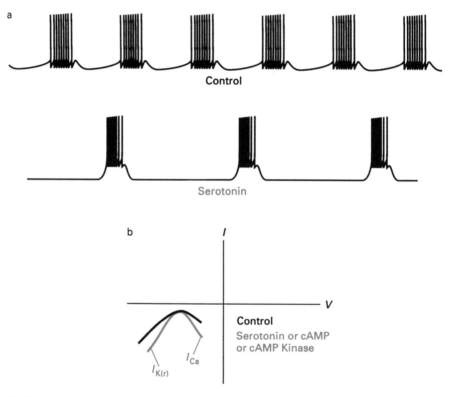

FIGURE 11-8. Serotonin modulates endogenous bursting activity in the *Aplysia* neuron R15. *a*: Serotonin causes an increase in the amplitude and duration of the interburst hyperpolarization as well as an increase in the frequency of action potentials during the burst in neuron R15. *b*: These changes result from increases in two voltage-dependent currents, an inwardly rectifying potassium current ($I_{K(r)}$) and a calcium current (I_{Ca}) (see Levitan and Levitan, 1988).

anism of its action is not understood. What is clear, however, is that complex interactions among a variety of different regulatory mechanisms are involved in the modulation of the endogenous firing pattern in the bag cell neurons.

Aplysia **neuron R15.** Bursting activity, the grouping of action potentials in bursts separated by periodic hyperpolarizations known as interbursts, is widespread in nervous systems. As we have mentioned in Chapter 2, bursting is used to drive rhythmic behaviors, as well as to increase the efficiency of secretion of peptide hormones. Although in some cases bursting activity is an emergent property of a multineuronal network (see Chapter 16), certain neurons generate bursts endogenously in the absence of synaptic or hormonal input. The archetypal endogenous burster is *Aplysia* neuron R15 (see Fig. 2-10c). The activity of R15 can be modulated for long periods of time by synaptic stimulation, or by the application of a number of different hormones or neurotransmitters. This is illustrated most dramatically by the fact that even though R15 is an endogenous burster, it is under tonic synaptic inhibition and only rarely is permitted to burst in the intact animal.

We shall consider here one well-understood example of modulation of bursting activity, the regulation of the activity of neuron R15 by serotonin. Serotonin application alters the activity of R15 in complex ways (Fig. 11-8a). An increase in the amplitude and duration of the interburst hyperpolarization is observed, and occasionally this can become so pronounced that bursting is inhibited completely. At the same time the frequency of firing of action potentials within the burst is increased. Voltage clamp analysis reveals the ionic mechanisms that underlie these changes (Fig. 11-8b). Serotonin causes an increase in an inwardly rectifying potassium current, and also increases a calcium current that contributes to the negative slope region of the steady-state current–voltage relationship. It might be thought that simultaneous increases in two opposing currents, one an inward depolarizing current and the other an outward hyperpolarizing current, might simply cancel each other out. However, note that both currents are voltage dependent and are active over different voltage ranges (Fig. 11-8b). The inwardly rectifying potassium current is activated during the interburst hyperpolarization, and its modulation by serotonin thus leads to a more pronounced interburst. In contrast the calcium current is active only at more depolarized voltages, and, accordingly, its enhancement by serotonin provides more depolarizing drive and an increase in spike frequency during the burst. The net effect of these concerted changes in two ionic currents is a more vigorous burst, which modulates the release of R15s neurosecretory peptide. This peptide is involved in regulation of water balance in *Aplysia,* and thus modulation of the bursting activity of R15 plays an essential role in osmoregulation.

Both of these actions of serotonin are mediated by cyclic AMP and cyclic AMP-dependent protein phosphorylation. The neuropeptide egg-laying hormone (ELH; see Fig. 6-6), which is released from the bag cell neurons during their afterdischarge, increases these same currents in neuron R15 (see Fig. 16-9a) via cyclic AMP. Thus two distinct first messengers, serotonin and ELH, acting via a single second messenger, cyclic AMP, have *divergent* actions on two different ion channels (Fig. 11-9a). Furthermore calcium also modulates the activity of these same two ion channels, and cyclic

a **Divergence**

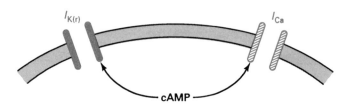

b **Convergence**

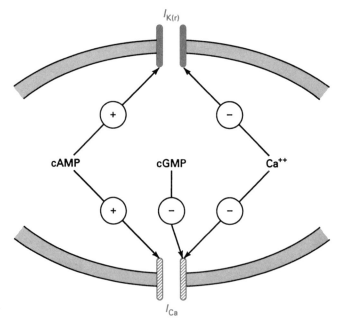

FIGURE 11-9. Divergence and convergence in second messenger actions. *a:* Cyclic AMP can have divergent actions on at least two different ion channels in neuron R15. *b:* The actions of several second messengers may converge on a single ion channel. The + and − symbols indicate activation and inhibition, respectively.

GMP regulates the calcium (but not the potassium) channel. In other words, there are also *convergent* actions of several different second messenger systems on a single class of ion channel (Fig. 11-9b). These findings, in neuron R15 and the bag cell neurons, emphasize that an understanding of how various modulatory systems *interact* is essential for a complete description of the regulation of neuronal firing patterns.

Modulation of Synaptic Efficacy

It is widely believed that modulation of chemical synaptic efficacy may underlie many important behavioral phenomena, including learning and memory. Modulation of electrical synapses is also important, but it has been less thoroughly investigated and we consider it only briefly at the end of this chapter. A change in the properties of ion channels can alter synaptic efficacy without changing action potential amplitude and duration or spontaneous neuronal activity. We shall discuss these ideas in the context of behavior in Chapter 17, and will restrict ourselves here to a brief description of several examples that are reasonably well understood at the cellular level, but whose behavioral consequences have not been investigated.

Aplysia neuron R15. For our first example let us return briefly to the actions of serotonin on neuron R15. We have seen that serotonin modulates the firing pattern in neuron R15 by increasing two opposing voltage-dependent ion currents. However, there is also a more subtle consequence of these actions of serotonin. As shown in Figure 11-10 (top), small hyperpolarizing stimuli, such as those that might be produced by rapid synaptic inputs, normally produce only small, rapidly reversible changes in the activity of R15. However after the voltage-dependent inward and outward currents have been enhanced by serotonin (Fig. 11-10, bottom), the small hyperpolarizing stimulus produces larger and longer lasting modulation of bursting activity, because there is more of the voltage-dependent current to be turned off or on by the small voltage changes. Thus the very obvious actions of serotonin on neuronal firing pattern are accompanied by a less conspicuous but nonetheless profound alteration in the sensitivity to synaptic input.

Of bullfrogs and rats. A similar subtle action of a transmitter on neuronal excitability has been described in at least two kinds of vertebrate neurons, the large B cells of the bullfrog sympathetic ganglion and rat hippocampal pyramidal cells. In each of these cell types a neurotransmitter alters neuronal excitability by inhibiting a voltage-dependent potassium current that is not active at the resting potential. In the sympathetic ganglion neurons a potassium current called the *M-current* (I_M) can be inhibited by a variety of neurotransmitters, including the peptides substance P and luteinizing hormone-releasing hormone (LHRH), and muscarinic cholinergic agonists. It is, in fact, inhibition of the M-current by LHRH that is responsible for the late slow inhibitory postsynaptic potential that we discussed in Chapter 8. Because the M-current is not very active at hyperpolarized potentials, application of one of these agonists has only a small effect on the cell's resting potential. However the response to a depolarizing stimulus, for example, an excitatory synaptic potential, will be very much enhanced because the M-current is not available to oppose the depolarization (Fig. 11-11a).

A similar phenomenon is observed in hippocampal pyramidal neurons, in which norepinephrine blocks a potassium current, probably a voltage- and calcium-dependent potassium current, that is responsible for the after-hyperpolarization (AHP) that follows an action potential. This current (I_{AHP}) is not active at the resting potential, and thus norepinephrine has little or no effect on the cell's resting potential. However a depolarizing stimulus that produces action potentials will be more effective in the presence of norepinephrine (Fig. 11-11b), because the afterhyperpolarization,

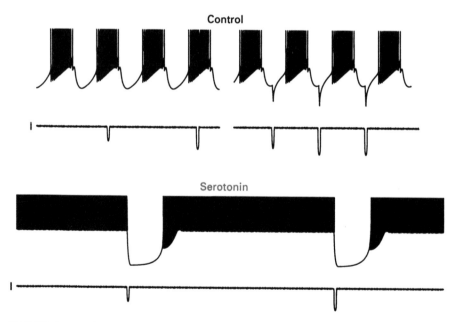

FIGURE 11-10. Serotonin modulates the response of neuron R15 to other stimuli. Serotonin changes the way the cell responds to small hyperpolarizing stimuli, such as those that might be provided by a synaptic input (modified from Levitan and Levitan, 1988).

which normally follows the action potentials and dampens the excitatory response, has been suppressed.

These few examples suffice to emphasize that the action of one transmitter may be modulated profoundly by that of another. Although the effects of a single transmitter on resting membrane properties may appear to be minor, its true influence on the excitability of a neuron may be evident only when it is coupled with stimulation of another synaptic or hormonal input.

Phosphorylation of Ion Channels: A Common Mechanism in Neuromodulation

Many of the modulatory phenomena that we have discussed above occur through the activation of protein kinases. One question that immediately arises concerns the identity of the proteins that are phosphorylated by the kinases. By analogy to alterations in the activities of some enzymes, which

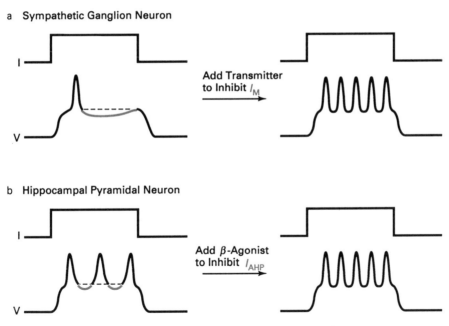

FIGURE 11-11. Modulation of responses to stimuli in vertebrate neurons. a: In experiments with sympathetic ganglion neurons of the bullfrog, addition of a transmitter that blocks the M potassium current (I_M) alters the cell's response to depolarizing stimuli. b: Similarly norepinephrine modulates the response of rat hippocampal pyramidal neurons to depolarizing stimuli. I_{AHP} refers to the current responsible for the spike afterhyperpolarization in these neurons (see Jones and Adams, 1987).

are known to result from direct phosphorylation of the enzyme molecules themselves, it seems possible that ion channels are targets for protein kinases, and that direct phosphorylation of ion channel proteins alters their functional properties. However, the experiments described above do not speak to this question. They involve injecting a kinase (or kinase inhibitor) into the cell, and subsequently measuring membrane properties. Such experiments demonstrate clearly that phosphorylation is both *necessary* and *sufficient* for the modulatory responses. However, it is conceivable that the phosphorylation target is some nonchannel membrane protein, or even a cytoplasmic protein, whose phosphorylation sets in motion a sequence of events that culminates in ion channel modulation, without phosphorylation of the channel itself.

Phosphorylation modulates channel activity in detached membrane patches. One experimental approach that can be used to explore this question is to investigate channel modulation in detached membrane patches. As detailed in Chapter 3, one of the modes of patch recording is the so-called *inside-out* patch, in which the former cytoplasmic surface of the patch membrane is exposed to the bathing medium (see Fig. 3-5). If one suspects that a channel under study might be modulated by phosphorylation, it is a simple matter to add a protein kinase, together with the magnesium ions and ATP that are necessary for the phosphorylation reaction, directly to the bathing medium while observing channel activity (Fig. 11-12a). This was first done for two molluscan channels, the S channel in *Aplysia* sensory neurons (Fig. 11-12b) that we have discussed above, and a calcium-dependent potassium channel in neurons from the garden snail *Helix* (Fig. 11-12c). In both cases addition of the catalytic subunit of the cyclic AMP-dependent protein kinase modulates the channel activity, as had been suspected from intracellular injections of the kinase. Note that these two types of channel are modulated in different directions by the same kinase, consistent with the diversity observed in the whole cell experiments. Furthermore the kinase induces a decrease in the *number* (N_f) of functional S channels, but does not alter the open probability (P_o) of those channels that are functional. In contrast, it is the *open probability* of the calcium-dependent potassium channels that is increased by the kinase, without any change in the number of functional channels. These experiments, which have now been done with several other kinds of channels, demonstrate that the phosphorylation target cannot be a cytoplasmic protein, but must be some channel regulatory component that comes away with the membrane patch when it is detached from the cell. It might be the ion channel itself, but it might also be some cytoskeletal component or other element, which scanning electron micrographs have shown are associated with these detached patches.

Phosphorylation modulates the activity of reconstituted ion channels. A complementary approach to the detached patch experiment is to examine the modulation of single ion channels *reconstituted* into artificial phospholipid bilayers. This is similar to the reconstitution of acetylcholine-dependent ion flux in liposomes, which we discussed in Chapter 9, but the experimental conditions are designed to allow single channel openings and closings to be observed. One way of doing this is to make plasma membrane

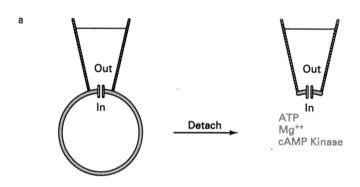

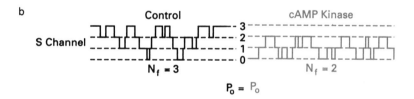

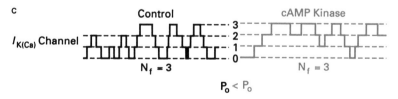

FIGURE 11-12. Phosphorylation modulates the activity of ion channels in detached membrane patches. *a:* Experimental configuration for testing the effects of a protein kinase on channel activity. *b:* Steven Siegelbaum and Eric Kandel found that the cyclic AMP-dependent protein kinase changes the number of functional S potassium channels in detached patches from *Aplysia* sensory neurons. *c:* This kinase also can increase the open probability of calcium-dependent potassium ($I_{K(Ca)}$) channels from *Helix* neurons. These findings are reviewed by Levitan (1985, 1988).

a

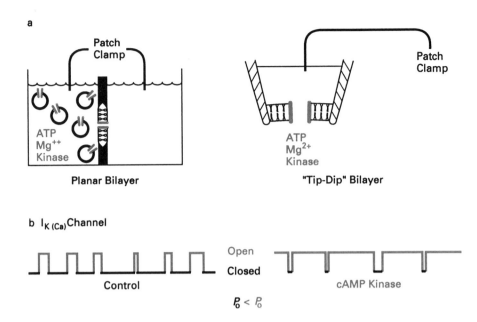

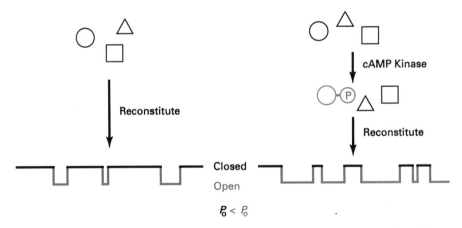

FIGURE 11-13. Phosphorylation modulates the activity of reconstituted ion channels. *a*: Two experimental configurations for ion channel reconstitution. *b*: The cyclic AMP-dependent protein kinase increases the activity of calcium-dependent potassium channels from *Helix* neurons in a bilayer. *c*: Phosphorylation modulates a purified reconstituted calcium channel (summarized in Levitan, 1988).

vesicles from the cells under investigation, and allow them to fuse with a bilayer that occludes a small hole in a partition separating two aqueous solutions (Fig. 11-13a left). When a vesicle fuses it dumps its membrane proteins into the bilayer, and if an ion channel is among these proteins, it is possible to measure single channel currents using relatively simple electronics. Another technique, which is in essence a marriage between the patch and bilayer approaches, involves the formation of a bilayer, containing an ion channel, at the tip of a patch electrode (Fig. 11-13a right). This permits single channel recordings and provides several other technical advantages. Figure 11-13b illustrates the modulation, by cyclic AMP-dependent protein kinase, of a calcium-dependent potassium channel extracted from a *Helix* neuron and fused into a bilayer at the tip of a patch electrode. Such reconstitution experiments allow the fundamental conclusion that, for at least some ion channels, the phosphorylation site is part of either the ion channel protein itself, or of some regulatory element that is so intimately associated with the ion channel that it swims with it in the bilayer.

Phosphorylation modulates the activity of purified ion channels. This approach has been extended in a particularly elegant way by examining the activity of ion channels that are reconstituted after they have been *purified* to homogeneity. One example of this is a voltage-dependent calcium channel from skeletal muscle, which was first purified on the basis of its ability to bind certain *dihydropyridines,* which are calcium channel modulators. The complex consists of three subunits (see Chapter 5), at least two of which are substrates for the cyclic AMP-dependent protein kinase. Calcium channel activity can be reconstituted in artificial bilayers from these three purified subunits. The gating of the channel is very different if the subunits are phosphorylated in the test tube prior to reconstitution (Fig. 11-13c). Another voltage-dependent ion channel, the sodium channel purified from brain, is also a good substrate for cyclic AMP-dependent protein kinase, but the functional consequences of this phosphorylation are not known.

The most comprehensive use of this approach is represented by work with the nicotinic acetylcholine receptor from *Torpedo*. The membrane of the *Torpedo* electroplaque contains protein kinase activity, including a cyclic AMP-dependent kinase, a protein kinase C, and a tyrosine kinase, all of which can phosphorylate the nicotinic receptor. As shown in Figure 11-14a, the pattern of phosphorylation of the various nicotinic receptor subunits differs for the three kinases. The locations of the specific phosphorylation sites for each kinase have been identified tentatively from examination of the primary amino acid sequences of the subunits, and in several cases have been confirmed directly by protein sequence analysis.

Protein
Stain

Autoradiogram

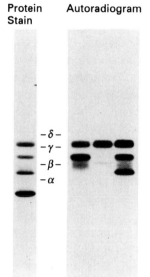

$-\delta -$
$-\gamma -$
$-\beta -$
$-\alpha -$

Cyclic AMP-Dependent Kinase	+	−	−
Protein Kinase C	−	+	−
Tyrosine Kinase	−	−	+

b

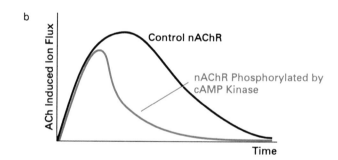

ACh Induced Ion Flux

Control nAChR

nAChR Phosphorylated by
cAMP Kinase

Time

c

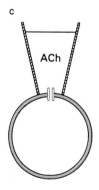

ACh

C
O

———— Desensitization ————→

Control nAChR

C
O

——— More Rapid Desensitization ———→

nAChR Phosphorylated
by Tyrosine Kinase

What are the functional consequences of phosphorylaton of the nicotinic receptor by the different kinases? The effects of the cyclic AMP-dependent kinase have been assessed by phosphorylating *Torpedo* postsynaptic membrane preparations with the catalytic subunit of the kinase, then purifying the phosphorylated nicotinic receptor and reconstituting it into phospholipid vesicles (liposomes). Acetylcholine-dependent ion flow out of the liposomes can be measured, using rapid flux techniques as described in Chapter 9 (see Fig. 9-12a). These experiments reveal that although the initial rate of ion flow is not altered by cyclic AMP-dependent phosphorylation, this treatment markedly increases the rate of agonist-induced receptor desensitization (Fig. 11-14b). Since it is known that the phosphorylation sites for this kinase are on intracellular loops of the γ- and δ-subunits, these regions must be involved in the desensitization process.

Phosphorylation by the tyrosine kinase also influences receptor desensitization. When the purified nicotinic receptor is reconstituted into liposomes, patch recordings on the liposome membrane reveal single channel openings and closings when acetylcholine is present in the patch electrode (this "liposome-attached" recording mode is analogous to a cell-attached patch). Agonist-dependent receptor desensitization is observed, even in the absence of any phosphorylation (Fig. 11-14c). When, however, the receptor is phosphorylated to different stoichiometries on tyrosine residues, the rate of this desensitization is dependent on the number of moles of phosphotyrosine per mole of receptor (Fig. 11-14c). Because the tyrosine phosphate is not only on the γ- and δ- but also on the β-subunit (Fig. 11-14a), this finding suggests that the latter subunit may also contribute to desensitization.

What is the significance of these findings? Recall from the discussion in Chapter 10 that the desensitization of the β-adrenergic receptor and rhodopsin can also be regulated by phosphorylation at intracellular sites. The use of this mechanism by a member of a different receptor family suggests that phosphorylation may be involved in desensitization in a wide variety of systems. The findings on the nicotinic receptor have not yet been set

FIGURE 11-14. Desensitization of the *Torpedo* nicotinic acetylcholine receptor is regulated by protein phosphorylation. *a*: Autoradiogram of a polyacrylamide gel showing the pattern of phosphorylation of the purified nicotinic acetylcholine receptor by different protein kinases. (courtesy of R. Huganir) *b*: Studies by Richard Huganir and Paul Greengard (summarized by Huganir and Miles, 1989) demonstrated that phosphorylation of the nicotinic acetylcholine receptor (nAChR) by the cyclic AMP-dependent protein kinase does not alter the initial rate of ion transport, but does enhance the rate of desensitization. *c*: Desensitization is also observed at the single channel level, and is enhanced by tyrosine phosphorylation (see Hopfield et al., 1988).

firmly in a physiological context, although there are some indications that elevations of cyclic AMP levels increase the rate of nicotinic receptor desensitization in mammalian skeletal muscle. However, the results do have important implications for our understanding of molecular mechanisms of ion channel modulation. They leave no doubt that, in this particular case, the regulatory element that is modulated by phosphorylation is actually part of the ion channel protein itself.

Phosphorylation of gap junction channels. As we discussed in Chapter 6, electrical synapses between neurons have their counterparts in the *gap junctions* that connect cells in many other tissues. It is known that phosphorylation by several different protein kinases can regulate gap junctional permeability in a variety of cell types. Furthermore the *connexins,* which come together to form the cell-to-cell gap junction channel (see Chapter 6), are good substrates for the cyclic AMP-dependent protein kinase. These findings suggest that phosphorylation of connexins may modulate gap junctional permeability. However, it is not known whether such phosphorylation of connexins contributes to modulation of the efficacy of electrical synapses in neurons.

Summary

Different neurons exhibit different patterns of endogenous electrical activity: some cells are normally electrically silent, others may fire action potentials in an irregular manner, and still others display regular and often complex firing patterns. These patterns of activity are not fixed, but are subject to modulation by stimuli from the neuron's environment. Such neuromodulation underlies short- and long-term changes in nervous system function, and thus is of fundamental importance for survival of the organism.

The mechanisms of neuromodulation are also diverse. We have discussed modulation of action potential amplitude and duration, neuronal firing patterns, and responses to synaptic input. These modulations often involve changes in the properties of one or another membrane ion channel, but the identity of the ion channel whose activity is modulated may be different from one neuron to the next, or even within the same neuron in response to different modulatory stimuli. We have therefore presented a number of examples of modulation, to provide a sense of the diversity and to emphasize that few unifying themes have emerged as yet in this rapidly changing field. One feature common to many of these examples is regulation via protein phosphorylation. At least in some cases, phosphorylation of the ion channel protein itself underlies the modulation of neuronal electrical properties.

Sensory receptor neurons

Although most neurons receive input from other neurons, the business of the brain is to act on information from the outside world. Specialized cells have evolved for the receipt of such information. These include cells that are responsible for sight, hearing, touch, taste, and smell, as well as those that signal to the brain the state of internal organs. Most external and internal stimuli are received by three classes of sensory cells: (1) those that respond to mechanical stimulation, (2) those that are activated by light, and (3) those that sense changes in their chemical environment. In addition, some cells respond to other signals such as changes in temperature. In all cases, the stimulus alters the activity of ion channels, sometimes through the agency of second messengers. Sensory cells therefore provide clear examples of the principles encountered in previous chapters. We will now discuss briefly a few of the varied transduction mechanisms that have been found in each of the major classes of sensory cells.

Generator Potentials

First let us consider in general terms the kind of output signal that these specialized sensory receptors send to the central nervous system. Some sensory cells are true neurons, with axons that travel from the sense organ to other parts of the nervous system. An external stimulus produces a depolarization or hyperpolarization of the membrane that is known as the *generator potential*. As is shown in Figure 12-1, the time course and amplitude of the generator potential generally mirror those of the stimulus. If a depolarizing generator potential is large enough to exceed the action potential threshold, the neuron will fire action potentials at a frequency that reflects the size of the stimulus. Thus we can see that information about stimulus strength is now encoded in action potential frequency. At the first synapse made by the sensory receptor neuron, the action potential frequency will determine the amount of transmitter released, and this in turn will control the magnitude of the depolarizing postsynaptic response. The latter will of course be translated back into a particular firing frequency in the postsynaptic cell.

Other sensory cells do not have axons. Although action potentials can sometimes be evoked in such cells, a sensory stimulus normally causes a depolarizing or hyperpolarizing generator potential that does not cause firing. Instead, the change in membrane potential alters the rate of spontaneous neurotransmitter release onto a postsynaptic neuron, and action potential frequency coding enters the picture only after this first synapse. As we shall see later, the classic example of this is the hyperpolarization of vertebrate photoreceptors in response to light.

Mechanoreceptors

Many sensory cells that respond to physical movement, termed *mechanoreceptors*, are found in or under the skin. These are true neurons, with their

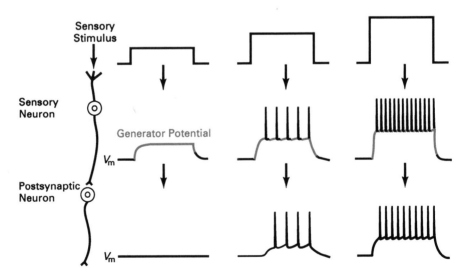

FIGURE 12-1. Generator potentials and frequency coding. The size of a sensory stimulus determines the size of the generator potential. This, in turn, is encoded in the frequency of firing of the sensory cell and/or its postsynaptic neuron.

cell bodies in the dorsal root ganglion. Recall that we have already considered the excitability of dorsal root ganglion neurons in the previous chapter. The business end of these mechanoreceptors is at the peripheral endings of their axons in the skin. Figure 12-2 shows that the morphology of such endings is very varied. One type of mechanoreceptor ending, located deep

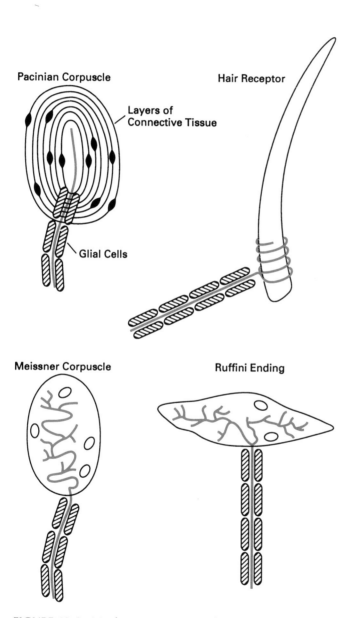

FIGURE 12-2. Mechanoreceptors in the skin (modified from Martin, 1981).

below the surface, is the *Pacinian corpuscle*. This consists of the bare ending of an axon, surrounded by "onion skin" layers of connective tissue. *Meissner corpuscles* and *Ruffini endings* are found closer to the surface of the skin and consist of more elaborate branchings of a nerve fiber, also enclosed in connective tissue. Other receptors comprise bare nerve fibers wrapped around the base of a hair follicle. Although in Figure 12-1 we suggested that frequency of firing generally mirrors the amplitude of a sensory stimulus, this is complicated by the fact that many sensory cells *adapt* their response to a *maintained* stimulus. This is particularly true of these different types of mechanoreceptors, whose electrical responses differ in that some fire only transiently when pressure is applied to the skin whereas others fire persistently throughout a maintained tactile stimulus. These and other receptors account for some of the varied sensations of touch and pressure on the skin. Mechanoreceptors are also found in many other parts of the body. For example, *Golgi tendon organs* and *muscle spindles* are found in joints and muscles, and provide information on the position and length of muscles.

The means by which a mechanical movement is transduced into the pattern of firing of the axon of a mechanoreceptor is not yet known. It is worth mentioning, however, that many cells possess *stretch-activated* ion channels that are normally closed, but are induced to open when physical pressure is applied to the plasma membrane. Such pressure probably distorts some direct link between the channel and a component of the cytoskeleton. Stretch-activated channels are found in a diverse group of cells and their function is not known. It is possible, however, that such channels underlie the transduction of movement into the electrical impulses of mechanoreceptors.

Hair cells of the cochlea. *Hair cells* respond to a specialized form of mechanical stimulation and are found in the inner ear of vertebrates. (These cells are entirely unrelated to the cells that innervate hair follicles as in Fig. 12-2.) They are found both in the vestibular organs and in the *cochlea*. In the former they are responsible for transducing information about gravity and movements of the head whereas in the cochlea, hair cells are the sensory cells of the auditory system. What is particularly interesting about these cells that are responsible for hearing is that they must not only provide the brain with information about the *intensity* of a sound, but must also respond selectively to different *frequencies* of sound waves.

Figure 12-3 shows a typical hair cell. It is elongated in shape with a distinctive arrangement of hairs, termed *cilia*, located at one end. The cilia are of two types. All hair cells have one long *kinocilium*, which, in an electron microscope, closely resembles moving cilia such as those in a sperm tail. Its

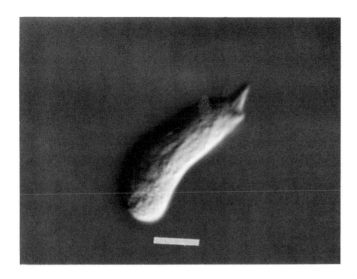

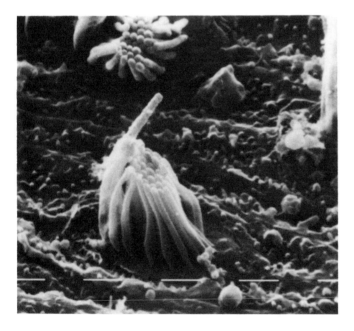

FIGURE 12-3. Hair cells. The upper micrograph is of a single hair cell isolated from the saccule of a toadfish. The tuft of cilia is at upper end of the cell (scale bar = 8 μm). The lower scanning electron micrograph shows the cilia of such cells in more detail (scale bars = 1 μm) (courtesy of Dr. Antoinette Steinacker).

structure is maintained by microtubules that extend the length of the kinocilium, with two central microtubules surrounded by a ring of nine others (Fig. 12-4). The remaining 30–100 cilia, which are arranged to one side of the kinocilium and are of varying lengths, are termed *stereocilia*. These do not contain microtubules but are filled with filaments of actin and proteins that cross-link actin filaments. The entire collection of cilia is termed the *hair bundle*.

Physical movements of the hair bundle cause a rapid change in the membrane potential of a hair cell. When the bundle is moved towards the kinocilium, the membrane depolarizes by 10–20 mV. In contrast, displacement of the bundle away from the kinocilium hyperpolarizes the cell (Fig. 12-4). The changes in membrane potential are caused by the opening and closing of channels located in the plasma membrane of the stereocilia themselves. These channels are relatively nonselective for cations such as sodium, potassium, and calcium ions. The response of the channels following mechanical displacement of the cilia occurs extremely rapidly, within 20–100 μsec. This means that opening of the channels is probably linked *directly* to mechanical deformation of the cilia, rather than through a second messenger system. The mechanism of the mechanical coupling to channels is not yet known. As we shall now describe, however, a striking feature of the response of hair cells in the cochlea is that their responses are specifically tuned to different frequencies of sound.

Hair cells are tuned by position in the cochlea and by electrical resonance.
Figure 12-5 shows that different cells respond optimally to different sound frequencies. This is illustrated by the *tuning curves* for several different hair cells, which represent the intensity of sound of different frequencies that must be applied to produce a fixed change in membrane potential. A major factor that determines the tuning curve for an individual cell is its position in the cochlea. The cell bodies of hair cells, together with their supporting cells, form a sheet of cells in the cochlea. The tips of the hair bundles are normally in contact with a stiff, carbohydrate-containing sheet, known as the *tectorial membrane,* that lies over the layer of cell bodies. Vibrations caused by sound waves entering the cochlea set up lateral movement of the tectorial membrane relative to the underlying cells. This, in turn, bends the hair bundles and transforms the mechanical vibration into an electrical oscillation of the membrane potential in the hair cells.

The mammalian cochlea is an elongated structure, folded into a spiral that resembles the shell of a snail. The sheets of hair cells extend from the base of the spiral to its apex. This mechanical design for the cochlea, coupled with the fact that changes in the thickness of the tectorial membrane occur along its length, provides a mechanism for the selective response of hair cells to different frequencies of sound. As was first suggested by the

a

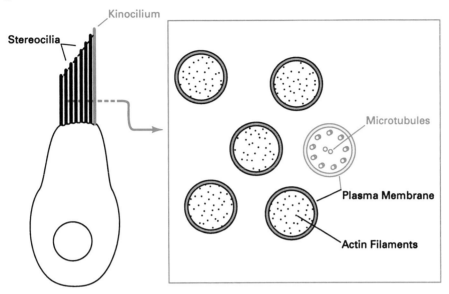

b

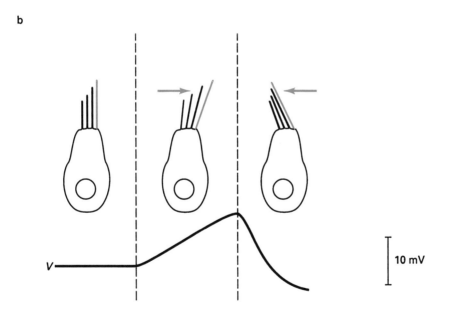

FIGURE 12-4. Hair cell cilia. *a*: Each hair cell has a single kinocilium and many stereocilia. *b*: Movement of the hair bundle towards the kinocilium depolarizes the cell. Movement in the opposite direction produces a hyperpolarization.

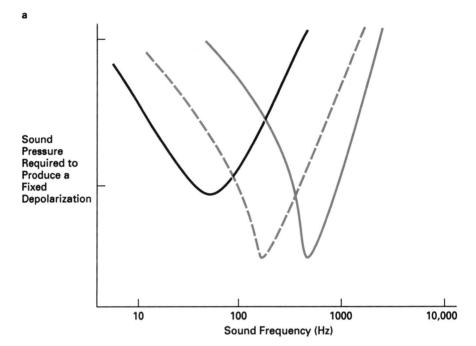

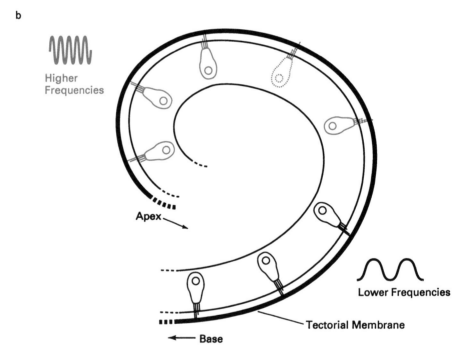

FIGURE 12-5. Tuning of hair cells. *a*: Tuning curves for three different cells. *b*: Spatial localization of cells with different frequency responses along the cochlea.

German physicist Helmholtz more than a hundred years ago, and elaborated by Georg von Bekesy in the 1950s, low-frequency sounds cause the greatest vibrations to occur at the base of the cochlea. In contrast high-frequency sounds maximally deflect the hair bundles in cells at the apex. The frequency to which a hair cell responds optimally is therefore determined by its physical position along the coiled cochlea.

In birds, and in amphibians such as bullfrogs and turtles, another mechanism for tuning has been discovered. The membrane potential of hair cells in these animals undergoes spontaneous oscillations, or can be induced to oscillate after a transient depolarization (Fig. 12-6a). Although the fre-

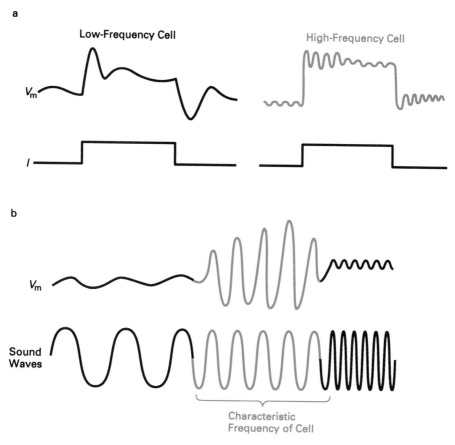

FIGURE 12-6. Membrane potential oscillations in hair cells. a: Endogenous oscillations and oscillations evoked by depolarizing currents were revealed in recordings by Fettiplace and his colleagues. b: Sound waves at the characteristic frequency of a cell cause the largest fluctuation in membrane potential. Experiments of this kind, using mechanical displacement of the hair bundles as a stimulus, have also been carried out by Lewis and Hudspeth (1983).

quency of oscillation can be increased or decreased by applying depolarizing or hyperpolarizing current, each cell has a characteristic frequency at which it oscillates around its resting potential. The characteristic frequency of different hair cells varies from tens to many hundreds of cycles per second. When a cell is stimulated with sound waves at its characteristic frequency, a maximal fluctuation of the membrane potential is evoked (Fig. 12-6b). Higher or lower frequency sounds are much less effective. The oscillations can largely be explained by the activity of only two types of ion channels, calcium channels and calcium-activated potassium channels. The opening of calcium channels causes the depolarizing phase of an oscillation. As calcium enters the cell, calcium-dependent potassium channels begin to activate. When a sufficient number of these have been opened, the membrane hyperpolarizes and calcium entry decreases. As intracellular calcium falls, the potassium channels close, and calcium channels again activate, renewing the cycle.

How is tuning to different frequencies achieved? Calcium current does not differ much from one hair cell to the next. In contrast, the rate at which the calcium-dependent potassium current activates varies enormously between cells (Fig. 12-7). It is these differences in the kinetics of potassium current that account for the fact that hair cells have different characteristic frequencies of oscillation. A particularly intriguing question that has yet to be answered is what determines the kinetics of the potassium channel in different cells. It could be that the channel proteins themselves differ, that they are modulated in different ways in different cells, or that their placement with respect to the sites of calcium entry varies. Such tuning of indi-

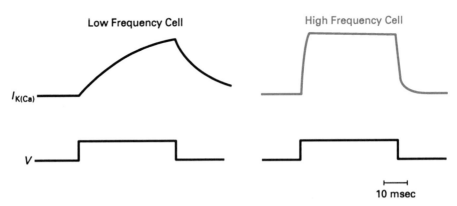

FIGURE 12-7. Calcium-dependent potassium current in hair cells. Voltage clamp recordings by Art and Fettiplace (1987) found rapidly activating currents in cells with high-frequency responses.

vidual hair cells by electrical resonance accounts for the selective response of cells in the lower range of auditory frequencies, in species such as birds and amphibians. Although it is more difficult to record from hair cells in mammals, there is at present no evidence that such tuning occurs in mammalian species. It would be surprising, however, if electrical tuning of cells to specific frequencies of stimulation did not occur, at some level, in many nervous systems.

Photoreceptors

The way that visual stimuli are translated into electrical activity of a photoreceptor cell in the retina is better understood than any other sensory event. The process termed *phototransduction* actually occurs in a variety of cells. For example, birds possess *extraretinal receptors,* which are found in the pineal gland within the brain itself. These allow a bird to sense changes in the light of its environment even in the absence of a functioning retina. In the retina of vertebrates there exist two cell types, the *rods* and the *cones,* which have different sensitivities and respond to different frequencies of light. The cone cells can further be subdivided into cells that preferentially sense different colors. However it is rods, from a variety of species, that have been the favored cell type for studying visual transduction.

Rods, rhodopsin, and cyclic GMP. Figure 12-8 shows the structure of a rod photoreceptor. The cell has two parts. The rod *outer segment* is elongated and contains a stack of flattened discs made from internal membranes. This is connected by a thin bridge to the remainder of the cell, the *inner segment,* that contains the nucleus, the mitochondria, and the presynaptic terminal that synapses onto other neurons in the retina. It is the outer segment that is the business end for visual transduction. Within the internal membranous discs is found the light-sensitive protein *rhodopsin.* This is made up of an *opsin* protein, bound to a light-sensitive molecule or *chromophore* termed retinal. The latter molecule may exist in a number of different forms, of which 11-*cis*-retinal and all-*trans*-retinal are the two major isomers (Fig. 12-9). On its own, neither opsin nor retinal absorbs visible light. In combination, however, absorption of a photon of light causes an isomerization of retinal from the 11-*cis* form to the all-*trans* form.

This light-dependent isomerization of retinal then causes a structural rearrangement of the protein. Rhodopsin that has been activated in this way is termed *meta*-rhodopsin. For all of the subsequent steps in visual transduction, it is useful to think of this molecule as analogous to a receptor that has just bound its neurotransmitter. In fact the structure of the opsin protein is very similar to that of neurotransmitter receptors such as the β-adrenergic receptors. Recall that receptors such as β-adrenergic receptor act through GTP-binding proteins. Not surprisingly, therefore, the steps that follow the production of *meta*-rhodopsin involve the production of a second messenger through the agency of a GTP-binding protein. We have already encountered this protein, transducin (G_T), in Chapter 10.

When *meta*-rhodopsin binds to transducin, GDP is replaced by GTP, and the α_T-subunit of transducin is liberated from its complex with the $\beta\gamma$-subunits. The target of the newly liberated α_T is an enzyme in the membranous discs, a *phosphodiesterase,* that cleaves the second messenger cyclic

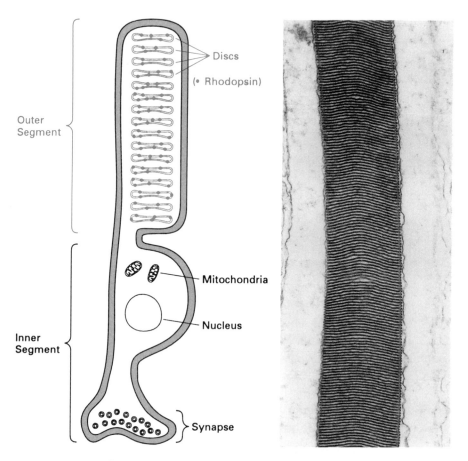

FIGURE 12-8. A photoreceptor. The drawing *(left)* is of an entire rod photoreceptor. The micrograph *(right)* shows only the outer segment of a salamander cone (courtesy of Dr. John E. Dowling).

GMP to 5′-GMP. Even in the dark, the levels of cyclic GMP in the outer segments are maintained by a balance between its rate of synthesis through guanylate cyclase and degradation by the phosphodiesterase. The action of α_T, formed after exposure to light, is to stimulate the phosphodiesterase, producing a *drop* in the levels of cyclic GMP. This drop occurs within about 100 msec of the onset of a light flash, sufficiently fast to account for a visual response. In many respects, photoreceptors are built backward. When excited by light, they respond by *dropping*, rather than raising, their concentration of the second messenger cyclic GMP. As we shall see, this results in the closure of channels that are normally kept open by this molecule. The channel *closure* in turn causes a *hyperpolarization* of the cell, and a *reduction* in the rate of transmitter release at the terminal.

This cascade of reactions that follows the formation of *meta*-rhodopsin produces a very significant amplification of the signal generated by light. It has been estimated that a single molecule of *meta*-rhodopsin, which is formed by the action of a single photon of light, diffuses in the membrane and activates several hundred transducin molecules before it is rendered inactive (see below). The subsequent stimulation of the phosphodiesterase by α_T provides further amplification such that a single photon of light can lead to the destruction of more than 10^5 molecules of cyclic GMP.

Cyclic GMP-gated channels. How is the change in cyclic GMP levels translated into an electrical response in the photoreceptor? To understand this we must first consider the electrical properties of a rod at rest in the dark. The dominant type of ion channel in the plasma membrane of the outer

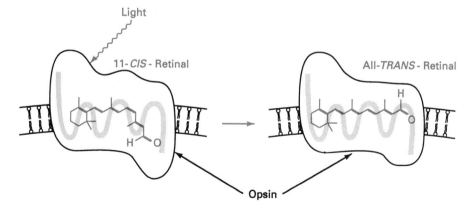

FIGURE 12-9. Rhodopsin. Rhodopsin is made up of an opsin protein, bound to the light-sensitive molecule, retinal. The light-induced isomerization of retinal was discovered by George Wald.

segments is a species of channel that allows sodium and calcium to enter the cell. Because of the abundance of sodium ions in the extracellular fluid, the major ion that enters the outer segments through these channels is sodium. The opening and closing of these channels is not very dependent on membrane potential. From the discussion in Chapter 3, we know that a cell with a preponderance of such sodium channels would be expected to have a very positive resting potential. The effect of the rod sodium channels is, however, counterbalanced by potassium channels. The interesting thing about these potassium channels is that they are found in a very different part of the cell, the membrane of the *inner* segment that includes the nucleus and synaptic terminal. Because there is good electrical continuity between the inner and outer segments, the mean membrane potential is kept fairly negative as a result of the open potassium channels. This spatial distribution of channels, however, creates a circulating current, termed the *dark current,* which flows in through the outer segment sodium channels, through the bridge into the inner segment, and out through the potassium channels (Fig. 12-10).

The effect of shining light on a rod is to shut down many of the sodium channels in the outer segment. This produces a marked decrease in the dark current (Fig. 12-10). As a result the potassium channels, which remain open in the inner segment, hyperpolarize the cell toward E_K, reducing the spontaneous release of neurotransmitter from the synaptic terminal. The closure of the sodium channels can be attributed directly to the drop in cyclic GMP in the cytoplasm of the outer segment.

Although in the previous chapter we stated that second messengers such as cyclic AMP and cyclic GMP frequently act by engaging the services of a protein kinase, this is not the case in the rod outer segment. Instead, it appears that the sodium channels bind cyclic GMP directly, and remain open only when cyclic GMP is bound. This can be demonstrated by making an inside-out patch recording on membrane from the outer segments. When cyclic GMP is added to the cytoplasmic face of the patch, a large increase in conductance, attributable to the opening of the sodium channels, can be measured (Fig. 12-11a). This occurs in the absence of ATP, which would be required for the activity of any protein kinase.

One interesting feature of the sodium channels is that, under normal conditions, the conductance of a single channel is extremely low. We stated in Chapter 3 that the conductance of open channels in biological membranes typically ranges from about 5 to 400 pS. In contrast, the conductance of these rod channels is only about 0.1 pS. This is, in fact, too low to resolve individual openings and closings of the channel using patch clamp equipment, and this value for the conductance had to be obtained by another technique (known as *noise analysis*). An explanation for the low

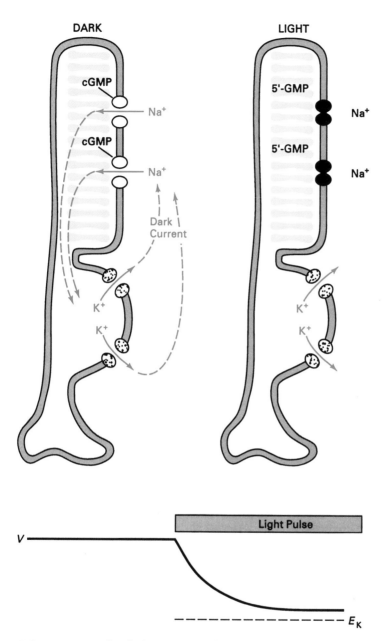

FIGURE 12-10. The dark current. In the dark, current flows through sodium channels in the outer segment of a rod photoreceptor. A pulse of light closes these channels, resulting in hyperpolarization of the rod.

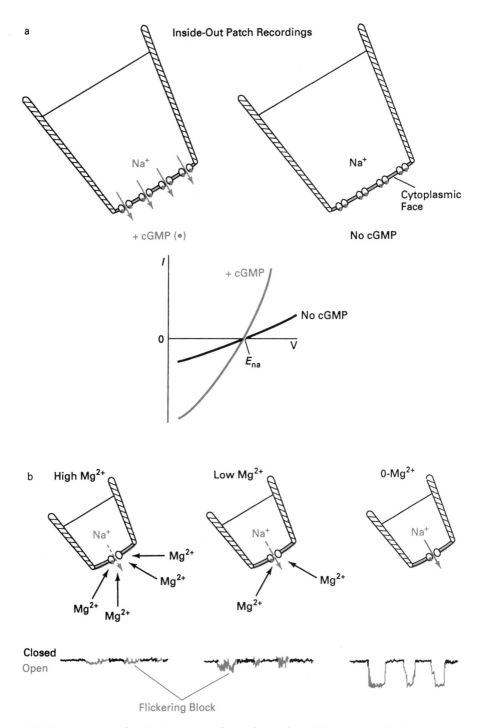

FIGURE 12-11. Cyclic GMP opens sodium channels. *a*: When cyclic GMP is added to the cytoplasmic face of an inside-out patch of membrane from a rod photore-ceptor, the I–V curve changes to reflect an increase in sodium conductance. This finding was obtained by Evgeniy Fesenko and colleagues in the Soviet Union. *b*: "Flickering" block of the cyclic GMP-regulated channel is relieved when divalent cations are omitted from the recording solutions (Haynes et al., 1986).

conductance is that the channels are partially blocked by calcium and magnesium ions, which are normally present in physiological solutions. The persistent barrage of these ions into the mouth of the channel produces a "flickering" block, effectively reducing the amount of current that is measured when the channel is open. This block is relieved when calcium and magnesium are omitted from the solutions, and individual opening and closing of the channel, which now has a conductance of ~25 pS, can be measured (Fig. 12-11b).

Terminating the light response. The analogies between visual transduction and neurotransmitter action can be taken further, when one considers how the response to a flash of light is terminated. In Chapter 10 we saw that rhodopsin can be phosphorylated by a rhodopsin kinase that resembes β-ARK, the kinase that phosphorylates and down regulates the β-adrenergic receptor. Rhodopsin kinase preferentially phosphorylates *meta*-rhodopsin, making it relatively ineffective at activating transducin, and thus terminating the light response. After phosphorylation, the all-*trans*-retinal dissociates from rhodopsin, leaving the opsin protein, which must bind another 11-*cis*-retinal before it can again be activated by light.

Figure 12-12 shows a simplified scheme representing the cyclic reactions that we have covered. Many features of visual transduction have yet to be

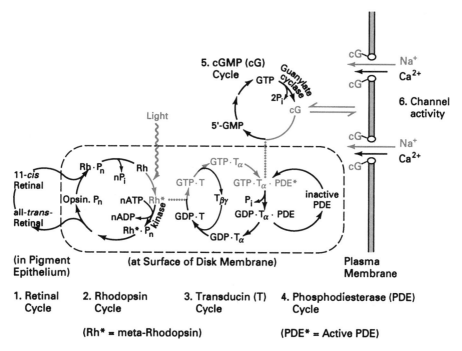

FIGURE 12-12. Cyclic reactions in visual transduction. The active messengers in this simplified scheme are highlighted in red (adapted from Lamb, 1986).

explained. For example, calcium ions can also enter through the outer segment cation channels, and changes in intracellular calcium are suspected to play an important role in adaptation of the rods to bright light. Furthermore, some invertebrate photoreceptors use mechanisms that differ from the cyclic GMP cascade. Although the transduction mechanism in these cells is not fully understood, light stimulates the formation of IP$_3$ (see Chapter 10), which, through the release of calcium, regulates the response to light. Again, this emphasizes the similarity of visual transduction to neurotransmitter action.

Before we leave the topic of photoreceptors, it should be mentioned that the structure of the outer segments is a dynamic one. Cells in the epithelium that overlie the tips of the outer segments continually engulf and digest these outermost tips, together with the enclosed rhodopsin-containing discs. New membranous discs containing the transduction machinery are continually being assembled near the base of the segment to replace the degraded discs (Fig. 12-13). The turnover of discs can be rapid, many discs being replaced within 1 hour. It may be that for reasons we do not under-

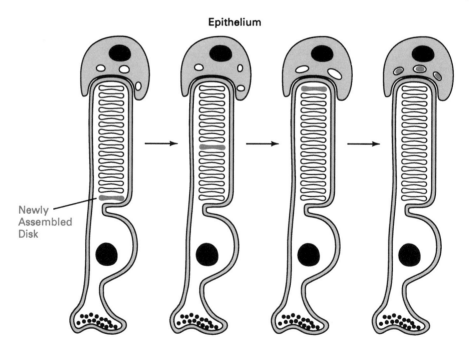

FIGURE 12-13. Turnover of photoreceptor discs. New discs are assembled near the base of the outer segment and travel to the tip of the cell where they are degraded.

stand, "newer" means "better" for the function of the components in visual transduction.

Chemoreceptors

The olfactory system. The task of a *chemoreceptor* cell is to signal to the nervous system a change in its chemical environment. Major use of chemoreceptors is made in those parts of the body specialized for taste (the *gustatory sense*) and smell *(olfaction)*. The latter sense is particularly remarkable in the specificity with which it can distinguish different odorant molecules. Figure 12-14a shows the structures of three molecules that, at first glance, appear generally similar in shape and size. The sensations that they elicit, however, leave no doubt that they stimulate different patterns of activity in neurons of the olfactory system.

Figure 12-14b shows the anatomy of olfactory chemoreceptors in a vertebrate. These are neurons that are aligned in sheets within the nasal epithelium. Their axons travel to the brain where they synapse within the olfactory bulb. Each receptor neuron has a dendrite that extends toward a layer of mucus lining the nasal cavity. There it forms a dendritic "knob" from which fine cilia extend into the mucus. It is this part of the cell that responds to odorant molecules that are borne in with the air circulating over the mucus.

At first, one might suggest that the response of the olfactory system to different odorants results from a process directly analogous to neurotransmission. The cilia of each chemoreceptor cell could bear a specific receptor protein that responds to one specific odorant. In some cases this is undoubtedly true. For example, air-borne or water-borne pheromones are used by many species to influence the behavior of sexual partners. The structure of these molecules is generally fixed, and it is probable that specific receptors for these molecules activate a direct line from the olfactory neurons to neurons controlling reproductive behaviors. Nevertheless, it is not reasonable to think that there exists a specific protein receptor for each of the vast number of odorants that we can distinguish. It seems more likely that a smaller number of receptors on different cells react to different aspects of an olfactory molecule. However nothing is known as yet about the nature of these receptors.

Although we know nothing of the receptors themselves, parallels between chemoreceptor cells and better understood sensory cells such as photoreceptors are beginning to emerge. The electrical response of the olfactory cells, like that of rods, is relatively slow, inward currents developing over several hundred milliseconds following application of odorants.

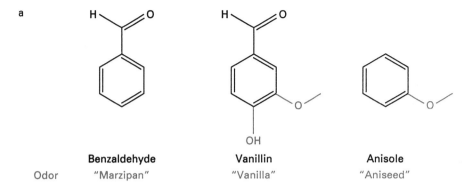

	Benzaldehyde	Vanillin	Anisole
Odor	"Marzipan"	"Vanilla"	"Aniseed"

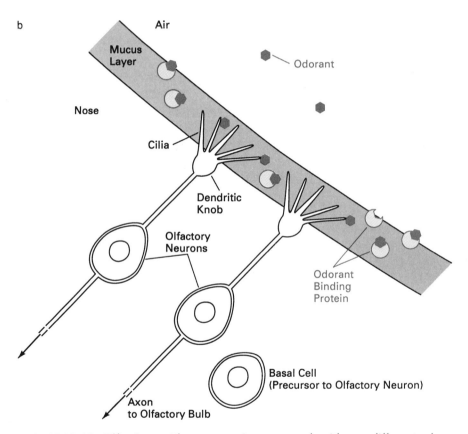

FIGURE 12-14. Olfaction. *a*: Three aromatic compounds with very different odors. *b*: Transport of odorants to olfactory neurons.

A cyclic nucleotide, cyclic AMP, may be involved in transduction. The application of odorant molecules stimulates adenylate cyclase in the membranes of the cilia, via a species of G_s protein that is specific to the olfactory cells. A cation channel, whose opening can be induced by cyclic AMP, exists in the plasma membrane. In many respects this channel resembles the cation channel in photoreceptors. In inside-out patch recordings, the channel can be opened by cyclic AMP in the absence of ATP. This indicates that cyclic AMP acts either by binding to the channel itself or to a molecule closely associated with the channel, and that the activation of a protein kinase is not required for opening. The olfactory channel differs from that in rods, however, in that it can be opened by both cyclic AMP and cyclic GMP; that in rods is highly specific for cyclic GMP.

Another point of similarity between olfactory and visual transduction is found in the way that the light-sensitive substance retinal and the odorant molecules are brought to the receptor cells. In the visual system, retinal is synthesized from the related molecule retinol, a highly lipophilic molecule. This is transported to the retina in a form that is bound to a carrier protein termed *retinol-binding protein.* This protein is closely related in its structure to *odorant-binding protein,* a dimer of two identical 19-kDa subunits found in the mucus around the olfactory cilia. The major role of odorant-binding protein is to bind and to concentrate airborne odorants and then to ferry them to the tips of the cilia (Fig. 12-14b).

One particularly interesting aspect of olfactory neurons is that, unlike most other neurons, they are not permanent. In the retina, we saw that photoreceptors are continually renewing their outer segments. Renewal in the adult olfactory system is accomplished by destroying *entire* olfactory neurons and replacing them by new neurons. The new neurons arise from precursor cells in a manner similar to the formation of neurons during embryonic development (Chapter 13). The axons of the newly formed olfactory neurons must then navigate their way to their postsynaptic targets in the adult olfactory bulb.

Other chemoreceptors. Another important class of chemoreceptive cells is found in the lingual epithelium of the tongue. These do not need to distinguish among as varied a selection of chemical stimuli as the olfactory neurons. Taste receptors are responsible for sour, sweet, salty, and bitter sensations in food applied to the tongue. Figure 12-15a shows a picture of a taste bud, with its receptor cells. In contrast to the olfactory cells, the taste receptors do not have axons but are innervated by afferent fibers that relay gustatory information to the brain. The sensing of stimuli occurs in finger-like projections or *microvilli* at the surface of the taste buds. Shown in Fig-

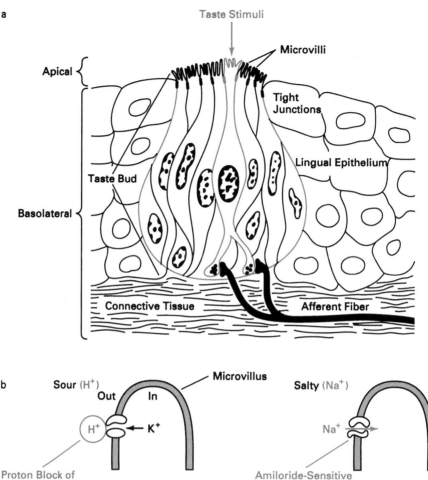

Taste Stimuli

Microvilli

Apical

Tight
Junctions

Lingual Epithelium

Taste Bud

Basolateral

Connective Tissue

Afferent Fiber

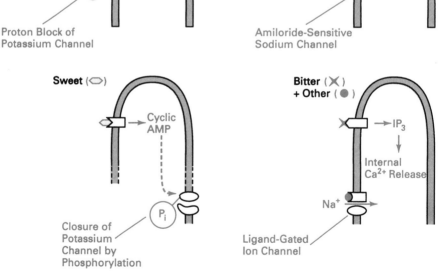

b

Microvillus

Sour (H$^+$)

Out   In

H$^+$ ← K$^+$

Proton Block of
Potassium Channel

Salty (Na$^+$)

Na$^+$

Amiloride-Sensitive
Sodium Channel

Sweet (⬯)

Cyclic
AMP

P$_i$

Closure of
Potassium
Channel by
Phosphorylation

Bitter (✕)
+ Other (●)

→IP$_3$

Internal
Ca^{2+} Release

Na$^+$

Ligand-Gated
Ion Channel

FIGURE 12-15. Taste transduction. *a*: Taste receptor cells within a taste bud. *b*: Some mechanisms that may transduce different taste stimuli (adapted from Kinnamon, 1988).

ure 12-15 are some of the mechanisms that have been proposed to account for transduction of taste stimuli. Sourness depends primarily on the acidity of a chemical stimulus, and salty sensations are evoked by solutions with a high sodium concentration. Interestingly, responses to such substances may not occur through specific membrane receptors. Instead, transduction of acidic sourness may occur through the direct action of protons on potassium channels in the apical region of the cell. Similarly, salt may be sensed by the direct entry of sodium ions through a class of sodium channels that are blocked by the drug *amiloride*. Sweetness and bitterness, on the other hand, are probably transduced by specific membrane receptors for sugars, amino acids, and other chemicals. Although proposed transduction mechanisms are illustrated in Figure 12-15b, the full details of these are yet to be discovered.

We must not forget that chemoreceptors are used for more than just the senses of smell and taste; they also transmit information about the chemical composition of various internal fluids. The majority of these are not normally considered to be sensory systems. However, one special case that deserves mention here is the role of chemical sensitivity in what is usually ascribed to the sense of touch. Painful insect bites, or agents that cause painful local inflammation of tissue, act by stimulating nerve endings in the skin. However, much of their effect is *indirect*. The immediate stimulus causes the release of active substances from nonneuronal cells in the skin. These substances include peptides such as *bradykinin* and lipid molecules such as the arachidonic acid metabolic products called prostaglandins (Fig. 12-16a). These then act directly on the sensory neurons (Fig. 12-16b). Like the skin mechanoreceptors we considered earlier, these neurons have their cell bodies in the dorsal root ganglia. Figure 12-16c illustrates the effects of bradykinin on a type of tumor cell that, although it is not a true neuron, seems to retain many features of the neurons that respond to these locally released agents in the skin. In these cells bradykinin evokes a transient hyperpolarization followed by a brisk neuronal discharge, which can be attributed largely to the activation of the IP_3/protein kinase C second messenger pathway.

Summary

Sensory cells have evolved pathways that allow ion channels to be regulated by external stimuli such as movement, light, or chemicals. Only in a few cases, such as in photoreceptors, is the means by which the external stimulus is transduced thought to be reasonably well understood. However, sensory cells appear to handle information in ways similar to those used by neurons that deal with information coming from a presynaptic pathway.

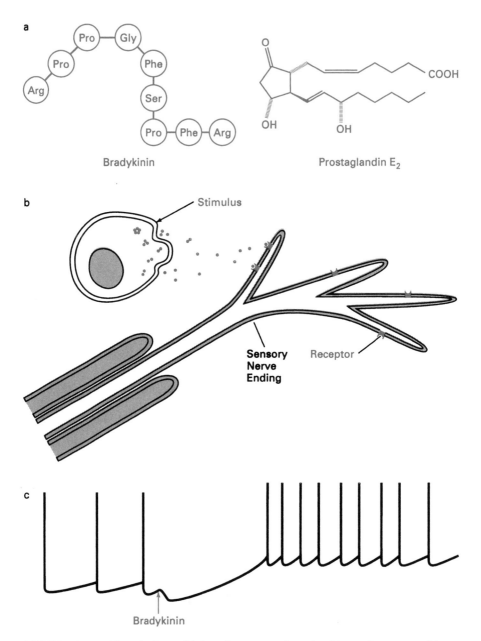

a Bradykinin   Prostaglandin E$_2$

b Stimulus

Sensory Nerve Ending   Receptor

c

Bradykinin

FIGURE 12-16. Chemical sensitivity of nerve endings in skin. *a*: Structure of bradykinin and one of the prostaglandins. *b*: Such substances may be released by nonneuronal cells to act locally on nerve endings. *c*: Actions of bradykinin on a hybrid neuroblastoma-glioma cell (termed an NG-108 cell). These actions have been studied by M. Nirenberg, H. Higashida, and their colleagues.

In particular, in many sensory cells stimulation alters the levels of second messengers such as cyclic nucleotides, which then open or close ion channels in the plasma membrane.

The study of sensory receptors has also uncovered some unexpected twists in signaling pathways. Ion channels that are opened *directly* by cyclic GMP and cyclic AMP have been found in photoreceptors and olfactory receptors. Some auditory receptors have been found to be exquisitely tuned to specific *frequencies* of stimulation, rather than solely to the amplitude of a sound wave. It seems possible that such mechanisms were discovered first in sensory cells because of the intense interest neurobiologists have in these cells that tell us about the outside world. On further search some of these mechanisms may also be found in nonsensory neurons.

Behavior and plasticity

The first two sections of this book dealt with the mechanisms that neurons use for intracellular and intercellular information transfer. In this final section we will discuss *behavior,* the output of the nervous system. We will also cover plasticity, changes in properties of individual neurons and the patterns of connections among them that lead to changes in behavior. Our discussion of plasticity will focus first on *development,* a time of rapid and dramatic changes in neuronal properties. Although our emphasis throughout this book has been on adult neurons, we begin with development because many of the cellular and molecular mechanisms that contribute to developmental plasticity are also relevant to plasticity in the adult.

Chapter 13 deals with early steps in the development of the nervous system, the *growth* and *survival* of neuronal precursors and their *differentiation* into adult nerve cells. Once an immature neuron begins to extend an axon, this axon must find its way, often over very long distances, to its synaptic target. The role of extracellular *adhesion molecules* in this process of *pathfinding* is summarized in Chapter 14. Such adhesion molecules play an important role in guiding neuronal processes through the extracellular matrix, and over the surfaces of other cells. Specific cell surface molecules also tell neurons when they have reached an appropriate synaptic target and can stop migrating. In Chapter 15 we discuss *synaptogenesis,* the formation of the complex chemical synapse that occurs when an axon reaches a suitable target. During development and also in the adult, synapses can be continually formed and eliminated. The mechanisms that contribute to the *selective maintenance* of some synapses therefore are also presented in this chapter.

The ways in which *networks of interconnected neurons* can generate behavior is the subject of Chapter 16. Certain small groups of neurons that can mediate surprisingly complex behavioral outputs have been analyzed in detail. These analyses have provided insights into the way individual neurons are uniquely tailored to their role in a network and the way changes in their cellular properties alter the output of the network. *Computational models* based on biological data can be used to investigate this organization and to make quantitative predictions that can be tested by further biological experiments. Finally, in Chapter 17 we end the book with a discussion of *learning and memory*. These fundamental features of animal behavior, the ability to modify behavior as a result of experience (learning) and to maintain the new behavior, often for as long as the animal lives (memory), have fascinated scientists for many years. Recent advances in our understanding of neuronal properties and their modulation (see Chapters 9 through 11), and of the selective stabilization of synaptic pathways (see Chapter 15), are providing new insights into the cellular and molecular mechanisms of learning and memory.

Growth, survival, and differentiation of neurons

Radical changes in the structure and connections of neurons occur during the *development* of the nervous system. Immature neurons are subject to the actions of chemical and mechanical influences that cause them to migrate to various locations in the nervous system, to extend axonal and dendritic processes toward other cells, and then to make and break synaptic connections with these cells before a final pattern of branching and connections is established. A full account of the different stages of neuronal development is beyond the scope of this book, which focuses on the properties of adult nerve cells. Nevertheless, neuronal *plasticity* in the adult animal may utilize mechanisms that are active during development, a period of profound plastic changes in neuronal structure and function. Furthermore, many of these mechanisms may also contribute to the *maintenance* of neuronal form and connections in the adult. We shall therefore now give a very brief and general account of the normal course of neuronal development. In this chapter we shall deal with factors that determine the existence and properties of select groups of neurons. The following two chapters will cover the growth of axons, and the formation of synaptic contacts.

Cell Determination

Early in the development of an embryo, there exist three layers of cells. The cells that are destined to become neurons are found in the most external layer termed the *ectoderm*. Immediately under the ectoderm lies a layer of cells termed the *mesoderm*. The first step in neuronal development is the *determination* of cells in the ectoderm to become neuronal precursors. This is often called *neural induction* (Fig. 13-1). In the nervous system of vertebrates, this appears to result from the action of factors—diffusible molecules—released from nearby cells in the mesoderm. After this stage the cells in the ectoderm can develop only into neurons, glial cells, or a limited number of other cell types.

The type of cells that a given immature cell can eventually give rise to also depends on its *position* in the developing nervous system. In the early development of the vertebrate nervous system, cells are specified to form the types of cells appropriate to structures such as the forebrain or the spinal cord. If a region of cells specified to form forebrain structures is, for example, rotated by 180° during development, then it may eventually form inappropriate forebrain structures at more posterior locations. It is likely that this specification is also brought about by factors released from mesodermal cells. Gradients of such factors have been postulated to exist in developing structures. The amount of a given factor to which a cell is

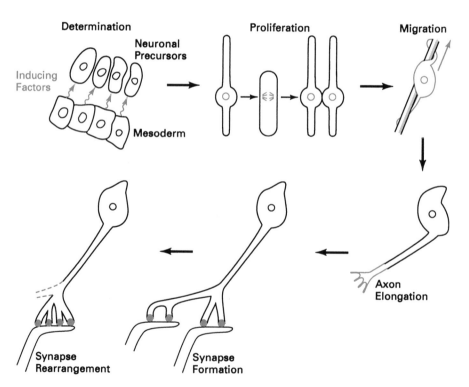

FIGURE 13-1. Some of the stages in the development of neurons that are discussed in this and the next two chapters.

exposed will depend on its position in the gradient, and this may be one way position controls the course of development. In addition, direct interactions with neighboring cells can induce neurons to follow different developmental paths.

Cell Proliferation

During and following the actions of factors that determine the developmental fate of a particular group of neuronal precursors, the cells begin to divide (proliferation in Fig.13-1). During development of the vertebrate central nervous system, the embryo invaginates such that the location in which the neuronal precursor cells proliferate is adjacent to internal fluid-filled ventricles. This location is termed the *germinal* zone. As the cells divide, they undergo a series of changes in shape that are characteristic of dividing cells in most epithelia (Fig. 13-2a). These are, in sequence:

1. The cells send out processes that span the thickness of the germinal zone (G_1 phase).
2. As the cells synthesize DNA, the nuclei of the cells move along the processes away from the ventricles (S phase).
3. The nuclei return along the processes toward the ventricular surface.
4. The processes of the cells retract.
5. The cells divide.
6. The two daughter cells each reextends processes to span the germinal zone, and the cycle begins again.

After several such cycles the cells either lose the ability to divide further and become *postmitotic,* or they migrate to another region of the developing nervous system where they undergo several more divisions before ceasing to proliferate. In the vertebrate nervous system, an immature cell that has undergone its final division, and is destined to become a neuron, is termed a *neuroblast.* There is some semantic confusion here in that, in *invertebrates,* the same term "neuroblast" is used to describe cells that *can* proliferate but whose *progeny* are destined to become neurons.

Cell Migration

After their final division, a few cells may already be situated close to their eventual location in the nervous system. The majority, however, *migrate* (Fig. 13-1) considerable distances to arrive at their final destination. During

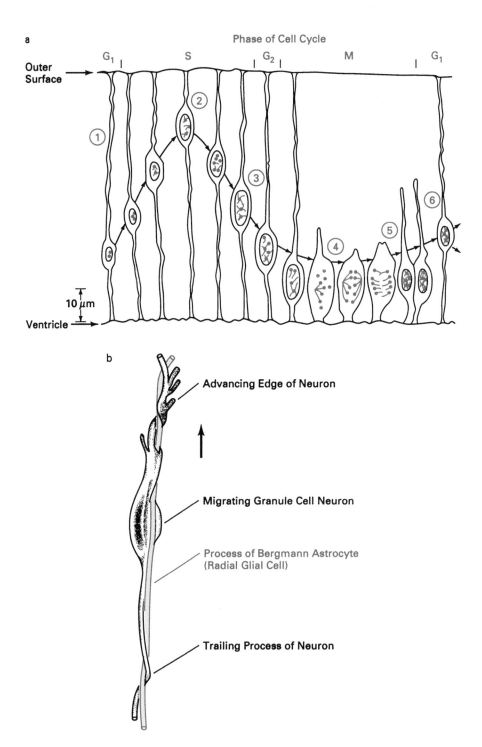

FIGURE 13-2. Proliferation and migration. *a*: Movements of cell nuclei during proliferation (Jacobson, 1978). *b*: The anatomical work of Pasco Rakic (1971) demonstrated the migration of neurons in the cerebellum along radial glial fibers.

such migration, a cell is exposed to factors released from other cells, including nonneuronal cells. It appears that such factors, in addition to direct physical interactions with other cells, play a central role in determining whether the cell will differentiate into a neuron, a glial cell, or some other cell type. Moreover, if a cell develops into a neuron, factors released by other cells influence the type of neurotransmitter it synthesizes and the specific mixture of receptors, ion channels, and other proteins that determines the characteristics of the fully differentiated neuron.

Another class of cell-to-cell interactions that occur during the migration of immature neurons determines the *direction* of cell migration. A clear example of this can be observed in the interaction of glial cells with developing *granule cell neurons* in the immature cerebellum of vertebrates (Fig. 13-2b). A class of glial cells known as *Bergmann astrocytes* is generated in the germinal zone of the developing cerebellar cortex. They differentiate into cells that extend cytoplasmic processes across the full thickness of the cortex. These radial glial fibers provide a pathway for the subsequent migration of neurons from the external surface (the external granule layer) down into the developing cortex. The cells that travel along these glial fibers eventually form a layer of granule cell interneurons in the mature cerebellar cortex.

Cell Death during Development

Entire populations of neurons may arise during development only to die before a stable mature nervous system is formed. Very elegant examples of how cell death is used to build a simple nervous system have been described in invertebrates such as leeches, nematodes, and grasshoppers. For example, the nervous system of the grasshopper comprises a series of segmental ganglia along the length of the body. These ganglia are not identical and contain different numbers of neurons. Initially the pattern of cell division is similar in all of the segments. Later, the "unwanted" cells in a given segment simply die, leaving the remaining cells to establish the appropriate pattern of connections. This phenomenon has been termed *programmed cell death.*

Neurons also die during development of vertebrate nervous systems. As many as three-quarters of the neurons destined for a specific neuronal pathway may die during early embryonic development. In some cases, survival of a neuron may require exposure to some diffusible *trophic factor* released by its postsynaptic or presynaptic target. Thus removal of a target, for example, a limb bud in a chick embryo, causes the atrophy of sensory neurons that normally would innervate the limb. Conversely, implanting an extra limb partially prevents the loss of cells in the sensory ganglia. In other cases, the survival of a group of neurons may depend on exposure to a specific factor at a critical time in development.

Peptide Growth Factors and Related Molecules

It is safe to say that we are only beginning to understand the multiplicity of signals that determine whether a cell will differentiate into a neuron, and whether it survives or dies before development is complete. Work with many nonneuronal cells has demonstrated the existence of peptide molecules that stimulate cell division, and that have therefore been termed *growth factors*. It now appears that families of molecules related to growth factors and their *tyrosine kinase receptors* play an important role in shaping developmental decisions in neurons as well as other cells.

Epidermal growth factor. One of the first of these growth factors to be fully characterized was *epidermal growth factor* (EGF). This 53 amino acid peptide, which was isolated by Stanley Cohen from salivary glands of mice, stimulates the proliferation of a variety of cell types, including astrocytes. When added to cells, EGF acts through a receptor that spans the plasma membrane. The intracellular domain of this receptor contains an enzyme activity that is a *tyrosine kinase*. One of the events that follow EGF binding to its receptor is that the tyrosine kinase activity is stimulated, resulting in the transfer of phosphate from ATP to tyrosine residues in specific internal proteins that are substrates for the kinase (see Fig. 10-16).

A number of additional protein factors related to EGF have now been identified, and their receptors have been found to resemble those for EGF. Stimulation of phosphorylation at tyrosine residues is not the only biochemical event that follows activation of these receptors. For example, activation of the phosphoinositide second messenger system on the cytoplasmic side of the membrane (see Chapter 10) may also occur, as a result of direct interactions of the EGF receptor with phospholipase C. An important aspect of the action of these factors is that they often do more than simply stimulate cell division. They may have complex effects on various properties of cells without influencing cell proliferation directly. Although it is unclear whether EGF *itself* is involved in regulating the properties of neurons, there is accumulating evidence that molecules related to EGF and its receptor play an important role in the early events that determine whether a cell will become a neuron rather than differentiating into another type of cell. Later in this chapter we shall describe some of this evidence, which is evolving from the study of neuronal development in *Drosophila*.

Nerve growth factor. There are also peptide growth factors whose importance for *neuronal* survival during development, and maintenance in adult

life, is not in doubt. The first example and the most thoroughly characterized of these factors is *nerve growth factor* (NGF).

The first clues to the existence of NGF came from experiments by E.
Bueker in 1948 that demonstrated that if a muscle tumor is implanted into
the body wall of an embryo, the tumor becomes innervated by neurons
from the sensory and sympathetic ganglia along the spinal cord. This innervation is accompanied by a great increase in the size of the ganglia that
project to the tumor. These experiments were followed up by Rita Levi-
Montalcini and Victor Hamburger, who provided evidence for the existence of a soluble substance that promotes the growth of neurons in these
ganglia. To characterize this substance, Levi-Montalcini developed a simple
bioassay. She placed small pieces of tissue containing sensory neurons or
neurons from the sympathetic nervous system into a culture dish. The addition of cells from the muscle tumor was then found to produce a very dramatic stimulation of the growth of *neurites* out of the explant (Fig. 13-3a).
(The term *neurite,* or *neuritic branch,* is used to describe either an axon or
a dendrite, and is a particularly useful term when it has not been established
whether the growing process is in fact an axon or a dendrite.)

Subsequent experiments by Levi-Montalcini and Stanley Cohen found
that certain tissues other than muscle tumors could also provide the neu-
rite-inducing factor. In an attempt to characterize the molecular properties
of NGF, Cohen treated a crude preparation of the factor with snake venom,
which contains a phosphodiesterase that should degrade nucleic acids but
leave proteins intact. Surprisingly, the snake venom alone was found to be
active in inducing neurite outgrowth, and was shown subsequently to contain high amounts of NGF. In mice, the submaxillary salivary gland, an
anatomical homolog of the snake gland that secretes venom, was also found
to be very rich in NGF. The salivary gland was the source from which the
protein was eventually purified.

Structure of NGF. As we shall describe below, NGF is a complex of three
different types of protein. However, its biological activity on sympathetic
and sensory neurons resides entirely in a complex of two identical peptide
chains that have a molecular weight of 13,259 each (Fig. 13-3b). This hom-
odimer has been termed the β-subunit of NGF. Although there is only one
gene for the β-subunit, the RNA that is transcribed from this gene may be
spliced in four alternative ways to yield four different messenger RNA species. The proteins produced by these messenger RNAs differ by up to 20
amino acids. Different tissues use different splicing alternatives, and thus
the precise structure of the β-subunit can vary from tissue to tissue.

NGF in the submaxillary gland of mice exists as a complex of three different proteins, in which the β-subunit is combined with two α-subunits
and two γ-subunits (Fig. 13-3b). (This large complex has a sedimentation

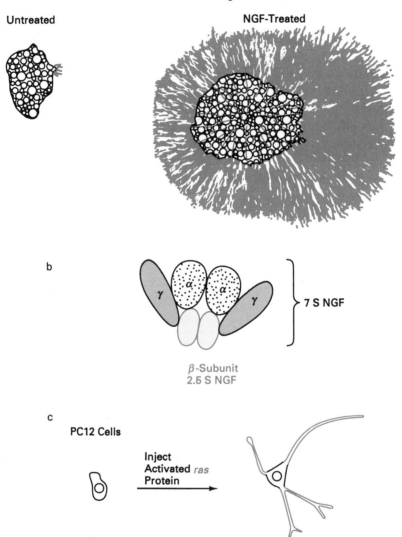

FIGURE 13-3. Nerve growth factor (NGF). *a*: Growth of neurites from sensory ganglia in culture is induced by NGF. *b*: The structure of 7 S NGF. *c*: An experiment by Bar-Sagi and Feramisco (1985) showed that injection of activated *ras* protein into PC12 cells induces neurite outgrowth.

coefficient of 7 Svedberg (S) units and is therefore sometimes termed 7 S NGF, in contrast to the β-subunit alone which has been called 2.5 S NGF.) The γ-subunits are peptidases that appear to play a role in the formation of mature β-subunits, which are cleaved from a slightly larger precursor protein. The role of the α-subunits is unknown. It is also not known whether NGF exists as this three protein complex in all of the tissues in which the biologically active β-subunit has been detected.

Direct measurements of the levels of NGF in various tissues, as well as the use of DNA probes to detect messenger RNA coding for the β-subunit, have shown that NGF is synthesized at a low rate by a variety of tissues that are innervated by sympathetic neurons. The reason for the atypically high levels in salivary glands and in the saliva of mice is not known. NGF levels may also become elevated on injury to a nerve, when it appears to be synthesized by Schwann cells.

Actions of NGF during development. Despite its name, NGF is not a true growth factor. It does not appear to cause the mitosis of neurons or their precursor cells. It is clear, however, that NGF plays a role in the normal development of sympathetic neurons and in their maintenance in adult animals. For example, injection of antibodies to NGF into embryonic or newborn mice results in the complete abolition of the sympathetic nervous system. Injection of antisera to other growth factors does not have as specific an effect. Destruction of the sympathetic nervous system also occurs in older animals that are deprived of biologically active NGF by repeated injections of anti-NGF antisera, although a more prolonged period of application is required than in the embryo or newborn. Conversely, injection of NGF itself into immature animals causes an enlargement of the sympathetic ganglia. Generally similar results are obtained with sensory neurons, although it appears that antisera to NGF are able to induce a depletion of sensory neurons only when applied early in embryonic development.

On the basis of these sorts of experiments, the hypothesis has emerged that target cells destined to be innervated by sympathetic or sensory neurons provide NGF to ensure the survival of their synaptic inputs. It may be that NGF acts to inhibit the mechanisms that would normally lead to cell death in this population of cells. At one time, it was thought that a localized source of NGF might also provide a chemical signal for the directional growth of axons during development of the sympathetic nervous system. Indeed, experiments with isolated cells in culture, as well as with the intact nervous system, have demonstrated that neurites will grow in the direction of an increasing gradient of NGF. However, studies using cDNA probes to measure the amounts of messenger RNA for NGF during development have shown that synthesis of NGF begins to occur only *after* sympathetic neurons have reached their targets. Dependence on NGF for survival also begins only at this time. Thus gradients of NGF are unlikely to play a role in the guidance of neurites during development.

Cells that respond to NGF. Although sympathetic and sensory neurons are the major neuronal types that are sensitive to NGF, a small number of other types of neurons also respond to this factor. For example, there exists a population of cholinergic neurons in the septum that projects to the hippocampus. If the axons of these neurons are cut, the cells normally die. Application of NGF prevents the death of these cholinergic neurons but does not affect the survival of other types of neurons in the septum. NGF also acts on certain nonneuronal cells such as chromaffin cells in the adrenal medulla, which can be induced to synthesize catecholamines and to extend neurites on treatment with NGF. Like sympathetic and sensory neurons, chromaffin cells are derived from the *neural crest* (see below). A cell line that also has these properties and that has been particularly useful in the study of NGF was derived from a tumor of adrenal chromaffin cells, termed a *pheochromocytoma.* We shall see below that these *PC12* cells have played an important role in studies of receptors for NGF and of its mechanism of action.

PC12 cells and the mechanism of NGF action. The way that NGF exerts its actions on sensitive cells is not fully understood. In part, this is because NGF may act at several different levels. For example, the effects on cell survival and on neurite outgrowth could involve very different mechanisms. Clues to its mode of action have, however, come from studies with PC12 cells. Receptors for NGF have been characterized in the plasma membrane of PC12 cells, and are found in other cells that respond to NGF. The receptor appears to be a single protein that crosses the plasma membrane once, with its N-terminal in the extracellular space and its C-terminal in the cytoplasm.

There is evidence that following binding of NGF to its receptor, the subsequent effects of NGF on neurite growth in PC12 cells may involve a protein known as *ras.* The *ras* protein is a 21-kDa phosphoprotein that binds GTP and is homologous to other GTP-binding proteins, such as the α-subunits that couple neurotransmitter receptors to the formation of second messengers (Chapter 10). It is found in most eukaryotic cells, and was first discovered as a protein encoded by a *proto-oncogene,* a cellular homolog of a transforming viral gene (for a more detailed discussion of proto-oncogenes, see Chapter 15). By analogy with the G protein α-subunits, it would be expected that a mutant form of *ras,* that is diminished in its ability to hydrolyze GTP, may be chronically activated. In fact, injection of such a mutant *ras* protein into PC12 cells induces morphological differentiation similar to that induced by NGF (Fig. 13-3c). Injection of the unmutated, and therefore unactivated, normal *ras* protein does not produce this effect.

The injection of antibodies to the normal *ras* protein has been found to block the effect of NGF on PC12 cells.

These results suggest that NGF could exert its effects on neurite out-growth by binding to its receptor, and activating a second messenger path-way that requires the *ras* protein for its normal function. There is still, however, much that we do not understand about the action of NGF. For example, NGF is taken up into sensitive cells by endocytosis. This usually occurs at the distal tips of axons, presumably those closest to the source of NGF in the target tissue. The NGF is then transported toward the cell body in the endocytosed vesicles. It is possible that this *retrograde transport* of NGF-containing vesicles to the soma of the cell is required for the regula-tion of genes necessary for the differentiation and survival of sensitive cells.

Cholinergic factor. One of the best studied examples of the way factors released from nonneuronal cells determine the eventual properties of migrating neurons has come from work with cells that form the sympa-thetic and parasympathetic nervous systems. These cells migrate from the *neural crest*, a column of undifferentiated cells at the dorsal margin of the neural tube (Fig. 13-4). Neurons of the sympathetic nervous system, which

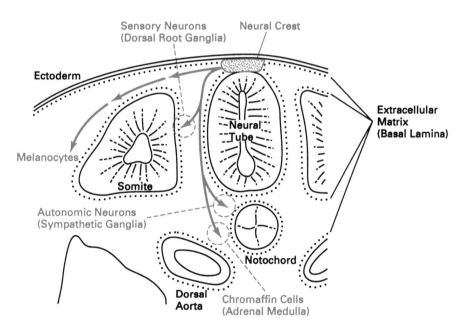

FIGURE 13-4. The neural crest. A cross section through the trunk region of a chick embryo. Cells that migrate from the neural crest region develop into sensory, sym-pathetic, and parasympathetic neurons, as well as other cell types. Also shown is the extracellular matrix of the basal lamina (see Chapter 14) (modified from Sanes, 1983).

arise from the central portion of the developing spinal cord, synthesize the enzymes required for the production of norepinephrine, which they use as a neurotransmitter. In contrast, neurons of the parasympathetic nervous system, which develop largely from the anterior developing spinal cord, synthesize and secrete acetylcholine as their major transmitter.

The development of the biochemical machinery to produce either the adrenergic or the cholinergic transmitter system is not determined by the position in the neural crest from which the cells migrate, but rather by a factor that the cells encounter during their migration away from the neural crest. When immature adrenergic neurons are isolated in culture they maintain the ability to make norepinephrine. However, if these neurons are cultured in the presence of nonneuronal cells such as heart cells, or other cell types that normally receive a cholinergic input, then the cells stop synthesizing tyrosine hydroxylase, the rate-limiting enzyme in the pathway for synthesis of norepinephrine (see Chapter 8). Instead, the neurons now synthesize choline acetyltransferase and make acetylcholine their transmitter. This results from the action of a glycoprotein that is secreted into the medium by the nonneuronal cells. This factor has been termed *cholinergic factor*.

Although the type of transmitter can be modified by the cholinergic factor in immature cells taken from young animals, adrenergic neurons from older animals are insensitive to its actions. Thus, there exists a critical period in development during which cholinergic factor is able to exert its effects. A particularly interesting aspect of the action of this factor is that its activity is strongly dependent on the amount of electrical activity in the neurons. Direct electrical stimulation of the cultures of neurons, or depolarization with solutions containing a high concentration of potassium ions, prevents the action of the cholinergic factor and maintains the adrenergic phenotype. This inhibition of the action of cholinergic factor by electrical activity is due to calcium entry through voltage-dependent calcium channels. Thus it is possible that the activity of calcium channels during development serves to fix the choice of transmitter in neurons of the autonomic nervous system.

The Use of Genetic Mutations to Investigate Neuronal Growth and Differentiation Factors

An approach that has become extremely powerful in studies of development is the analysis of mutants in which some aspect of normal development has been perturbed. A favored animal for such genetic experiments is the fruit fly *Drosophila*. We have already seen, in Chapter 5, how a study of mutant flies termed *Shaker* led to the characterization of a family of potassium channels. We shall now give a brief account of two mutations, termed *Notch* and *sevenless,* to illustrate the major insights that are currently being gained about protein factors that regulate neuronal differentiation.

The Notch *locus.* During normal development of *Drosophila,* a strip of cells along the ventral midline of the embryo comes to form the *neurogenic region* (Figure 13-5a). The cells in this region develop either into neuroblasts, which give rise to neurons along the ventral cord of the fly, or into dermoblasts, which eventually form the epidermis over the ventral cord. (Remember that *Drosophila* is an invertebrate, and that "neuroblasts" therefore divide to give rise to neurons.) Under normal conditions, about one-quarter of the cells in the neurogenic region form neuroblasts. However, in a class of mutant flies first noted early in this century, the cells that normally become dermoblasts turn into neuroblasts, resulting in hypertrophy of the nervous system and loss of the ventral epidermis.

The defect in these animals was localized to a stretch of DNA termed the *Notch* locus. A complete deletion of the DNA in this region produces

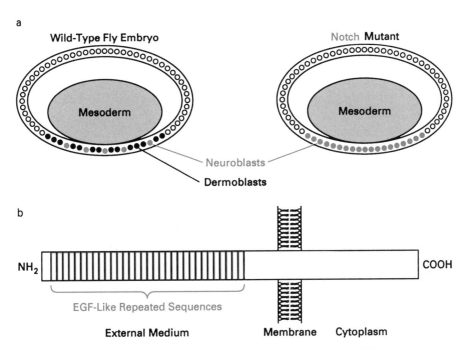

FIGURE 13-5. The *Notch* mutation. *a*: Cross section of a normal fly embryo and of a *Notch* mutant, with its overabundance of neuroblasts. *b*: The sequence of the *Notch* protein was deduced by Spyros Artavanis-Tsakonas (1988) and his colleagues.

the change in the fate of the precursor cells for the epidermis described above. This, in turn, causes the death of the embryo. Other mutations of the DNA in this region are not always lethal, but result in a variety of abnormalities of development particularly affecting the eyes and wings.

The *Notch* locus spans approximately 38 kilobases (kb) of DNA in the chromosome. This is transcribed into one large piece of RNA that is subsequently processed to form a messenger RNA species of 10.2 kb, encoding a protein of 2703 amino acids (Fig. 13-5b). The predicted structure of this protein is very interesting. It has one stretch of hydrophobic amino acids that is presumed to span the plasma membrane. Near the N terminal of the protein there exists a sequence of 38 amino acids that contains six cysteine residues. Variants of this 38 amino acid sequence, each containing six cysteines, are repeated in tandem throughout most of the 1700 amino acids that are predicted to lie on the extracellular side of the membrane. The repeated sequences are highly homologous to regions of EGF and related growth factors. It has therefore been suggested that these EGF-like repeats act as a form of membrane-immobilized growth factor. This factor may influence the properties of neighboring cells that come into direct contact with the plasma membrane of a cell that expresses the protein.

The exact role of the *Notch* protein is not yet known. It is found in the membrane of actively dividing cells and, although it is normally found in both the neural and epidermal precursor cells in the neurogenic region, it is not confined exclusively to this region. Moreover, *Notch* is only one of a number of mutations that produce abnormal development in this region. It is likely that the *Notch* protein, perhaps in concert with proteins derived from other neurogenic loci, is used in the cell to cell interactions that fine tune the development of neurons.

Sevenless. Another *Drosophila* mutation that is providing significant insight into the molecules that determine the formation of specific types of neurons is the *sevenless* mutation. This mutation alters the fate of a photoreceptor named R7 in the compound eye of the fly. The eye of a fruit fly comprises several hundred units termed *ommatidia*, each of which contains a precise arrangement of eight photoreceptor neurons, R1–R8 (Fig. 13-6). One of these, R7, is specialized to respond to ultraviolet light. As is typical of photoreceptors, these cells have elongated processes, termed *rhabdomeres*, that contain the photosensitive pigments. The elongated rhabdomeres are aligned around a central axis. Six of the cells, R1–R6, span the length of the ommatidium. The rhabdomeres of the other two cells, R7 and R8, span only half the length of the ommatidium each, with cell R7 sitting on top of R8. Situated directly above these sensory neurons is a group of four cone cells whose function is to synthesize nonneuronal components of the overlying eye structure including the cornea. (These cone cells are themselves nonneuronal and are not to be confused with cone photoreceptors of the vertebrate retina.)

During normal development, the cells that make up an ommatidium arise from an amorphous collection of precursor cells. Cells that can be recognized as R1–R8 and the cone cells come into being in a fixed sequence, with R7 being the last sensory neuron to form. In the *sevenless* mutant, however, the R7 cell does not develop. Instead, the cell that would normally have become R7 differentiates into one of the overlying cone cells (Fig. 13-6).

The *sevenless* locus codes for a large protein that is predicted to have two transmembrane regions. Again, as in the *Notch* system, a homology with the EGF growth factor system is found in this protein. In particular the domain near the C-terminus, believed to be on the cytoplasmic side of the membrane, strongly resembles the tyrosine kinase domain of the receptors for EGF and similar growth factors. Moreover, it has been shown that development is abnormal when this tyrosine kinase is defective in cell R7 itself, but not when it is defective in other cells in the eye. Although the

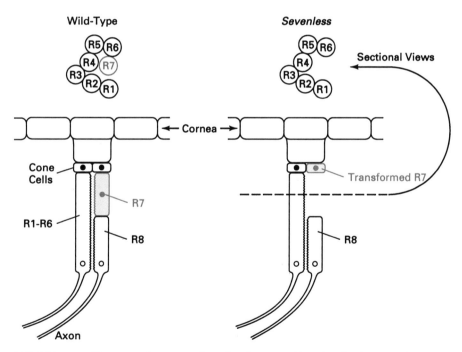

FIGURE 13-6. The *sevenless* mutation. Sectional and longitudinal views of normal and *sevenless* ommatidia (Palka and Schubiger, 1988).

size and structure of the remainder of the protein differ substantially from the EGF receptor, these findings, together with related genetic studies, indicate that the *sevenless* protein is likely to be a receptor for a growth factor-like signal generated by neighboring cells.

Further work has shown that cell R8, the cell that lies immediately below R7, generates the signal required to transform a precursor cell into sensory neuron R7. Another mutation has been found that also produces loss of R7. This mutation has been termed *bride of sevenless*. In contrast to *sevenless*, the protein encoded by *bride of sevenless* is defective in cell R8 rather than R7. This protein may therefore be the ligand that activates the *sevenless* receptor protein in R7, or it could be required for the production of such a ligand. According to this scheme, the precursor cell above R8 develops into a cone cell in the absence of the appropriate signal from R8. When the *sevenless* receptor is successfully activated, however, the appropriate proteins to build the ultraviolet-sensitive R7 sensory neuron begin to be synthesized.

Steroid Hormones

For many cells in the nervous system, appropriate development depends on the action of hormones secreted by other organs in the body. Some of these are listed in Table 13-1. In contrast to peptide hormones, these hormones are lipid soluble and readily enter the brain from the blood. They also cross cell membranes without the need for specific carriers or receptors in the plasma membrane. There is evidence that each of these hormones influences the nervous system, based primarily on the finding of receptor proteins for these hormones in the cytoplasm of neurons or glial cells. In most cases, the different receptor proteins are localized to specific groups of neurons rather than being uniformly distributed throughout the nervous system.

How do steroid hormones work? The best understood mode of action of these hormones is through a class of receptor proteins that, in the presence

TABLE 13-1 Steroid Hormones Important for Neuronal Development

Class	Example	Source
Androgens	Testosterone	Gonads
Estrogens	β-Estradiol (estrogen)	Gonads
Progestins	Progesterone	Gonads
Glucocorticoids	Corticosterone	Adrenal gland
Mineralocorticoids	Aldosterone	Adrenal gland

of the hormone, directly bind DNA in the nucleus, allowing the transcription of specific genes. The general structure of these receptor proteins is shown in Figure 13-7a. There is a high degree of homology in the regions of DNA binding and of hormone binding in the different receptors for this class of hormones. Figure 13-7b shows part of the sequence of the DNA-binding region of the receptors, which is rich in the amino acid cysteine. These are believed to form a coordination complex with zinc ions, producing two *zinc fingers*. A zinc finger is a structural feature that allows a protein to bind DNA.

When a cell is exposed to hormone, the receptors bind the hormone and undergo a structural change that allows the DNA-binding regions to interact with specific short sequences in the DNA of the cell. These DNA

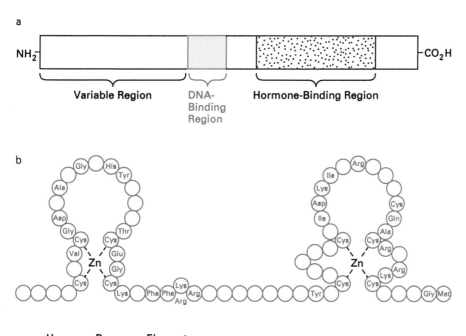

FIGURE 13-7. Steroid receptors. *a*: An outline of the three distinct regions found in these receptors. *b*: Amino acid sequence of a DNA-binding region. The named amino acids are found in nearly all receptors of this family. *c*: DNA sequence within the prolactin gene required for binding the estrogen receptor (Evans, 1988).

sequences have been termed *hormone response elements* (HREs). They are found in the proximity of genes coding for proteins whose expression is regulated by the steroid/thyroid family of hormones. Although the exact position of the HRE relative to the structural gene encoding the protein itself may vary in different genes, the sequence of the HRE is very similar for each of the genes that is regulated by a specific hormone. The sequence of the HRE in the gene for prolactin, a peptide whose synthesis is regulated by the steroid hormone estrogen, is shown in Figure 13-7c.

Organizational effects of steroid hormones. The best understood examples of the neuronal action of such hormones have come from studies of the role of the gonadal steroids in the control of reproductive behaviors. Differences exist between the brains of males and females in the size and synaptic connections of several well-defined groups of neurons. One such group in rat brain is the *sexually dimorphic nucleus of the preoptic area* (SDN-POA), which is five times larger in males than in females. The increased number of neurons in males results from the action of testosterone during development. Thus, if a female rat is treated with testosterone during late embryonic development and over the first 10 days after birth, the SDN-POA develops as in a male (Fig. 13-8). Interestingly, this action of testosterone is

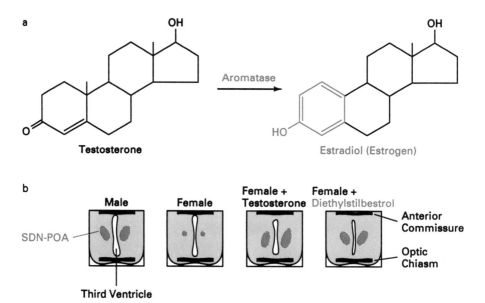

FIGURE 13-8. Steroid hormones. *a*: Aromatization of testosterone to estrogen. *b*: Sections through the sexually dimorphic nucleus of the preoptic area (SDN-POA) were prepared by Roger Gorski (1989) and his colleagues, using normal male and female rats and steroid-treated female rats.

not mediated by a receptor specific for testosterone. Enzymes exist within the neurons that convert testosterone into estrogen. This process is termed *aromatization* (Fig. 13-8). Thus estrogen is actually the active hormone within the cells, and treatment of immature females with compounds such as diethylstilbestrol, a highly active analog of estrogen, also causes the male pattern of development of the SDN-POA.

The action of a steroid hormone to alter this developmental pathway is an example of what have been termed *organizational* effects of steroid hormones. In particular, the action of the hormone is permanent and is restricted to a critical period during development. For example, treatment of female rats with testosterone later than six days after birth cannot alter the size of the SDN-POA. It is not yet known which specific steps in neuronal development (Fig. 13-1) are sensitive to testosterone and estrogen.

Activational effects of steroid hormones. The role of steroid hormones is not confined to the formation of specific neuronal pathways during development. The hormones also regulate the properties of many mature neurons and thereby influence the onset of specific animal behaviors. Again, such effects are particularly obvious in reproductive behaviors, and have been termed *activational* effects. For example, the morphology and extent of dendritic branching in neurons that innervate muscles involved in copulation remain sensitive to changes in testosterone in adult male rats. Another well-studied effect is the action of estrogen and progesterone in the onset of *lordosis* behavior in female rats. During this behavior, the female rat assumes a characteristic body posture that indicates sexual receptivity. Groups of neurons that control lordosis behavior are found in the *ventromedial nuclei* within the hypothalamus. An elevation of the levels of estrogen produces morphological changes in these neurons, increasing the size of their somata, and increasing the synthesis of a variety of proteins, including the receptors for progesterone. Subsequent elevation of progesterone levels, as would normally occur during the estrus cycle, induces the synthesis of further proteins that are essential for lordosis behavior to occur.

The nature of the changes in neuronal properties and possible synaptic remodeling that are induced by the consecutive actions of estrogen and progesterone is not yet understood. However, it is known that estrogen increases synthesis of the neuropeptide Met-enkephalin and of receptors for the peptide oxytocin in the ventromedial nuclei, and that receptors for other transmitters, such as GABA and acetylcholine, may decrease in number. The steroid hormones may be considered to be agents that prime neuronal pathways for the occurrence of stereotyped reproductive behaviors.

Regulation of neurons by testosterone in songbirds. One of the clearest examples of the influence of steroid hormones on the properties of a neuron is found in songbirds such as canaries. In this case the hormones produce profound plastic changes in the properties of adult neurons, emphasizing that there are commonalities in mechanisms of developmental and adult plasticity. During the breeding season, adult male canaries generate a song made up of a fixed pattern of individual sounds, termed syllables. This stereotyped pattern of syllables is termed a *stable song*. Although in any one bird the song pattern remains relatively fixed during the spring reproductive season, the specific pattern of syllables that comprises the stable song is lost in the months that follow. At this time the animals vocalize variable patterns of syllables, termed *plastic song*, and may generate new syllables that were not used in prior years. In the next breeding season, the bird acquires a brand new pattern of syllables in its stable song.

One group of neurons that controls singing in canaries is found in the forebrain, and is termed the *nucleus robustus archistriatalis* (RA, Fig. 13-9a). These neurons receive inputs from another nucleus in the forebrain, the *higher vocal center* (HVA). A number of RA neurons project in turn to motor neurons in the medulla. These send connections directly to muscles of the *syrinx* that produces the sounds comprising a song. The RA and HVA nuclei are several times larger in adult male canaries than in females, whose songs are much less elaborate than those of the males. In the RA nucleus, it has been shown that this difference arises from an increase in the number of neurons as well as in the extent of dendritic branching and synaptic connectivity in the male (Fig. 13-9b). Treatment of a female canary with testosterone increases the size of the RA and HVA nuclei, inducing the growth of dendrites and new synaptic connections. Restructuring of these neurons by testosterone also evokes a stereotyped pattern of singing comparable to that of the male.

Further evidence that testosterone plays an ongoing role in maintenance of the form and connections of these neurons comes from measurements of testosterone in male canaries throughout the year. During summer and early fall, levels of testosterone in the blood are low. At this time, variable plastic song predominates. Thereafter testosterone levels rise and the RA and HVA nuclei double in size. As this occurs the pattern of song changes to the stereotyped stable song, which is dominant in the spring.

One interesting aspect of the remodeling of neural form and connections in song birds is that not only can preexisting connections be modified by hormones, but new neurons may be added to a pathway. There are very few such examples of precursor cells dividing, differentiating, and being added to a neuronal pathway in an adult animal. When radiolabeled thymidine, which incorporates into the DNA of dividing cells, is injected into the brain of an adult primate, for example, no labeled neurons are found. In contrast, the same experiment in adult songbirds labels a variety of neurons, including those of the HVA nucleus, although not those of the RA nucleus. As in a normal developing brain, neurogenesis in adult birds occurs in a zone of precursor cells adjacent to the ventricle, after which the cells migrate along radial glial cells to their final destination and establish the form and connections of mature neurons (see Fig. 13-1). It appears that this neurogenesis represents a turnover of neurons, rather than a continual increase in cell number throughout the life of the animal. Because neurogenesis occurs at about the same rate in both males and females, and at

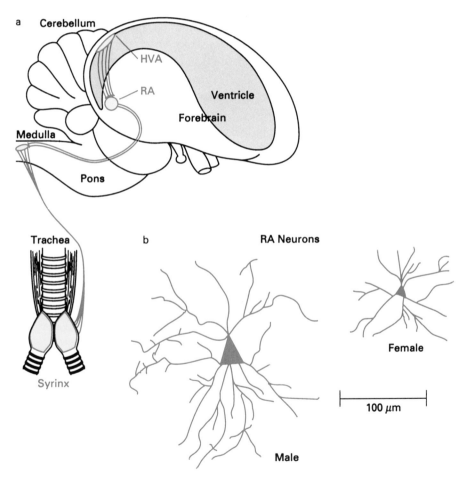

FIGURE 13-9. Steroid actions in songbirds. *a:* Canary brain circuits that control singing were identified by Nottebohm (1989) and his colleagues. *b:* Morphology of neurons in the nucleus robustus archistriatalis of male and female zebra finches (Gurney, 1981).

times when new patterns of song are not being elaborated, the significance of new neurons in the HVA for singing behaviors is not yet known.

What is the mechanism of steroid action in the nervous system? Studies such as those described above suggest that in many neurons, proteins that influence development and the eventual pattern of dendritic branching and synaptic connections may be encoded by genes coupled to HREs for steroid and related hormones. Nevertheless, some neurons that are highly sensitive to the steroid estrogen appear not to possess the estrogen receptor, indicating that the effects of the steroid must be indirect. Moreover the pharmacology of steroid responses does not always match that of the classic steroid receptors. In addition, it has been found that some steroids can act directly at the plasma membrane of a neuron, enhancing responses to the neurotransmitter GABA. Such findings demonstrate that there is much we do not yet understand about the mode of action of these hormones in the nervous system.

Summary

The development of the nervous system requires the participation of a variety of factors that influence neuronal determination, proliferation, migration, and differentiation. These factors include true growth factors such as proteins resembling epidermal growth factor, and survival factors such as nerve growth factor. Hormones released from remote organs, including steroid hormones, can also profoundly influence neuronal form and function in the developing as well as in the adult nervous system. Molecular genetic approaches using *Drosophila,* as well as other creatures whose genetics is well understood, are providing new insights into the mechanisms of action of some of these developmental factors.

Adhesion molecules
and axon pathfinding

We know that the electrical behavior of neurons can be transformed and modulated in ways that allow a neuron to control specific behaviors. Electrical behavior, however, is not the only aspect of neuronal activity that is subject to regulation by other cells or external stimuli. Structural features, such as the number, size, and type of synapses that a neuron makes, can also be modified. Furthermore, the entire shape of a mature neuron's dendritic branches may alter over time. This is illustrated in Figure 14-1, which shows tracings of the dendrites of a neuron of the superior cervical ganglion in the intact nervous system of a mouse. This cell was injected with a fluorescent dye and drawings were made immediately of the shape of its dendrites. After a period of several weeks in the intact animal, the ganglion was again exposed and drawings made of the fluorescent cell. Clear changes in dendritic branching pattern, and by inference in the synaptic connections on the dendrites, occur during this time. Although the factors that control such changes in cell form are largely unknown, these experiments lend support to the idea that long-term regulation of neuronal activity in the mature nervous system may be accompanied by rearrangements of neuronal structure and connections. To examine such rearrangements, we must first understand the factors that lead to specific patterns of branching during formation of the nervous system.

Axon Outgrowth during Development

Once an immature neuron has reached its final location in the nervous system, it must establish contacts with its appropriate synaptic partners by extending axonal and dendritic branches toward these partner neurons (see Fig. 13-1). In some cases, a neurite may have been established even during cell migration. For example, in the case of the developing granule cells of the cerebellum that we discussed in the previous chapter, some neurite extension occurs before the cells migrate along the Bergmann glial cell. A trailing neurite is then left along the length of the glial fiber and eventually becomes the axon of the granule cell. In general, however, neurite extension occurs *after* cell migration.

The growth cone. The outgrowth of neurites is guided by a specialized region of the cell known as the *growth cone,* which is found at the leading tip of a neurite. Growth cones have been investigated mainly in cell culture, where neurite extension can be examined readily under the microscope. Figure 14-2a,b illustrates the major features of the growth cone region. The central core is an extension of the neurite process itself and is rich in microtubules that provide the structural support for axoplasmic transport (see Chapter 1). Using video-enhanced microscopy, bidirectional transport of granules can be observed up to the central core. In addition, the core of the growth cone is rich in mitochondria, endoplasmic reticulum, and vesicular structures.

Surrounding the central core is a region that is generally devoid of organelles but very enriched in the contractile protein *actin.* Time-lapse pictures of these regions, which are known as *lamellipodia,* reveal the existence of undulating waves of movement that have been termed ruffling. Finally, very thin straight processes known as *microspikes* or *filopodia* are found at the extremities of the lamellipodia. These, like the lamellipodia, are rich in

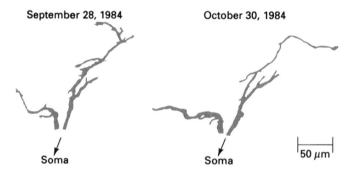

FIGURE 14-1. Remodeling of dendrites. A neuron in the superior cervical ganglion of a mature mouse was injected with a fluorescent dye by Purves and Hadley (1985). The two drawings show the change in shape of its dendrites during a period of 1 month.

actin. The microspikes are in constant motion, extending from and retracting back into the lamellipodia. Growth of the neurite occurs when a microspike extends and then, instead of retracting, remains in place while the lamellipodium advances toward the end of the microspike (Fig. 14-2c). New microspikes then extend from the newly advanced border of the lamellipodium.

Axonal pathfinding. The growth of an axon in the nervous system does not proceed randomly but follows a relatively precise pathway toward its target. Moreover the pathway that an axon follows is specific to the cell itself, and may be very different from that of neighboring cells. An example of this is shown in Figure 14-3, which illustrates the growth of the axons

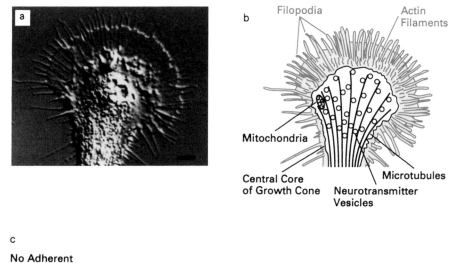

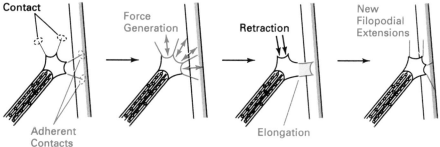

FIGURE 14-2. The growth cone. *a*: Photograph of the growth cone of an *Aplysia* neuron in cell culture (Forscher and Smith, 1988). *b*: Components of a growth cone. *c*: Scheme showing how selective adhesion of a growth cone to its substrate may guide the direction of growth.

of motor neurons in the embryonic spinal cord of a zebrafish. Three such motor neurons, termed RoP, MiP, and CaP (for rostral-, middle-, and caudal-primary motor neurons), are found in each segment of the body of a zebrafish. Initially, the growth cones of all three cells extend ventrally in the same direction away from the cell bodies. After a short period of growth, however, the axon of the MiP cell turns sharply and continues to extend dorsally between the dorsal muscle and the spinal cord. The other two axons continue ventrally until the junction of the dorsal and ventral muscles is encountered. At this point the axon of the RoP cell turns laterally while the CaP cell continues to grow ventrally. The three neurons come to innervate different muscles in each body segment of the zebrafish.

Another clear example of stereotyped branching patterns is in the grasshopper embryo, whose relatively simple nervous system has allowed the mapping of the precise pattern of navigation of the axons of several different identified neurons. On encountering a specific feature of the environment, such as another neuron or a glial cell, the growth cone of an axon may be forced to follow one pathway while that of a neighboring neuron may continue in another direction toward its appropriate target. It has been observed that when a growth cone reaches such a *landmark* cell, it may actually extend its filopodia deep into that cell. It is possible that such inter-

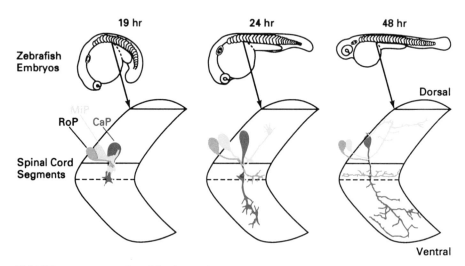

FIGURE 14-3. Axonal pathfinding. The embryonic development of the spinal cord of a zebrafish was studied by Westerfield and Eisen (1988) by following the paths of motor neurons that were filled with fluorescent dyes. The branching of each of these three neurons follows a stereotyped pattern to innervate a different set of muscles.

actions produce biochemical changes that alter the adhesivity of membranes and thereby change the subsequent migration of axons.

In some cases, as in the growth of nerve tracts, a large number of axons initially may all grow in the same direction. A bundle of closely associated axons is called a *fascicle* and the formation of such bundles is known as *fasciculation*. In the growth of a fascicle of axons, the first neuron to enter the pathway follows cues from the environment. This neuron is frequently called the *pioneer* cell. The growth cones of the subsequent axons may extend along the axon of the pioneer neuron to form the fascicle. It appears, however, that these follower neurons sometimes have the ability to make the appropriate navigational decisions in the absence of the fibers of pioneer cells.

Adhesion is important for directed growth. Such stereotyped patterns of branching and directed growth of axons have been described in a variety of systems in both vertebrates and invertebrates. What mechanisms might give rise to such stereotyped behavior? Of primary importance among the factors that control the direction a growth cone follows is the *differential adhesion* of different growth cones to other cells and surfaces that are encountered along the way. The general problem of how a cell adheres to, and interacts mechanically with, its environment is not specific to the nervous system. Neurons however, are remarkable in the degree of specificity that is required to control both the differential navigation of different axons and dendrites and the eventual establishment of synaptic contacts.

We will therefore give an account of the various adhesion molecules that are known to exist in or on neurons. It should be pointed out that in addition to their role in axonal navigation, the molecules to be described may also influence the migration, differentiation, and morphogenesis of neurons. A possible distinction that can be made between an adhesion molecule and a growth factor–receptor complex is that adhesion molecules need to be present in sufficient abundance to provide mechanical adherence, rather than simply to relay messages from the external environment to a cell.

Molecules that guide the extension of neurites in a particular direction operate through one of three general mechanisms (Fig. 14-4):

1. Cell–substrate adhesion, via molecules that allow neurons and other cells to adhere to noncellular components in the extracellular space.
2. Cell–cell communication through direct membrane contact. Many interactions of neurons or other cells may require the direct binding of a molecule on the surface of one cell with a complementary receptor in the membrane of a neighboring cell. A special case are molecules involved in cell-to-cell adhesion, known as *cell adhesion molecules* (CAMs).

3. Cell–cell communication via soluble factors, some of which we discussed in the last chapter. These include factors such as growth factors, hormones, and neurotransmitters that can be secreted either by other neurons or by nonneuronal cells.

We shall now consider each of these mechanisms in turn.

Cell–Substrate Adhesion

What is extracellular matrix? We have stated that the ability of a neuron to adhere to specific surfaces that it encounters plays a major role in its migration to the appropriate site in the nervous system, and in its extension of neurites toward the appropriate targets. For some types of neurons, much of their migration and axon elongation occurs not over the surface of other cells but through an *extracellular matrix* that is relatively devoid of cells. This matrix is required not only as a physical substrate for migration, but can have a profound influence on the properties of cells in contact with it. For example, the normal division and differentiation of many cells, such as Schwann cells, require interaction with this matrix. Because the fixed components of this extracellular space are relatively simple compared to the chemical composition of cell membranes, much of what has been learned about neuronal adhesion and neurite outgrowth has come from the study of cell–substrate interactions.

Substrate adhesion molecules. The space surrounding many nonneuronal cells is filled with a loose latticework of glycoproteins and sugars. Close to the membrane of cells that are not migrating through the space, this latticework becomes denser and forms a *basement membrane* (Fig. 14-5a). The

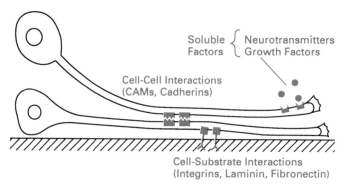

Soluble Factors { Neurotransmitters Growth Factors

Cell-Cell Interactions (CAMs, Cadherins)

Cell-Substrate Interactions (Integrins, Laminin, Fibronectin)

FIGURE 14-4. Factors that determine the amount and direction of neuritic growth. These include the interaction of a neuron with other cells, with the substrate, and with soluble molecules released by other cells.

a

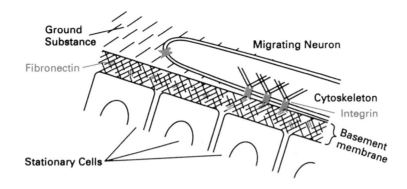

Ground
Substance

Migrating Neuron

Fibronectin

Cytoskeleton

Integrin

Basement
membrane

Stationary Cells

b

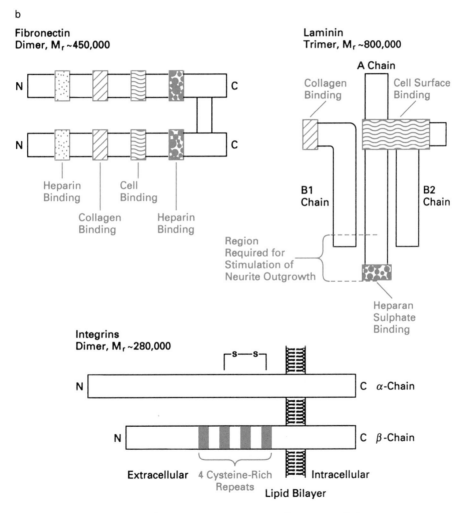

Fibronectin
Dimer, $M_r \sim 450,000$

N
C
N
C

Heparin
Binding

Cell
Binding

Collagen
Binding

Heparin
Binding

Laminin
Trimer, $M_r \sim 800,000$

A Chain

Collagen
Binding

Cell Surface
Binding

B1
Chain

B2
Chain

Region
Required for
Stimulation of
Neurite Outgrowth

Heparan
Sulphate
Binding

Integrins
Dimer, $M_r \sim 280,000$

S—S

N
C α-Chain
N
C β-Chain

Extracellular   4 Cysteine-Rich
Repeats

Intracellular

Lipid Bilayer

FIGURE 14-5. Neuron–substrate interactions. *a:* The extracellular matrix. *b:* The structure of molecules involved in cell–substrate interactions.

major components of this latticework are listed in Table 14-1. Of particular interest are the proteins *fibronectin* and *laminin*. These can be obtained in purified form, and their effects on cell adhesion and neurite outgrowth can be assessed by culturing isolated neurons in tissue culture dishes whose surface has been coated with these proteins. Both substances readily promote the adhesion and extension of neurites by several types of neurons. Figure 14-5b illustrates the structures of fibronectin and laminin. Both are very large glycoprotein complexes that contain several distinct domains, each of which appears to have a specific role in binding other components of the extracellular matrix, or in binding to cell membranes and promoting neurite outgrowth. Thus they may be considered a major part of the "glue" that attaches cells to the matrix.

The attachment of fibronectin and laminin to cells is mediated by receptor proteins termed *integrins*, which are located in the plasma membrane of many neurons and other cells (Fig. 14-5b). For example, the extracellular domain of the integrin fibronectin receptor binds to the sequence of amino acids Arg-Gly-Asp-Ser that is found in the fibronectin molecule. Similar sequences are found in many molecules that bind to other integrins in a wide variety of nonneuronal cells. The short cytoplasmic domain of the integrins appears to connect directly to the cytoskeleton, thus providing a direct link between the intracellular scaffold of the cells and the external latticework (Fig. 14-5a). There is a tyrosine residue in the cytoplasmic domain that can be phosphorylated by a tyrosine kinase, and it has been proposed that such phosphorylation uncouples the link from the extracellular matrix to the cytoskeleton. It is not yet known, however, if all of the effects of extracellular matrix molecules on parameters such as neurite outgrowth are mediated through the integrin family of receptors.

Role of extracellular matrix in migration of neural crest cells. One of the preparations most favored by developmental neurobiologists is the *neural*

TABLE 14-1 Some Components of the Extracellular Matrix

Collagens	A family of glycoproteins rich in proline
Fibronectin Laminin Chondronectin	Elongated glycoproteins that bind to receptors on cell membranes and also to other components of extracellular matrix such as the collagens
Hyaluronic acid Chondroitin sulfate Heparan sulfate	Glycosaminoglycans—unbranched disaccharide polymers

crest, a collection of cells on the dorsal edge of the developing neural tube. We encountered these cells in Chapter 13 during our discussion of the actions of cholinergic factor (Fig. 13-4). Cells that are destined to become neurons of the sympathetic or sensory ganglia arise from the neural crest. In addition to sympathetic and sensory neurons, neural crest cells may develop into melanocytes, Schwann cells, and cells of the adrenal medulla. The analysis of this system of cells has been central to the identification of many of the molecules involved in neural differentiation. Figure 13-4 also shows the directions in which cells from the neural crest migrate to their final destination. The pathways through which these cells travel are relatively devoid of cells but are filled with extracellular matrix containing fibronectin. Levels of fibronectin in this matrix are high during the period of migration but drop after migration has ceased. Moreover, the ability of neural crest cells to adhere to a collagen matrix has been shown to depend directly on fibonectin.

Cell–Cell Adhesion

Although the extracellular matrix plays an important role in the migration of cells and the extension of neurites in the periphery, the central nervous system does not possess a well-defined extracellular matrix. Basement membranes are found only along the cerebral blood vessels and lining the fluid-filled cerebral ventricles, and thus the movement of neurons and their axons in the central nervous system occurs largely over other cells. Moreover, even in the peripheral nervous system, much of the growth of axons occurs over the surface of epithelial cells and other axons. In contrast to the molecules described above, which have been termed *substrate adhesion molecules* (SAMs), the physical association of the membranes of two cells occurs through *cell adhesion molecules* (CAMs).

Cell adhesion molecules. Cell adhesion molecules were initially discovered by Gerald Edelman and his colleagues using suspensions of cells from chick retina. When such cells are dissociated in culture, they reaggregate readily into clumps of cells. Antibodies that can specifically prevent this reaggregation were made, and then used to isolate a membrane glycoprotein that binds to these antibodies. The first protein to be isolated in this way was termed N-CAM for neuronal-CAM.

  N-CAM is one of a family of cell adhesion molecules that has now been found to be expressed on neurons and on a wide variety of other cells. Figure 14-6 shows the similarity of N-CAM to three other CAMs, which represent only a fraction of the CAMs that are likely to exist. Ng-CAM is an adhesion molecule found on specific axonal tracts. (Ng-CAM appears to be the same as molecules that have been termed L1, G4, and NILE in different species). Myelin-associated glycoprotein (MAG) is a glial cell adhesion molecule. Fasciclin II, as will be described below, is a glycoprotein expressed on a subset of grasshopper neurons. The extracellular part of these molecules has a series of domains that each contains about 50 amino acids between two cysteine residues. These, by forming −S−S− bridges, can form the domains into loops. Very similar structures (termed C2 type domains) are found in immunoglobulins, molecules involved in the mounting of immune responses by lymphocytes. The CAMs are therefore considered to belong to the *immunoglobulin superfamily.*

Most of the CAMs are believed to cause cell–cell adhesion by binding to the same CAM on an adjacent cell. Thus N-CAM molecules on one cell

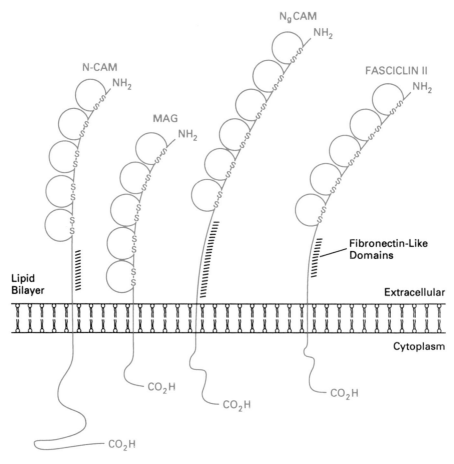

FIGURE 14-6. Cell adhesion molecules. Examples of the structure of four CAMs.

bind directly to N-CAM molecules on its neighbor. However, as shown in Figure 14-6, some of these molecules also contain regions of homology to fibronectin. These latter regions (termed type III domains) include the sequence Arg-Gly-Asp-, which is used in binding to the integrins, suggesting that these molecules may interact with cell surface proteins in more than one way. Another important family of related cell adhesion molecules is the *cadherins*. N (for neuronal)-cadherin is the best-studied example in the nervous system, although it is also found in nonneuronal cells. Like the CAMs, an N-cadherin molecule binds to another N-cadherin on an adjacent cell. In contrast to the CAMs, however, this binding requires the presence of external calcium ions.

A central test for the involvement of the CAMs and cadherins in normal development has been to expose developing cells to antibodies raised against specific adhesion molecules. For example, antibodies that block the homophilic binding of N-CAM or N-cadherin have been shown to disrupt the normal development of the retina into sharply defined cell layers (Fig. 14-7). Another test for the role of N-cadherin in the extension of axons from the retina has been to place fragments of retina onto a single layer of cells that do not normally express N-cadherin on their surface. In this condition, although retinal cells in the explant possess N-cadherin, no neurites extend out from the retinal fragments onto the monolayer of cells. When, however, the cells constituting the monolayer are induced to make N-cadherin,

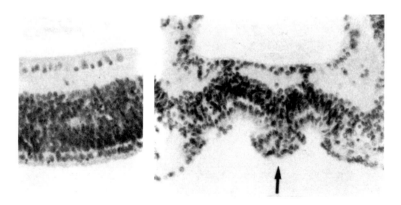

FIGURE 14-7. Disruption of retinal fragments by antibodies to N-cadherin. The effects of antibodies on the structural integrity of embryonic chick retina were tested by M. Takeichi and his colleagues. The photograph on the *left* shows a stained section of a normal fragment incubated with a control antibody for 4 days. The fragment on the *right* was incubated with the antibody to N-cadherin. The arrow points to the broken photoreceptor layer (Matsunaga et al., 1988b).

by introduction of an active gene for this adhesion protein into the cells, vigorous neurite outgrowth is observed (Fig. 14-8).

Do CAMs and cadherins account for specificity of axon guidance? Earlier in this chapter we illustrated that growing axons make specific decisions about the direction in which they extend, and that these decisions may differ from those of neighboring axons (Fig. 14-3). The differential adhesion of the growing neurite to specific molecules on the surface of nearby cells appears to be central in the choice of the pathway that an axon will follow, and there is evidence that cell adhesion molecules may be involved in these choices. Because the prototype molecules N-CAM and N-cadherin are widely distributed throughout the nervous system, they are generally viewed as acting as a relatively nonspecific glue that may aid cells in migrating and extending processes through surrounding tissue. In contrast, some of the other CAMs, such as Ng-CAM and fasciclin II, are found only in specific neurons and in specific tracts of axons. Moreover, such CAMs may appear on the surface of neurons transiently, at specific times during devel-

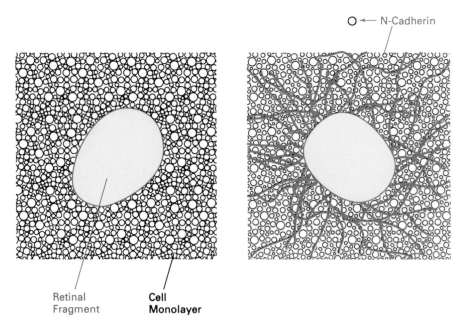

O ◀— N-Cadherin

Retinal
Fragment

**Cell
Monolayer**

FIGURE 14-8. N-Cadherin guides optic nerve fibers over the surface of other cells. This shows another experiment by M. Takeichi and his colleagues (see Fig. 14-7). Fragments of retina, placed over a layer of cells that lack N-cadherin, do not extend neurites from the explant. When the cells forming the layer are modified to express N-cadherin, outgrowth occurs (Matsunaga et al., 1988a).

opment, and their appearance may be localized to areas of membrane in contact with certain other cells.

An instructive example of the possible role of a CAM, fasciclin II, in the selective guidance of axons is provided by work on grasshopper embryos. In very early embryos, fasciclin II is found on the surface of all ectodermal cells. At a later stage, however, it can be detected only on a subset of cells and neurons within each segment of the developing nervous system. Figure 14-9 illustrates the disposition of fasciclin II in three neurons, termed MP1, dMP2, and vMP2, which adhere to the mesectodermal cells at the midline. Axon outgrowth from these three cells occurs initially over the surface of glial cells and along a basement membrane. At this stage the axons and growth cones do not contain fasciclin II, although the cell bodies do (Fig. 14-9a). In time the axons of two of these neurons, MP1 and dMP2, which grow in a posterior direction, begin to approach the axons of other MP1 and dMP2 neurons that are located in adjacent segments. At this time, these two neurons begin to express fasciclin II over their *entire* membrane, including that of the growth cones (Fig. 14-9b). When the axons reach the next segment, they cease to navigate along the basement membrane, and now adhere to the axons of MP1 and dMP2 neurons in this segment, forming a fascicle of axons (Fig. 14-9c). When the developing embryo is incubated with antibodies to fasciclin II, the ability of these axons to recognize each other, and to form this MP1/dMP2 fascicle, is selectively impaired.

In contrast to the MP1 and dMP2 neurons, the vMP2 neuron extends its axon in an anterior direction. This axon eventually joins the axons of other vMP2 neurons to form an independent fascicle of vMP2 neurons (Fig. 14-9c). Although the soma of vMP2 expresses fasciclin II, at no stage does the axon or growth cone appear to bear this adhesion molecule. Still later in development, the fasciclin II at the soma of all three neurons is lost, while that in the MP1/dMP2 axons remains (Fig. 14-9d).

The picture that emerges from these and similar results is that adhesion molecules arise on the surface of growing neurons, at specific times and places, to provide an adhesive "road surface" that routes the axonal traffic in appropriate directions. Thus, for the vMP2 neuron described above, fasciclin II at the soma provides transient adhesion to the mesectodermal cells, while another CAM, as yet unidentified, may provide adhesion to the axons of other vMP2 neurons. For the MP1 and dMP2 neurons, the same CAM, fasciclin II, contributes to adhesion both to the mesectodermal cells and to the other homologous axons.

Inhibition of axonal adhesion. The focus of this section has been on molecules that may *promote* the selective adhesion of membranes. The direction of extension of an axon may, however, also be driven by molecules

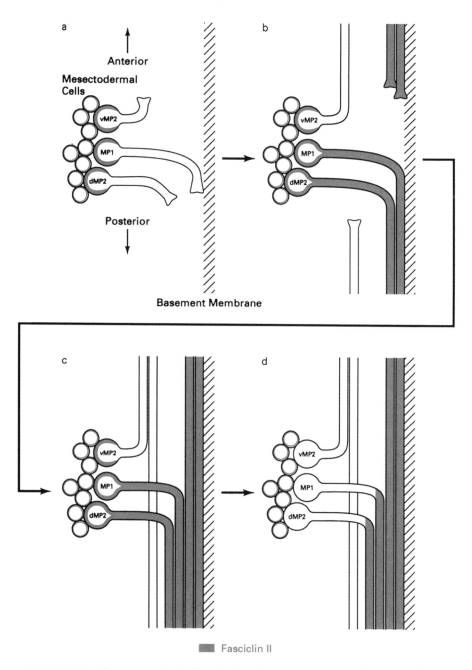

Fasciclin II

FIGURE 14-9. Changes in CAM distribution during axonal growth. Corey Goodman and his colleagues used immunocytochemistry to study progressive changes (*a–d*) in the localization of the cell adhesion molecule fasciclin II during axonal outgrowth from three grasshopper neurons (Harrelson and Goodman, 1988).

that *inhibit* adhesion. Some neurons appear actively to *retract* their growth cones after contacting the axons of certain other neurons or glial cells. For example, neurons do not normally extend neurites over the surfaces of oligodendroglia or myelin membranes. This appears to be due to the presence of two proteins on the surface of the glial cells that selectively prevent adhesion of neuronal growth cones. The application of antibodies to these cell surface proteins, instead of hindering neurite growth as in the case of the CAMs, neutralizes the inhibitory effect and allows axons to extend actively over the glial membranes. Investigation of the nature of the proteins that render specific membranes nonpermissive for axon growth is likely to be a profitable field of research in the future.

Diversity of cell adhesion molecules. A relatively large number of specific cell adhesion molecules must exist, if they are to account for the many different paths that growth cones of different axons take as they move toward their final destination. Some that have been described to date have been given such poetic names as amalgam, neurofascin, and contactin. Many more, perhaps some with structures unrelated to the immunoglobulin superfamily, are certain to be discovered. In addition, considerable diversity could in principle be generated by expressing different forms of the same basic molecule. For example, the N-CAM protein exists in three closely related forms of different length. These all share the same extracellular domain but differ in the length of the C-terminus region. Two of these cross the plasma membrane as shown in Figure 14-6, whereas the third lacks the membrane spanning and cytoplasmic domains and is anchored to the cell membrane through an inositol-containing lipid.

It is likely that, in addition to changes in the *expression* of different CAMs, *modifications* of a CAM that is already expressed may produce a progressive change in the adherent properties of a cell or a neurite. For example, much of the sugar that is bound to the N-CAM protein consists of an oligosaccharide termed *polysialic acid*. During development of the chick brain, there is conversion of an embryonic form of N-CAM to an adult form. This occurs due to a progressive loss of polysialic acid, from about 26% of the weight of the N-CAM molecule in the embryo to only 9% in the adult. This reduction in bound oligosaccharide produces a marked increase in the adhesive power of the N-CAM molecule, perhaps because the polysialic acid hinders the normal adhesive interactions.

Cell–Cell Communication via Soluble Factors

The recognition and binding of different immobilized extracellular and cell-membrane proteins by receptors on the surface of a neuron have many analogies with the interaction of a neurotransmitter or hormone with its receptor. Thus it is not surprising that neurotransmitters themselves, and also the occurrence of neuronal electrical activity, may influence neurite outgrowth. Figure 14-10 illustrates the actions of the neurotransmitter serotonin on an identified neuron of the mollusc *Helisoma*. This neuron, termed B19, may be placed in cell culture where it normally extends neuritic branches. Application of serotonin to the growing cell, however, abruptly terminates the elongation of the neurites and causes loss of lamellipodia in the growth cones. In contrast, the growth of cells that do not bear serotonin receptors is unaffected by this transmitter.

This regulation of neurite growth by neurotransmitters occurs locally through receptors on the growth cones themselves, and does not require signals from the cell body. By severing a neurite just before the region of the growth cone, it is possible to isolate growth cones from the remainder of the cell (Fig. 14-10b). Interestingly, such isolated growth cones continue to extend along the substrate for a considerable time after they have been severed from their cell bodies. Isolated growth cones of neuron B19 respond to serotonin in the same way as those on intact B19 cells, indicating that inhibition by serotonin is a local event occurring at the growth cone itself.

The action of neurotransmitters on growth cones need not always be inhibitory. Application of the peptide substance P, or cyclic AMP analogs, stimulates axon outgrowth in some cultured mammalian neurons. One line of thinking, based on experiments with the calcium indicator dye fura-2 (see Chapter 7), is that normal extension of growth cones occurs only when the calcium concentration in the growth cone is within a range of concentrations permissive for growth. Neurotransmitters that act to change intracellular calcium levels, moving them into or out of this permissive range, stimulate or inhibit growth. In the same way, electrical activity itself can also influence neurite extension, by allowing calcium entry through voltage-dependent calcium channels. For example, direct electrical stimulation of neurons such as B19 prevents axon elongation. Because most of these studies have been carried out using neurons in cell culture, the relevance of such effects of neurotransmitters and electrical stimulation to axonal growth in the intact nervous system is not yet established. It does seem likely, however, that these influences shape the final branching patterns at times when synaptic contacts are being established.

Soluble growth factors also influence both the amount and the direction of axon outgrowth. For example, as we saw in Chapter 13, neurons from the sympathetic ganglia and sensory neurons from dorsal root ganglia are sensitive to NGF. When a pipette containing NGF is placed into a culture dish containing sensory neurons, a gradient of NGF is established in the dish as NGF slowly leaks into the medium. The axons of the cells are then

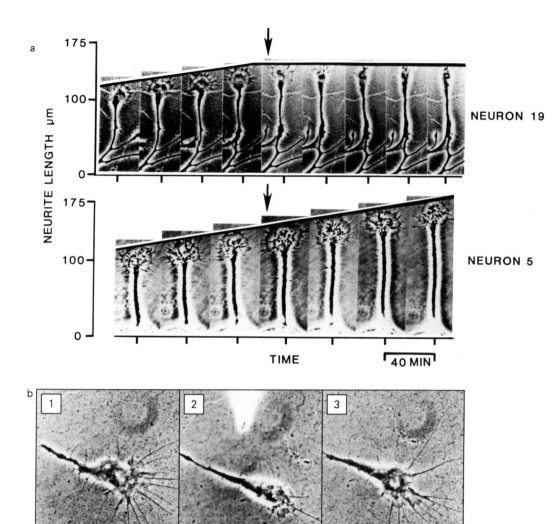

FIGURE 14-10. Inhibition of growth cones by neurotransmitter. *a*: Experiments by Stanley Kater and his colleagues showed that application of serotonin (5-HT) to *Helisoma* neuron B19 inhibits growth of its neurites. Neuron B5 is unaffected by serotonin. *b*: Serotonin also inhibits an isolated B19 growth cone, when applied from an external pipette (*2*). Growth recovers when serotonin is removed (*3*) (Haydon et al., 1984).

observed to change their direction of growth and extend toward the tip of the pipette. When injections of NGF are made into brains of newborn rats, aberrant growth of neurites from sympathetic ganglia toward the injection site occurs. There is evidence that the axons of other kinds of neurons can also be guided by soluble factors, although the chemical nature of these factors is not known. It is unlikely that the high directional selectivity of axonal growth can be accounted for by soluble factors alone, although molecules that diffuse through the extracellular space could exert an important influence on the routing of axons.

Biochemical Properties of Growing Axons

Secretion of proteases. To reach their targets in intact developing tissues, growing neurites must penetrate other tissues and forge their way through extracellular matrix. To aid their progress, growth cones appear to secrete *proteases,* enzymes that partially digest proteins in the extracellular matrix. One way this can be demonstrated readily is by plating neurons in cell culture on a dish covered with a layer of protein, such as fibronectin, laminin, or gelatin, that has been modified chemically so as to fluoresce. Degradation of the proteins is detected as loss of fluorescence near the cell and the growth cone. The release of proteases must, of course, be selective and controlled because, as we have seen, many extracellular proteins are important in cell adhesion. In addition to aiding the mechanical progress of a growth cone, it is possible that such extracellular proteolysis serves as a cell-to-cell signal to trigger responses that follow contact of one cell by the growth cone of another.

Synthesis of GAP-43. An interesting protein that is strongly suspected of influencing the ability of axons to grow is *GAP-43,* which is short for 43 kDa *growth-associated protein.* A number of names have been given to GAP-43 by different investigators, working in different systems and species. These include F-1, B-50, pp46, and GAP-48. Current evidence suggests that each of these represents the same, or a very closely related, protein.

What is the evidence that GAP-43 is associated with neuronal growth? During the extension of axonal branches, the pattern of proteins that is synthesized at the soma and then transported down to the tip of a developing axon differs from that in fully formed nerves. Figure 14-11 illustrates the pattern of newly synthesized proteins transported along the axons of neurons in the retina of a hamster. These proteins were labeled by injecting a radioactive amino acid into the retina. Four hours later, they were extracted from the *superior colliculus,* a brain area that receives extensive synaptic input from the retina. The pathway from the retina to the collic-

ulus in mammals closely resembles the *retinotectal pathway* in lower vertebrates that is described in the next chapter. The two-dimensional separation of proteins by gel electrophoresis shows that several proteins are
either enriched or depleted in a 2-day-old animal relative to adult animals.
Among these, one protein in particular, GAP-43, is present in large amount
in the growing cells. However, it undergoes a decrease of 90% or more in
the several weeks that follow the period of maximal growth.

GAP-43 is a highly acidic protein. Its synthesis and transport occur at a
high rate during periods of axonal extension in many different neuronal
pathways. When axonal outgrowth is stimulated in cultured cells, for
example by NGF, GAP-43 is one of the proteins whose synthesis is induced.
Moreover, when a pathway in the mammalian nervous system is severed,
the synthesis of GAP-43 may be markedly stimulated. This occurs only in
those cases in which the pathway regenerates, but is not seen in tracts in
which no regeneration occurs. Such findings suggest that GAP-43 may be
an important component of the neurite outgrowth process.

The growth cones of developing axons also contain GAP-43. Highly
purified growth cones can be prepared by centrifugation techniques, and
GAP-43 is a major component of such preparations. The primary structure
of GAP-43, which is known from the cloning of the DNA encoding the
protein, does not have the hydrophobic sequences characteristic of integral
membrane proteins. However, GAP-43 may be closely associated with the
internal surface of the plasma membrane of growth cones. Moreover, the
introduction of GAP-43 into nonneuronal cells promotes the extension of

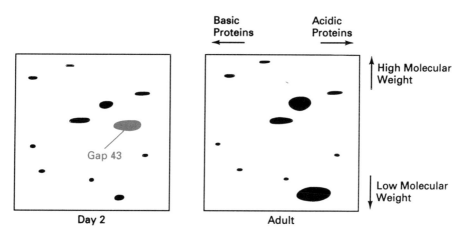

FIGURE 14-11. GAP-43. Representation of a two-dimensional separation of newly
synthesized proteins transported along axons of retinal ganglion cells in a 2-day-
old hamster and in an adult (Benowitz and Routtenberg, 1987).

filopodia, and of processes resembling neurites. GAP-43 does not disappear altogether during development and can be detected readily in certain regions of adult brain. It is a phosphoprotein and can be phosphorylated by the enzyme protein kinase C. It is therefore possible that this intriguing protein plays a role in plastic changes in the adult nervous system (see Chapter 17).

Summary

Developing neurons extend neurites, which become the axons and dendrites of the adult neuron. These neurites follow specific paths and branch in characteristic ways. The leading tip of the neurite, the growth cone, appears to sample the extracellular environment and contribute to decisions about the direction of neurite extension.

Molecules of various types are essential for appropriate pathfinding by growing neurites. For example, neurites grow selectively toward or away from certain soluble factors, including growth factors and neurotransmitters. In addition, a variety of adhesion molecules play a primary role. These include protein molecules in the membrane of the neurite and in the extracellular matrix that mediate specific adhesion of the neurite to the substrate over which it is growing. Other membrane proteins promote the adhesion of neurites of different cells to each other in specific patterns. Some of these molecules and mechanisms that regulate neuronal development and differentiation may also participate in neurite outgrowth in adult nervous systems, either during recovery from injury or in response to novel stimuli from the environment.

15

Formation, maintenance, and plasticity of chemical synapses

Following cellular determination and neurite elongation, developing neurons form the specific synaptic connections that are essential for brain function (Fig. 13-1). One particularly striking aspect of the formation of synapses is its extraordinary sensitivity to patterns of electrical activity in the developing pathways. We will now describe synaptogenesis during development and its guidance by the pattern of electrical stimulation to which an immature neuron is exposed. Synaptogenesis is not, however, restricted to the developing nervous system. We have hinted that reorganization of neuronal form and function occurs in the adult nervous system, and in fact mature neurons retain most of the machinery necessary for restructuring their synaptic connections by mechanisms similar to those operating in development. We will therefore provide some examples of synaptic plasticity in adult neurons. Finally, we will describe some of the changes that occur in the properties of cells once synaptic contacts have been established.

Synaptogenesis during Development

Morphological changes during synapse formation. When growing axons reach the cell on which they will finally make synaptic contacts, changes occur in the shape of their growth cones (Fig. 15-1). The lamellipodia that are characteristic of rapid growth shrink in size and the filopodia extend from the tip of the neurite in an irregular pattern. This change in the appearance of growth cones is associated with a slower rate of elongation. When the growth cone finally contacts a cell with which it forms a synapse,

343

a further change takes place in its structure. The lamellipodium and filopodia disappear as neurotransmitter vesicles from the central core region advance into the tip of the neurite. The point of contact is then transformed into a full-fledged synapse with the accumulation of material in the synaptic cleft and the thickening of the postsynaptic membrane to form a *postsynaptic density*. Functional synaptic communication, however, can occur as soon as the neurite contacts the postsynaptic cell. Indeed, experiments in cell culture have demonstrated that even extending growth cones are capable of releasing neurotransmitter.

As described in Chapter 8, some neurons do not contact their targets directly but release neurotransmitter locally to influence neighboring cells without making specialized synaptic contacts. Changes in the terminals of these cells must also occur when their axons have reached their final destinations. An example of how terminal morphology can change without synapse formation is given in Figure 15-2, which shows the change in structure of a growth cone of an *Aplysia* bag cell neuron in cell culture following elevation of cyclic AMP levels. The actin-rich lamellipodium is invaded by microtubules and other organelles from the central core region to produce a club-like ending that is packed with secretory granules. Although such changes can be induced experimentally in cell culture, it is not known whether second messengers such as cyclic AMP play a role in similar transformations of growth cones in the nervous system.

What determines the choice of postsynaptic target? Once synapses have formed on postsynaptic targets, they do not remain static, but may either

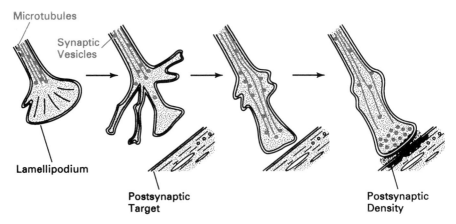

FIGURE 15-1. Synapse formation. Transformation of a growth cone into a presynaptic ending.

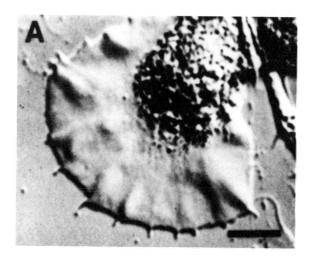

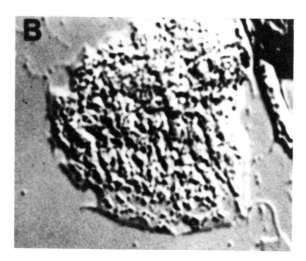

FIGURE 15-2. Changes in structure of a growth cone in cell culture. Photographs of the growth cone of an *Aplysia* neuron before (*A*) and after (*B*) treatment with drugs that elevate cyclic AMP levels (Forscher et al., 1987).

expand or retract and may withdraw altogether. A neuron whose process has retracted from one postsynaptic cell may subsequently form a more stable synapse on another cell. Synapse formation during development is therefore a relatively protracted process. For example, in the mammalian brain, the number of synapses formed may increase over a period of many weeks or months after birth. Thereafter the total number of synapses declines toward that of the adult. Moreover, in some cases synapses continue to be made and broken during adult life.

As will be described below, the fine-tuning of synaptic connections appears to result in large part from the competition of different axons for the same postsynaptic cells. A major question that has not yet been resolved is how a growing axon recognizes an appropriate postsynaptic target in the first place. An extending axon does not make synaptic contacts with every cell that it encounters on its path. When it reaches an appropriate postsynaptic target, therefore, some signal must be generated that instructs the growing cell to slow its growth, contact the postsynaptic cell, and form a synaptic terminal. The nature of this signal is not known, but it may perhaps be related to the molecules that regulate interactions through cell–cell contact (see Chapter 14).

How specific is this interaction of a growing axon with its target? Can the axon of a motor neuron, for example, form a synapse on any muscle cell, or is there a strict one-to-one recognition of a specific muscle cell or small group of cells? In many cases it appears that synapse formation can, initially, be rather indiscriminate. Synapses from a subset of motor neurons can be removed by severing their axons and allowing the synapses to degenerate. Under these conditions, the muscles can readily be reinnervated by other motor neurons. Moreover, when neurons are placed in cell culture, they may make synapses with other neurons or muscle cells with which they would not normally make contact in the intact nervous system.

Although the mature pattern of synaptic connections is influenced strongly by mechanisms such as the elimination of presynaptic terminals through competition (see below), some degree of specificity in the initial choice of postsynaptic target must exist both for the appropriate class of target cell and for the relative position of different cells within that class. One of the most thoroughly studied and clearest examples of this is to be found in the visual system of vertebrates such as frogs, fish, and birds.

Synapse Formation in the Visual System

The retinotectal system. In many lower vertebrates, the major neuronal projection from the *retina* extends from the optic nerve to part of the midbrain known as the *optic tectum*. This pathway controls many of the rapid visual reflexes of such animals. A feature of this pathway that has attracted much attention from developmental biologists is the particularly clear point-to-point mapping of the input from different parts of the retina to corresponding points on the surface of the tectum. Such mapping can readily be demonstrated by shining points of light on the retina and recording electrophysiological responses in the tectum. As shown in Figure 15-3a, stimulation of the dorsal visual field (ventral retina) triggers responses in the dorsal tectum, whereas stimulation of the ventral visual field produces responses in the ventral tectum. Similarly retinal neurons that respond to light in the temporal visual field (anterior retina) connect to neurons in the posterior tectum, while those responding to light in the nasal field project to the anterior part of the tectum.

The chemoaffinity hypothesis. Although the final connections to the tectum are arranged in an exquisitely precise array, the path that the axon of a retinal cell follows to reach its final destination may not be direct. An ingrowing axon may bypass many potential postsynaptic targets in the tectum before establishing its synaptic contacts (Fig. 15-3b). The major hypothesis to account for the specificity of these connections is the *chemoaffinity hypothesis* of Roger Sperry. The simplest form of this hypothesis states that there exist specific molecules in the presynaptic and postsynaptic cells (usually thought of as being on the surface of the cells) that differ either in their chemical identities or in their relative amounts in different regions of the tectum. These differences constitute a biochemical label for each cell. The fact that cells from one region of the retina connect only to the appropriate region of the tectum is explained by requiring correct matching of presynaptic and postsynaptic labels for synapse formation to occur.

Many experiments support the chemoaffinity hypothesis. Some of these involve surgical manipulation of the inputs to the tectum. For example, the optic nerve of a frog can be severed and the eye rotated through 180°. The axons of the retinal cells will in time regenerate and reinnervate the tectum. Under these conditions cells in the retina form new synaptic connections with the same part of the tectum that they innervated before, even though the region of visual space projected by a given region of retina to tectal sites is now 180° different. Moreover, if half of a retina is removed, the remaining cells in the retina will reextend axons that make synaptic contacts with the appropriate half of the tectum. Initially, these axons will not innervate the remainder of the tectum, which does not bear the appropriate label. (As will be described below, however, slower remodeling of the connections does occur, and the entire tectum is eventually innervated by the half-retina.)

Further evidence that molecules in the membrane of retinal cells may determine which region of the tectum they come to innervate has come

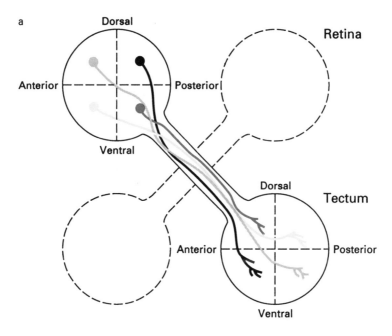

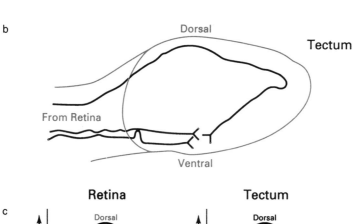

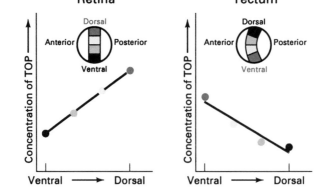

FIGURE 15-3. Retinotectal connections. *a*: The topology of connections. *b*: Drawings of the irregular growth of a retinal axon into the ventral tectum of an adult newt (Fujisawa, 1981). *c*: Results of an experiment by Trisler and Collins (1987) measuring the gradient of TOP in retina and tectum.

from experiments in which the physical adhesion of retinal cells to tectal cells has been tested in a dish. Retinal cells have been dissociated and then incubated with the tectum. Cells from the dorsal retina are found to bind preferentially to the ventral tectum, while cells in the ventral retina adhere more strongly to dorsal tectum. Such preferential stickiness of retinal cells matches the pattern of synaptic connections that eventually is formed by these cells. In other experiments, explants of retina have been placed over alternating strips of cell membranes prepared from posterior and anterior tectal cells. Cells in explants from the posterior retina preferentially grow on the membranes of anterior tectal cells. Interestingly this preference results from a *repulsive* factor in the membranes of the posterior tectal cells. Heating of the posterior tectal membranes neutralizes this repulsion and the retinal cells then fail to discriminate between posterior and anterior tectal membranes.

Finally there is direct evidence that gradients of proteins and other molecules exist in the retina. One protein was discovered by making monoclonal antibodies to different regions of the retina. One of these antibodies was then found to bind a 47-kDa membrane-bound glycoprotein that is present in high amounts in the dorsal retina but only in much lower amounts in the ventral retina (Fig. 15-3c). A corresponding gradient of the same protein was found in the tectum, where its concentration is greatest in ventral regions and lower in the dorsal tectum. This protein has been named TOP, because of its topographical distribution. Although TOP is a good candidate for a molecule that could label retinal cells from different regions, presently there is no evidence as to whether TOP plays a role in synapse formation.

Rearrangement of Synaptic Connections

The retinotectal system. Synapse formation is a two-stage process. Once an initial set of contacts is made by the incoming axons, perhaps guided by adhesive molecules, there follows a second prolonged period of restructuring or *sorting* of these synapses. At this time, some terminal branches and their synaptic contacts may be withdrawn from a tectal cell while the connections from other retinal neurons may be strengthened. Although such remodeling of connections is most obvious during the initial development of the retinotectal pathway, it is clear that in some instances this continues throughout adult life.

Particularly clear examples of the ongoing restructuring of synapses occur in frogs and goldfish. In these species, the retina and tectum continue to increase in size by the addition of new cells. In the eye, new neurons are added as a ring to the circumference of the retina. In contrast, in the tectum the new cells are added only to the posterior tectum (Fig. 15-4). Thus to maintain the correct mapping of connections from the retina to the tectum, the entire set of synapses from the retina continually retracts and reconnects to a new set of tectal neurons. In the goldfish, the retina and tectum continue to increase in size, and therefore to realign their connections, throughout adult life.

A shift in the pattern of synaptic connections must also occur in the experiment described above, in which the optic nerve is cut and part of the retina is removed. Initially only the area of the tectum that corresponds to the intact retina is reinnervated by the axons from the remaining retinal cells. In time, however, branches of these axons extend to make synapses over the entire tectal area. In this way, the map of the visual field represented by the remaining part of the retina spreads out over the whole tectum.

Electrical activity of neurons determines the final pattern of synaptic contacts. Experiments with the tectum, and with many other parts of the developing nervous system, suggest strongly that the fine-tuning of synaptic connections is determined by the pattern of electrical activity in the pre-

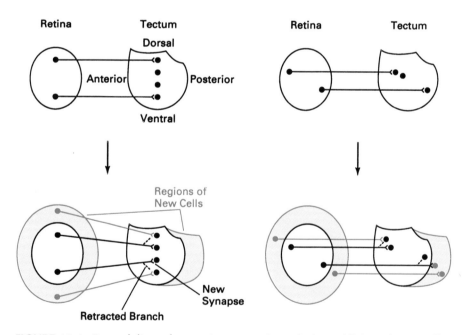

FIGURE 15-4. Remodeling of synaptic connections during addition of new cells. As the size of the retina and tectum increase, connections are broken and reformed to maintain the overall map of retina onto the tectum.

synaptic neurons. A given postsynaptic cell may initially be innervated by many different presynaptic cells. It appears, however, that only those inputs whose activity is correlated in time with postsynaptic activity are stabilized (Fig. 15-5). Fibers that generate weaker responses, or fail to influence the postsynaptic cell altogether, retract their synaptic contacts. Such fibers may, however, successfully stimulate and establish stable contacts with other postsynaptic cells.

The hypothesis that excitatory synapses are stabilized when they successfully trigger action potentials in the postsynaptic cell was proposed by the Canadian psychologist Donald Hebb in 1949. It has been used in models of both development and learning (see Chapter 17). A restatement of this hypothesis is that when a postsynaptic neuron becomes depolarized, it generates a biochemical reaction or a trophic factor that stabilizes the excitatory synapses that are firing at that time. An important aspect of this hypothesis is that a given presynaptic input to a cell need not, by itself, be of sufficient strength to induce a large depolarization in its target. If that input is fired at the same time as a number of other inputs, and their combined action depolarizes the cell, all of these inputs will tend to be stabilized. If, in contrast, a given input fires *asynchronously* with most of the other inputs onto that cell, this input will tend to be eliminated. Although the nature of the trophic factors involved is unknown, the general hypothesis that synapses are stabilized or eliminated based on their patterns of activity is able to explain many different findings about synapse formation and remodeling.

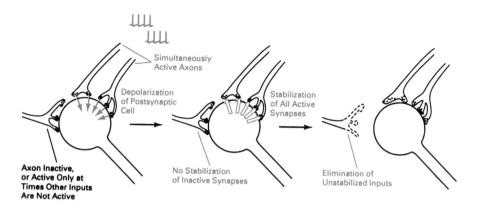

FIGURE 15-5. Hebb's rule. Excitatory synapses that successfully stimulate a postsynaptic neuron, or are active when the postsynaptic neuron is depolarized, are selectively stabilized (Hebb, 1949).

Ocular dominance columns. In lower vertebrates including frogs, all the fibers from one retina cross the midline to innervate the contralateral tectum. Thus, in the frog, there is normally no competition between inputs from the two eyes in one tectum. It is possible, however, to implant a third eye into a tadpole. The fibers from this third eye grow normally into one of the tecta, where they must compete with axons from the normal eye for synaptic space on the tectal neurons. The final pathways that are established can be measured by injection of a radiactive amino acid into one of the eyes. The radiolabel is taken up by retinal cells and transported to their terminals in the tectum. The amount of radioactivity in these terminals can then be visualized directly by placing a piece of X-ray film against slices made from the tectum (Fig. 15-6). It is found that connections have been established in a pattern of alternating columns, each of which contains inputs primarily from one eye. Under normal conditions, no such columns are observed in the tecta of frogs and other lower vertebrates.

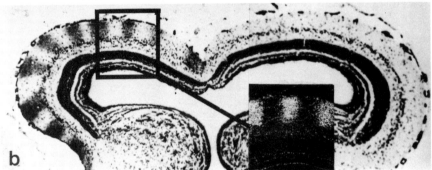

FIGURE 15-6. The three-eyed frog. *a*: Three-eyed frogs have been studied by Martha Constantine-Paton and her colleagues. *b*: An autoradiograph of the tectum shows the formation of stripes of inputs from the normal and the implanted eye. The inset shows an enlargement under dark-field illumination (Constantine-Paton and Law, 1978).

The formation of such columns, termed *ocular dominance columns,* can be understood in terms of resorting of synapses based on their electrical activity, a process that occurs during normal development (Fig. 15-7). Because all of the photoreceptor cells in one eye point toward one general region of visual space, the inputs in that one eye will tend to be activated approximately simultaneously. The other eye, however, covers a somewhat different visual field. Although the inputs from that eye generally will also be correlated with each other, they will tend not to be active at the same time as input from the competing eye. Thus a given small region of the tectum is innervated initially by inputs from both eyes. A small difference in the amount of input from one eye, however, tends to stabilize all of the

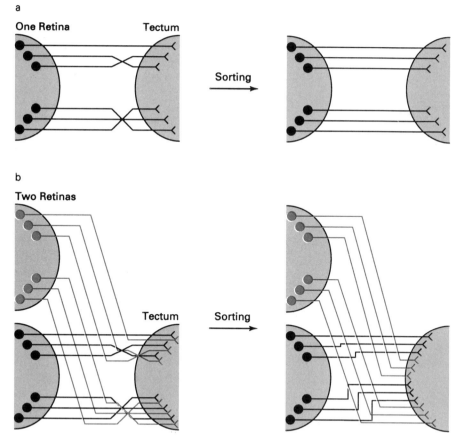

FIGURE 15-7. Segregation of retinal inputs to the tectum. *a:* Sorting of inputs from one eye only. *b:* Sorting in the presence of input from a second eye.

approximately synchronously active synaptic inputs from that eye, at the expense of inputs from the other eye. Such stabilization leads to the formation of areas or columns that are preferentially innervated by one eye or the other.

Segregation of inputs from two eyes also occurs *normally* in the visual systems of mammals, in which each retina sends some fibers to both hemispheres of the brain. Indeed, it was the experiments of David Hubel and Torsten Wiesel on kittens and monkeys that first led to the idea that patterns of electrical activity shape synapse rearrangement. Sheets of cells that respond preferentially either to one eye or the other are found in the *lateral geniculate,* a visual relay station, of adult cats and monkeys. In the visual cortex, inputs from the two eyes are segregated into ocular dominance columns. The sorting of these inputs occurs by mechanisms that are similar to those in the three-eyed frogs. At first, geniculate and cortical cells are innervated relatively uniformly by incoming fibers carrying information from both eyes. At this early time many cells can be stimulated by inputs from either eye. However, as development proceeds, synapses are broken, reformed, and stabilized to produce the alternating ocular dominance columns.

The formation of these segregated patterns of synaptic connections is not programmed innately but depends on the electrical activity of the presynaptic fibers. This can be demonstrated in a number of ways. For example, tetrodotoxin can be introduced to block action potentials. This treatment does not interfere with the initial innervation itself, but prevents the subsequent rearrangement into columns, leaving the presynaptic inputs from the two eyes relatively uniformly distributed. In other experiments the two eyes have been stimulated either synchronously or asynchronously. For example, if the two optic nerves of a cat are stimulated electrically such that activity from the two eyes occurs at different times, then ocular dominance columns develop. If, however, the two nerves are stimulated simultaneously, such that the input from the two sets of presynaptic fibers is identical, then no segregation occurs.

We emphasized earlier in this book that many neurons are capable of generating spontaneous patterns of electrical activity in the absence of external inputs. Interestingly, such spontaneous activity may play a role in the sorting of synapses before the visual system is fully functional. In three-eyed frogs, as well as in other species, inputs from the two eyes can still segregate even when animals are reared in the dark or before visual inputs are functional. The ocular dominance columns formed in these circumstances are still blocked or reversed by treatment with tetrodotoxin, indicating that ongoing electrical activity is required for the segregation.

There is now evidence that activation of the NMDA type of glutamate receptor during development may be involved in the selective stabilization of retinotectal projections in the frog, and in related phenomena in higher vertebrates. As will be seen in Chapter 17, this developmental stabilization of active synapses finds a parallel in schemes to account for plastic changes occurring in the adult brain. One piece of evidence that implicates NMDA receptors in development is that treatment of tectal cells with an NMDA receptor antagonist does not prevent the stimulation of the tectal cells by retinal afferents, but reverses the segregation into ocular dominance columns in the three-eyed frog. Discussion of possible mechanisms by which the NMDA receptor might control such long-term stabilization of synapses is deferred to Chapter 17.

Synaptic rearrangement at the neuromuscular junction and in other systems. This kind of remodeling of axonal and dendritic branches and synaptic connections occurs not only in visual pathways, but also in many other regions of the nervous system of both vertebrates and invertebrates, and may be a very general phenomenon. Another well-studied example of this is found at neuromuscular junctions of vertebrates. When the axons of motor neurons first contact their skeletal muscle targets, the different branches of an axon contact several muscle fibers. Moreover, each muscle fiber is contacted by the terminals of several different motor neurons. This situation is termed *polyneuronal innervation.* Over a period of a few weeks, however, many of these branches are withdrawn. Eventually, in the adult, each muscle fiber comes to be innervated by only one motor neuron. This elimination of synapses can be observed both morphologically and electrophysiologically. As shown in Figure 15-8a, at the neuromuscular junction of the neonate, graded stimulation of a presynaptic nerve produces graded postsynaptic potentials in the muscle. This is because the postsynaptic response is made up of responses to several presynaptic axons, and graded stimuli to the nerve trigger action potentials in a progressively larger proportion of these presynaptic fibers. In the adult, however, different intensities of stimuli either trigger an action potential in the axon of the one motor neuron contacting the muscle, or fail to excite this one axon, resulting in an all-or-none postsynaptic response.

Although the loss of polyneuronal innervation during development of the neuromuscular junction is produced by removal of axonal branches and their associated synapses from the muscle fiber, the number of individual synaptic terminals made by branches of the one axon whose inputs are stabilized actually increases (Fig. 15-8b). As in the retinotectal system, this restructuring of neuritic branches and synapses depends on the electrical activity of the presynaptic terminals. If electrical activity is abolished by pharmacological treatment of the presynaptic axons, for example, by using a local anesthetic, the elimination of synapses and the loss of polyneuronal innervation are greatly slowed. On the other hand, direct stimulation of the presynaptic fibers substantially enhances the rate of synapse elimination.

A similar restructuring of synapses has also been observed in many other parts of the nervous system, including the cerebellum, the cochlear nucleus, the superior cervical ganglion, the ciliary ganglion, and the submandibular gland of the rat. In each of these a progressive decrease in the number of axons that innervate a given postsynaptic cell is found during development, at times when little or no change occurs in the total number of pre- and postsynaptic cells. As in the retinotectal system, hypotheses can be invoked

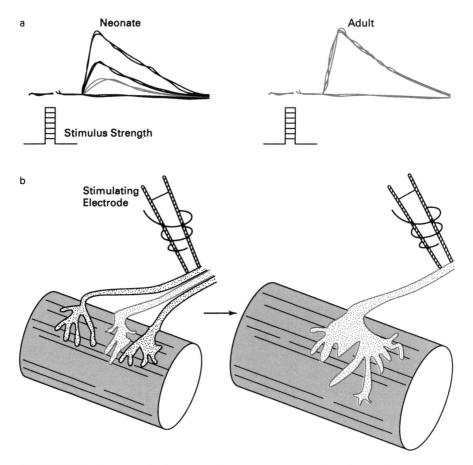

FIGURE 15-8. Synapse elimination at the neuromuscular junction. a: Postsynaptic potentials of varying sizes are recorded in the neonate, whereas stimulation of an adult junction gives an all-or-none postsynaptic potential. b: Elimination of synapses at the neuromuscular junction.

to explain these synaptic rearrangements based on the selective stabilization of active synapses. For example, it may be that some trophic factor is rapidly released from the postsynaptic cell when it is activated, and such a factor can act only on the most recently active presynaptic terminals to stabilize them and to stimulate their growth. There is no reason to expect, however, that mechanisms of synaptic rearrangement are the same in all parts of the nervous system. For example, the NMDA receptor cannot play a role at the neuromuscular junction because it is not found at this synapse.

Although, as a result of competition, the number of presynaptic neurons that innervate a single postsynaptic cell may decrease during development, not all such competition leads to a single "winning" presynaptic cell. The competition may be localized to relatively small regions on the postsynaptic membrane. For example, on a cell that has a complex pattern of dendritic branches, competition may lead one cell to establish its terminals on one branch, at the expense of synapses from other cells. On another branch or a different region of the dendritic tree, the terminals of a different cell may gain precedence, and the mature postsynaptic cell will come to be innervated in different regions by different axons. As a general rule it appears that cells that have limited dendritic branches, and in which all presynaptic inputs compete for postsynaptic membrane on the soma, become singly innervated. Cells with complex geometries remain multiply innervated.

Synaptic Plasticity in the Adult Nervous System

Sprouting in adult nerves. We have given an account of the extension of neurite branches and the making and breaking of synapses during the formation of the nervous system. It is clear that the adult nervous system retains nearly all of the machinery required for such synaptic plasticity. This can be demonstrated at the adult neuromuscular junction, where the stability of the synaptic connections depends on ongoing electrical activity in the muscles. Synaptic transmission can be blocked either by application of tetrodotoxin to the motor axons, or by local injection of an agent such as α-bungarotoxin, which blocks the response of muscle cells to acetylcholine released at the synapses. When this occurs, new branches are formed at the synaptic terminals. They extend over the muscle fiber, forming new areas of synaptic contact.

Further evidence that adult axons are capable of sprouting new processes and forming new synapses comes from studies of recovery from injury of motor nerves at the neuromuscular junction. Following partial denervation of a muscle by cutting some of its incoming axons, the remaining intact motor neurons form new branches that extend toward the denervated region and establish new synapses (Fig. 15-9). These new *sprouts* may extend either from the terminals of motor neurons (terminal sprouts) or from the axons at the nodes of Ranvier (nodal sprouts). The sprouts that establish synapses on the denervated muscles become stabilized, perhaps through the action of some trophic factor from the muscle cells (see below). Sprouts that fail to reach a target, however, are eventually retracted. Sprouting can also be evoked by manipulations that do not sever axons or block their electrical activity entirely. For example, when axoplasmic transport is blocked in some motor axons by application of agents such as the microtubule-disrupting drug colchicine, sprouting is induced in the terminals of adjacent axons.

In addition to being able to generate sprouts near a terminal region, mature neurons are capable of regenerating full axons. In the case of the neuromuscular junction, when the axon of a motor neuron is severed, the distal part of the axon degenerates. The remaining part of the axon, which is attached to the soma, forms a new growth cone at its distal end and

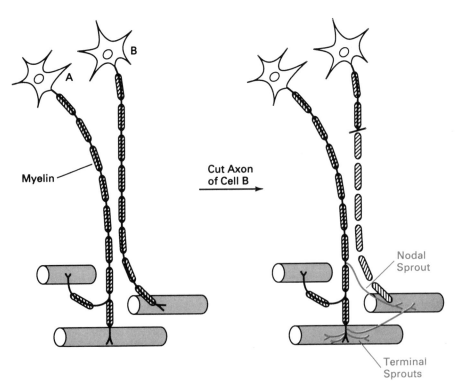

FIGURE 15-9. Axonal sprouting at the adult neuromuscular junction (modified from Brown, 1984).

grows back to the denervated muscle. Biochemical changes characteristic of developing axons can be detected in the regenerating axons. For example, the concentration of the protein GAP-43 (see Chapter 14) increases dramatically at this time. When the growth cone reaches the muscle, new synapses are formed. If the original sites of synaptic contact have come to be filled by branch sprouts from neighboring terminals whose axons were not severed, then these sprouted branches may retract as a result of the reinnervation. In many respects, therefore, this restructuring of synaptic branches in the adult nervous system resembles what occurs during development.

Does remodeling of synapses occur continually in the adult? Although the experiments described in this chapter indicate that the maintenance of synapses in mature neurons is dependent on ongoing electrical activity, and that such neurons are fully capable of retracting their processes or establishing new synapses, the extent to which these processes occur in the absence of experimental manipulations is not easy to assess. Nevertheless a variety of experiments have provided morphological evidence that such remodeling occurs normally, both at the neuromuscular junction and within the nervous system (e.g., Fig. 14-1). Some of these studies have examined changes in the structure of synapses as a function of the experiences to which an animal has been exposed. In addition, changes in the size of cell bodies, in the pattern of axonal branches, and in the number of synaptic contacts and of dendritic spines can be detected clearly in aged animals. The functional consequences of this substantial remodeling of neuronal structure in adult animals have yet to be discovered.

Sites of synapse formation are marked by proteins in the basal lamina. The process by which a neuromuscular synapse is established by a *regenerating* axon may differ somewhat from synapse formation by the first axons of embryonic motor neurons. In particular, the first synapses of the embryonic axons may form at random locations on the muscle. Regenerating axons, on the other hand, form synapses specifically at those sites where a synapse had previously existed. It appears that sites of former synaptic contact are marked by specific molecules that induce the growth cone to form a presynaptic structure. Interestingly, these molecules are not located on the surface of the muscle membrane. Instead they are in the *basal lamina*, the extracellular layer of proteins that is secreted by the muscle cells and surrounds the muscle (Fig. 15-10).

The existence in the basal lamina of molecules that mark the sites of former synaptic contact can be demonstrated by severing both the motor axons *and* the muscle fibers, causing the latter to degenerate (Fig. 15-10). In this condition, only the basal lamina remains. The sites of former synaptic contact can be identified both morphologically and by the presence of the enzyme acetylcholinesterase. It is found that when the regenerating motor axons reach the basal lamina they form apparently normal synaptic terminals specifically at these sites. Thus it appears that some time after an initial synapse is established, muscle cells secrete a marker protein into the basal lamina at the synaptic cleft. Antibodies have now been generated against various proteins of the basal lamina and one protein that is selectively localized to the synaptic part of the basal lamina has been identified. This protein, termed *s-laminin* (for synaptic laminin), turns out to be a homolog of the laminin protein that we discussed as a normal component of the extracellular matrix in Chapter 14. This, or a similar protein, may help persuade a growth cone to form into a synapse. Although synapses in the central nervous system do not possess a morphologically distinct basal lamina, it is possible that marking of points of synaptic contact by molecules derived from the postsynaptic neuron does occur, and plays a role in the restructuring of connections of such neurons.

Loss of presynaptic terminals after axotomy or interruption of axoplasmic transport. We have seen that cutting an axon, or interrupting transport in an axon, produces a restructuring of the synaptic contacts made *by* that axon and others that innervate the same target. These same procedures can also produce changes in the presynaptic contact *onto* the cell whose axon has been cut. For example, if the axons of neurons of the sympathetic ganglion are severed or exposed to colchicine, the synapses that they receive from other neurons retract (Fig. 15-11). As expected, this is accompanied by a substantial reduction in the synaptic potentials evoked in ganglionic

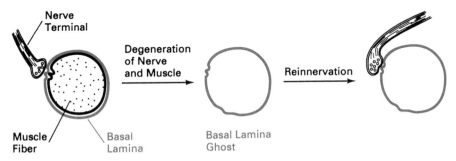

FIGURE 15-10. Basal lamina marks sites of synapse formation at the neuromuscular junction. This experiment, by Marshall et al. (1977), demonstrated that a regenerating motor axon specifically forms synapses at sites on basal lamina that had previously been occupied by a synapse.

neurons by stimulation of the preganglionic nerve. When transport is restored or the axons of the sympathetic ganglion cells are allowed to regrow and form new contacts, the branches of the presynaptic axons also extend to reestablish their full complement of synapses.

Such a loss of presynaptic terminals following interruption of an axon has been observed in many different pathways. It appears, therefore, that the normal maintenance of synaptic endings may depend on factors available only from an intact postsynaptic cell. In the majority of cases the putative factor and its mode of action are unknown. For sympathetic neurons, however, NGF may play an important role in this phenomenon. For example, application of NGF to the sympathetic ganglion following axotomy prevents the loss of presynaptic terminals illustrated in Figure 15-11. In addition, treatment of animals with an antiserum to NGF, which would be expected to bind to endogenous NGF and thereby prevent its uptake by cells, induces loss of synapses on sympathetic neurons even though their axons remain intact.

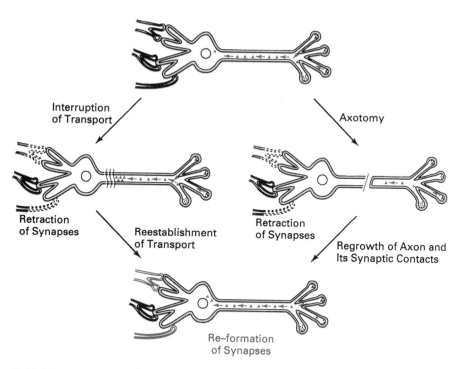

FIGURE 15-11. Loss of presynaptic terminals after interruption of axoplasmic transport or axotomy. Maintenance of presynaptic inputs may depend on a postsynaptic factor that is transported from the terminals back toward the soma.

Changes in Cell Properties Following Synapse Formation

The formation of a synaptic contact is usually followed by a sequence of changes in the properties of the postsynaptic cell. Receptors become reorganized in the postsynaptic membrane, and the types of proteins synthesized by the postsynaptic cell may alter dramatically. Some of these effects occur because the incoming axons induce new patterns of electrical activity in the postsynaptic cell. Other changes are the result of factors that do not depend directly on stimulation of the new input.

Receptor reorganization at the neuromuscular junction. Again, the neuromuscular junction of vertebrates has provided a classic preparation for the study of many of these effects. One of the first events observed following the arrival of the growth cone of a motor neuron at the muscle is the *clustering* of acetylcholine receptors under the newly formed presynaptic terminals. Even before the arrival of the motor neuron fibers, the immature muscle cells have an abundant concentration of acetylcholine receptors in the plasma membrane. These are distributed relatively uniformly over the surface of the cell, although some *hot spots* of clustered receptors can also be detected. When the nerve arrives, however, it induces the appearance of new clusters of receptors under the newly formed presynaptic terminals (Fig. 15-12). These clusters appear to arise both because newly synthesized receptors are inserted preferentially into the membrane under the terminal, and also because preexisting receptors may be induced to aggregate at these sites. The biochemical signals that induce the clusters are the subject of much current investigation.

The receptors under the terminals are termed *junctional* receptors while those in the uninnervated parts of the membrane are termed *extrajunctional.* The extrajunctional receptors are synthesized and degraded at a higher rate than the receptors under the synaptic cleft, and in time the extrajunctional receptors disappear altogether. The formation of junctional clusters appears not to depend on the electrical activity of the presynaptic nerve or of the muscle itself. Not only does the physical location of receptors and enzymes change during the maturation of the nerve–muscle contact, but the nature of the acetylcholine receptors themselves alters at this time. Patch clamp recordings have shown that the properties of the embryonic receptor channel are different from those of the adult. As we saw in Chapter 9, the fetal muscle receptor is made up of α-, β-, γ-, and δ-subunits. As the neuromuscular junction matures the synthesis of the γ-subunit ceases and synthesis of a new ϵ-subunit begins. The change in channel properties can be attributed to this switch from the γ- to the ϵ-subunit.

a 15 Day

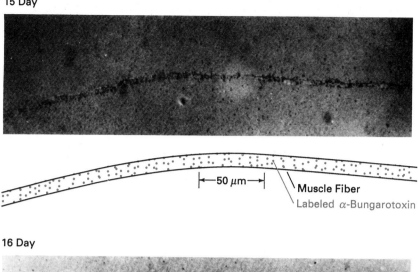

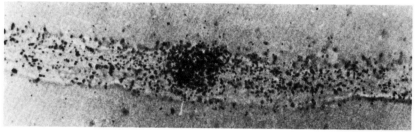

←—50 μm—→ \Muscle Fiber

Labeled α-Bungarotoxin

16 Day

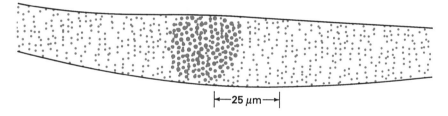

←25 μm→

b

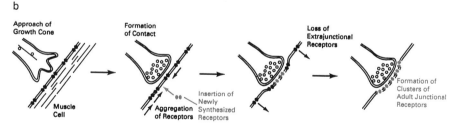

Approach of
Growth Cone

Formation
of Contact

Loss of
Extrajunctional
Receptors

Muscle
Cell

Aggregation
of Receptors

Insertion of
Newly
Synthesized
Receptors

Formation of
Clusters of
Adult Junctional
Receptors

FIGURE 15-12. Receptor clustering. *a*: Autoradiographs of neuromuscular junctions from 15- and 16-day rat embryos were made by Bevan and Steinbach (1977). The junctions were incubated with α-bungarotoxin, a ligand that binds to acetylcholine receptors. At the 16-day stage clustering is apparent. *b*: Scheme showing the onset of receptor clustering and the insertion of new species of receptors following innervation.

Effects of electrical activity on the properties of a neuron or muscle cell.
We have already encountered the fact that electrical activity within a neuron alters its response to growth factors and may, of itself, produce long-term changes in the properties of a neuron. In Chapter 13, for example, we saw that the action of cholinergic factor on neurons of the autonomic nervous system is blocked by electrical activity. The reader will not be surprised that it is again the neuromuscular junction that has provided a wealth of information on the role of electrical activity in shaping and maintaining the mature synaptic junction. The mechanisms by which electrical activity influences the properties of muscle are not well understood. In some cases, stimulated release of neurotransmitter may induce prolonged effects on protein synthesis and structural features of the muscle cell. In other cases, electrical activity in the postsynaptic cell alone, without release of neurotransmitter, is all that is required to influence the properties of a cell. In such cases, the effects of electrical activity may perhaps result from changes in intracellular calcium and the activation of calcium-sensitive enzymes.

When the activity of the presynaptic fibers is eliminated, either by denervation or by pharmacological block of the input to the muscle cells, a spectrum of changes is observed in the muscle. For example, there is a change in the properties of the voltage-dependent sodium channels, which revert to a form that is insensitive to the blocking agent tetrodotoxin. Such tetrodotoxin-insensitive sodium channels are normally found only early in development. In addition, although as described above receptor *clustering* does not depend on electrical activity, a block of activity does produce an increase in the *number* of extrajunctional acetylcholine receptors, leading to *supersensitivity* of the muscle to acetylcholine. Direct electrical stimulation of the muscle itself can largely prevent these changes, indicating that continued electrical activity in the muscle is required for its normal characteristics. Another example is provided by acetylcholinesterase, the enzyme that terminates the actions of acetylcholine. Like the acetylcholine receptor, this enzyme also undergoes clustering after the formation of synaptic contacts. Normal aggregation of the enzyme at the synapse appears to require ongoing electrical activity in the muscle, and is compromised by treatments that paralyze the muscle. Conversely, direct electrical stimulation of muscles promotes aggregation of the enzyme.

Fast and slow muscle fibers. A dramatic example of the way the pattern of neuronal activity can influence a postsynaptic target comes from a finding by Sir John Eccles and his colleagues. Mammals have two forms of skeletal muscle, fast and slow. The fibers that make up such *fast twitch* and *slow twitch* muscles differ in a number of characteristic ways. For example, the fast and slow fibers possess different forms of the muscle contractile protein myosin. Furthermore, the fast muscle fibers, which are pale in color, are used to produce rapid voluntary phasic movements. In contrast slow muscle fibers, which are rich in the protein myoglobin and hence are reddish in color, are used in maintaining a fixed posture. The neurons that innervate these two different types of muscle fiber also have very different electrophysiological properties. The motor neurons innervating fast muscle typically fire intermittent bursts of rapid trains of action potentials. These action potentials are followed by only small afterhyperpolarizations, allowing these cells to fire at frequencies as high as 30–60 spikes/sec. In contrast, the neurons that innervate the slow muscle fibers appear to have a slightly different set of ion channels that controls their pattern of firing. These cells have a larger afterhyperpolarization following an action potential and fire at a slower, sustained rate of only about 10–20 spikes/sec (Fig. 15-13a).

The Eccles group found that if a nerve containing the axons of fast motor neurons is forced to innervate a slow muscle, the muscle changes its properties to those of a fast muscle. Conversely, the innervation of a fast muscle by slow motor neurons causes the muscle to take on the properties of slow muscle (Fig. 15-13b). Subsequent work has confirmed that it is the change in pattern of electrical activity that is responsible for the shift in both the biochemical and contractile properties of the muscles. As shown in Figure 15-13c, for example, when the nerve to a fast muscle is forced to fire continually at a slow rate of about 10 action potentials per second for a period of several weeks, the muscle takes on the characteristics of a slow muscle. But the pattern of incoming electrical activity may not be the sole determinant of the response characteristics of a muscle. For example, very early in development some myotubes, the precursors of mature muscle fibers, differentiate as fast or slow fibers in the complete absence of neural input. Moreover, the role of patterned activity in the conversion of slow to fast muscles is less clear than that for the fast to slow conversion.

Experiments have also been carried out on the effects of stimulation on neurons in parts of the nervous system other than the neuromuscular junction. One example has been provided by neurons of the sympathetic ganglion. Electrical activity in these neurons is essential for maintaining levels of tyrosine hydroxylase, the rate-limiting enzyme in the synthesis of norepinephrine (see Chapter 8). Agents that block electrical activity cause levels of messenger RNA coding for this enzyme to fall, whereas stimulation of activity produces a prolonged increase in levels of this messenger RNA. Stimulation of the sympathetic ganglion also influences the levels of a neuropeptide, substance P. In this case, the effects are exactly the opposite of

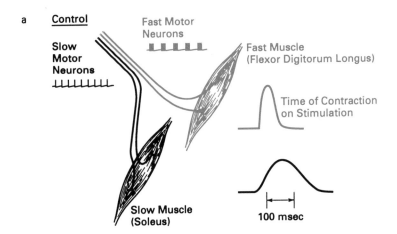

a   <u>Control</u>

Fast Motor Neurons

Slow Motor Neurons

Fast Muscle (Flexor Digitorum Longus)

Time of Contraction on Stimulation

Slow Muscle (Soleus)

100 msec

b   <u>Cross-Innervated</u>

Fast Motor Neurons

Slow Motor Neurons

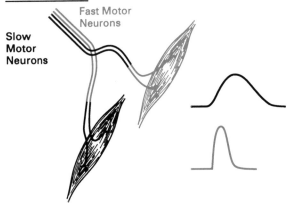

c   <u>Stimulated</u>

Slow Firing Pattern

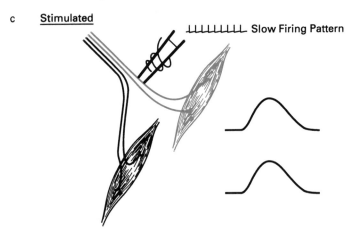

FIGURE 15-13. Properties of muscle depend on pattern of electrical input. Buller et al. (1960) showed that innervation of a fast muscle with a slow motor neuron (a,b) causes the muscle to take on the properties of a slow muscle. Imposition of a slow pattern of firing on the fast motor neuron (c) has the same effect (Salmons and Streter, 1976).

those on tyrosine hydroxylase. On blocking electrical activity, for example, by denervation of the ganglion, the levels of substance P rise, while stimulation prevents this rise.

Possible Roles of Proto-oncogenes in the Nervous System

How do changes in the electrical activity of neurons produce profound plastic effects such as those mentioned in the previous section? There must exist biochemical pathways that transduce the external signal, for example, calcium entry through calcium channels or the activation of a neurotransmitter receptor, into an intracellular message that alters the expression of specific genes in the nucleus of a neuron or muscle cell. Although some of these pathways may be related to the second messengers discussed in Chapter 10, it is becoming evident that *proto-oncogenes* are also important participants in these reactions.

An *oncogene* is an aberrant gene that causes uncontrolled growth in a population of cells. Some were discovered in viruses that infect cells and induce such abnormal growth. At first one might think that although oncogenes should be of great interest to cancer researchers, they may not be directly relevant to the study of neurons. However, it appears that most oncogenes code for proteins that have counterparts in normal cells, including neurons, and some of these normal gene products may have specific roles in neuronal growth and in the response of neurons to external stimulation. We have already seen that proteins closely related to growth factors play an important role in determining the characteristics of a mature neuron (see Chapter 13).

The term proto-oncogene is used to describe the normal cellular counterpart of the oncogene. Most of the proteins encoded by proto-oncogenes appear to be components of signaling pathways. One such proto-oncogene is called *erb-B*. It is now known that this is the gene that codes for the EGF receptor. The corresponding oncogene produces a truncated version of the receptor that, when it is expressed in cells, can produce uncontrolled growth in the absence of EGF. Some known proto-oncogenes, and the location of their protein products within a cell, are listed in Table 15-1. The properties and locations of the proteins clearly match those expected for components of biochemical pathways that could transduce external signals to changes in the nucleus of a cell. We have already encountered one of these proteins, *ras*, in Chapter 13 as a possible intracellular mediator of the effects of NGF on neurite outgrowth. We will now describe briefly some findings that suggest that the proto-oncogenes *src, fos,* and *jun* may also be important in determining the life-style of a neuron.

The src *proto-oncogene.* Like the EGF receptor and related receptors, the protein encoded by the *src* proto-oncogene is a tyrosine kinase. It differs from these receptors, however, in that it is not an integral membrane protein, and therefore is probably not a receptor itself. Because it is a phosphoprotein and its molecular weight is 60,000, the protein is sometimes termed $pp60^{c-src}$. This cellular *src* tyrosine kinase begins to be made within cells at the time that they start to differentiate into neurons. Moreover, the kinase persists at very high levels in fully mature neurons. Although a *src* tyrosine kinase is found in cells other than neurons, the levels in neurons are usually 6 to 20 times higher than in other cells. In addition, the *src* kinase in neurons differs from that in other cells in that it contains a stretch of six additional amino acids. Although this finding suggests that the neuronal-specific *src* protein kinase is likely to play a significant role in both developing and mature neurons, this role has yet to be discovered.

The fos *and* jun *proto-oncogenes.* Experience with molecular genetics has taught us that when external factors switch genes on and off, their action is coupled to the activity of a specific set of proteins in the nucleus. These nuclear proteins act as a cellular switch that controls the synthesis of new sets of proteins, thereby effecting a change in the properties of a cell. The proto-oncogenes *fos* and *jun* may participate in such initiation of the synthesis of new proteins in neurons in response to neurotransmitters or direct electrical stimulation.

The protein products encoded by the *fos* and *jun* proto-oncogenes belong to a family of closely related proteins. In many nonneuronal cells, exposure to growth factors that stimulate cell division causes a very rapid synthesis of some of these proteins (Fig. 15-14a), which then move to the cell nucleus. The *fos* and *jun* proteins each contains a helical region in which the amino

TABLE 15-1 Examples of Proto-oncogenes and Their Cellular Localization

Localization	Proto-oncogene	Function of Protein Encoded by Proto-oncogene
Secreted proteins	*sis*	Growth factor
Membrane proteins	*erb-B*	Receptor
	mas	Receptor
Cytoplasmic proteins	*ras*	GTP-binding protein
	src	Tyrosine kinase
Nuclear proteins	*jun*	Regulators of transcription
	myc	
	myb	
	fos	

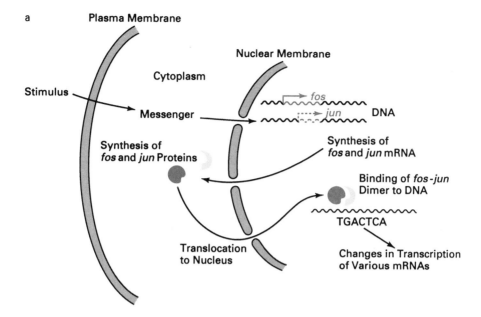

a

Plasma Membrane

Nuclear Membrane

Cytoplasm

Stimulus

Messenger

fos

jun

DNA

Synthesis of
fos and *jun* Proteins

Synthesis of
fos and *jun* mRNA

Binding of *fos-jun*
Dimer to DNA

TGACTCA

Translocation
to Nucleus

Changes in Transcription
of Various mRNAs

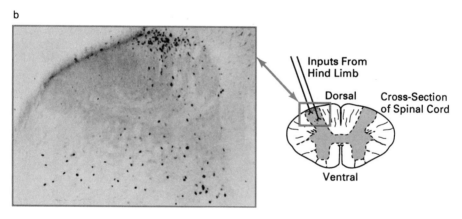

b

Inputs From
Hind Limb

Dorsal

Cross-Section
of Spinal Cord

Ventral

FIGURE 15-14. Proto-oncogenes *fos* and *jun*. *a*: Formation and nuclear translocation of the *fos* and *jun* proteins after stimulation of a cell (Morgan and Curran, 1989). *b*: Steven Hunt and his colleagues stained sections of spinal cord with antibodies to c-*fos*. This section came from an animal that had received noxious stimulation of a hind limb. Dark spots show the location of neuronal nuclei in which c-*fos* has been induced (Hunt et al., 1987).

acid leucine is found at every seventh position. This relatively common structural motif is known as a *leucine zipper*. It tends to be involved in protein–protein interactions, and its presence in the *fos* and *jun* proteins allows them to form a dimer. The *fos–jun* dimer then binds to certain sequences on DNA, and thereby regulates the ability of nearby genes to be transcribed into messenger RNA. The *fos–jun* complex can therefore be thought of as a messenger linking growth factor receptors to the synthesis of specific proteins. A variety of diverse factors including NGF, EGF, insulin, and acetylcholine have been shown to induce *fos* expression as much as 100-fold in PC12 cells.

What role *fos* may play in normal neuronal development or in synaptic plasticity is not clear. One particularly interesting aspect of the *fos* gene is that it can be induced by electrical activity alone, apparently as a result of calcium entry into cells. Even in certain adult neurons, synthesis of the *fos* protein can be evoked experimentally by manipulations such as direct electrical stimulation, application of convulsants or excitatory amino acids, and noxious sensory stimulation (Fig. 15-14b). It has been suggested, therefore, that the *fos–jun* pathway may allow short periods of neuronal stimulation to produce very long-term changes in gene expression. Because there exist *families* of regulatory proteins such as *fos* and *jun*, each family member may be evoked by a different stimulus to a neuron, and may in turn evoke a unique program of responses in the nucleus. It is interesting to speculate that such pathways might be implicated in the synaptic plasticity that contributes to long-term behavioral plasticity of the kinds we shall discuss in Chapter 17.

Summary

When the developing axon reaches its appropriate postsynaptic target, it stops elongating. A series of characteristic morphological changes, culminating in synapse formation, then occurs. Among the cues used by a neuron in choosing its correct postsynaptic partner are chemical labels that help to match appropriate pre- and postsynaptic cells. Not all synapses that form during development persist in the adult animal. Certain synapses are selectively stabilized, whereas others are lost. The pattern of electrical activity in the presynaptic neuron is important in the choice of synapses to be stabilized. It also regulates the properties of the postsynaptic cell.

Synapses may also be broken and reformed continually in the adult animal. Some of the same mechanisms that govern synapse formation and stabilization during development contribute to this adult synaptic plasticity. This may be observed experimentally by severing presynaptic axons and allowing the synaptic terminals to degenerate. Neighboring undamaged axons come to occupy the vacated synaptic sites until the regenerating axon grows back to take them over again. Such connections may regenerate with sufficient specificity to allow recovery of appropriate synaptic function.

Molecular mechanisms involved in synapse formation and synaptic plasticity, during development and in the adult, are being vigorously pursued. Neurobiologists are still in the very early stages of putting together a jigsaw puzzle in which growth factors, proto-oncogenes, tyrosine protein kinases, neurotransmitters, and patterns of electrical activity are among the pieces. When the picture is complete, it should be possible to view the path by which an undifferentiated cell becomes a mature neuron, the involvement of electrical activity in determining the characteristics of the mature cell, and the role of various factors in the plastic properties of the adult neuron.

Neural networks
and behavior

As cells go, neurons are not loners. Every function of the nervous system, from regulation of autonomic activities such as heartbeat to the control of complex animal behaviors such as dating and mating, reflects the coordinated action of a *network* of interacting neurons. A major challenge of neurobiology is to understand the nature of the interactions and computations that neural networks carry out. In this chapter, we describe a number of simple neural networks whose biological role is known. These representative examples have been chosen to illustrate how specific cellular properties of different neurons in a network are essential to the function of the network as a whole.

Models of Neural Networks

It is evident that a network comprising many neurons may generate patterns of activity that could not have been predicted by the study of a single cell in isolation. These properties of a network that can be attributed to interactions between cells are referred to as its *emergent properties*. Over many years, attempts have been made to understand such emergent properties by analyzing simple mathematical or computer models of interacting units. Figure 16-1a illustrates a typical model network with a set of input units, some internal units and a set of output units. The strength of a "synaptic" connection between one neuron and another, for example, neuron i and neuron j in the figure, is set by a parameter a_{ij} known as the *synaptic weight*.

To make numerical calculation of the behavior of networks a practical proposition, it has usually been necessary to make highly simplified assumptions about the properties of single units in the network. For example, the earliest models assumed that a neuron has only two states, *on* when it fires an impulse and *off* when it is silent, much like elements in a digital computer. Many other models now assume a sigmoidal *input–output function* such as that shown in Figure 16-1b. In such cases the neuron fires as long as the balance of excitatory and inhibitory "synaptic" inputs from other cells in the network exceeds the threshold for firing. The firing frequency depends on the sum of the inputs at any given time.

The major lesson to be learned from such studies is that complex patterns of activity and relatively sophisticated computations can be carried out by networks of very simple units. For example, some networks can recognize patterns of inputs corresponding to letters of the alphabet. When stimuli in the shape of a letter "A" are applied to a two-dimensional array of input neurons, one set of output neurons will fire. A different set of output neurons fire when the shape "B" is applied. Other networks generate patterns of firing that very closely mimic pathological states such as epileptic seizures (Fig. 16-1c). Still other networks tackle mathematical problems. In all cases the responses of a network are in some way encoded in the pattern of activity of a set of output units. When the synaptic weights at individual connections are allowed to alter as a result of predetermined rules, the outputs of model networks may display features of learning.

The study of model networks is likely to have a profound influence in some engineering fields, for example, the design of computers, in years to come. Application to the understanding of real neurons and their interactions has, however, been slow. This will no doubt change as the advent of faster and more sophisticated computers accelerates the pace of theoretical modeling, and allows the integration of what we learn about the properties of real neurons into the appropriate models. We shall therefore now give an account of several *real* neural networks. Our goal is not to give an exhaustive review, but rather to provide selected examples of how the intrinsic properties of neurons and their synaptic interactions shape the behavior of a network.

Networks Generating Rhythmic Movements

Central pattern generators. Many animal behaviors, such as walking or swimming, require the rhythmic contraction of muscles. We have already seen that a single neuron is capable of generating rhythmic bursts in the absence of external stimulation. However, most rhythmic behaviors require that opposing groups of muscles be contracted and relaxed in a coordinated

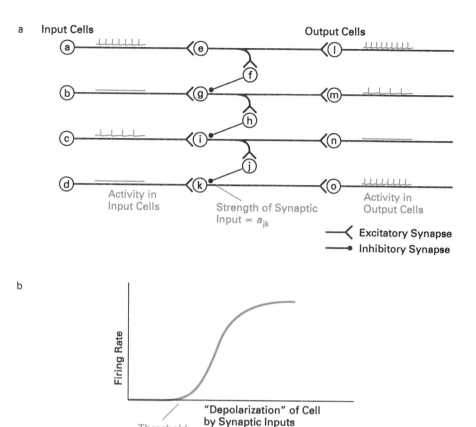

a

Input Cells   Output Cells

Activity in
Input Cells

Strength of Synaptic
Input $\propto a_{jk}$

Activity in
Output Cells

⟨― Excitatory Synapse
●― Inhibitory Synapse

b

Firing Rate

Threshold
for Firing

"Depolarization" of Cell
by Synaptic Inputs

c

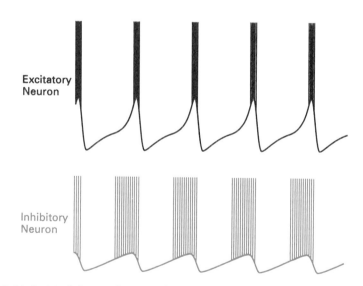

Excitatory
Neuron

Inhibitory
Neuron

FIGURE 16-1. Model neural networks. *a:* Model neural network with synaptic weights assigned to each connection. *b:* Sigmoidal input–output function in a model neuron. *c:* Pattern of firing in a model neural network generating epileptiform activity (modified from Kaczmarek, 1976).

manner. This coordination can be carried out only by a network in which different neurons innervate different muscles.

In theory, there are many ways to build a rhythmic network. For example, the generation and coordination of rhythmic movements could occur through a chain of reflexes in which receptors in the muscles signal the state of extension or contraction of each muscle to the remainder of the network, and this information would be essential for the network to function rhythmically. If this were the case, the muscles themselves would be integral components of the network. This, however, does not appear to be the case for most rhythmic networks that have been examined in detail. Rather, the pattern of outputs to different muscles is generated by a *central pattern generator,* a network of neurons that even in the absence of direct feedback from the muscles themselves is capable of generating the appropriate patterns of rhythmic excitation. (It is important to remember, however, that while sensory feedback is often not needed for the basic rhythmic movements, it *is* required to shape these movements to the needs of the animal in the real world.)

Rhythmic movements can be generated by networks with reciprocal inhibition. The very simplest circuit that can generate alternating contraction and relaxation in two different muscles consists of only two neurons. Each neuron makes an inhibitory synapse onto the other (Fig. 16-2a). For such a circuit to generate rhythmic output, it is not necessary that these neurons be endogenously active in the absence of other synaptic inputs. It is, however, necessary for the neurons to display *postinhibitory rebound.* This simply means that after the membrane potential of the cell has been hyperpolarized for a short period of time, the cell becomes more excitable than usual. When the membrane is then allowed to return toward its normal resting potential, one or more action potentials may result. This is a relatively common phenomenon in neurons, and, when it follows an experimentally applied hyperpolarizing current pulse as in Figure 16-2b, the action potential is often termed an *anode break spike.* In some cases, the explanation for postinhibitory rebound is that an inward current, such as a voltage-dependent sodium current or a T-type calcium current, is partly inactivated at the resting potential. Transient hyperpolarization, for example, by an inhibitory input, removes some of this inactivation so that the threshold for an action potential becomes more negative. As the cell depolarizes toward the resting potential, the increased inward current triggers an action potential before inactivation again develops.

Networks based on the simple two neuron circuit do indeed exist, and contribute to locomotion in some species. Figure 16-2c illustrates the activity of two neurons in *Clione,* a small marine mollusc. This animal swims

in the sea by moving a pair of wing-like structures (termed *parapodia*) that
are alternately flexed in a dorsal and ventral direction. A major component
of the central pattern generator for swimming appears to comprise four
swim interneurons (Fig. 16-2c). One upswing neuron and one downswing
neuron are found on each side of the nervous system. An action potential
in an upswing neuron generates an inhibitory postsynaptic potential (IPSP)

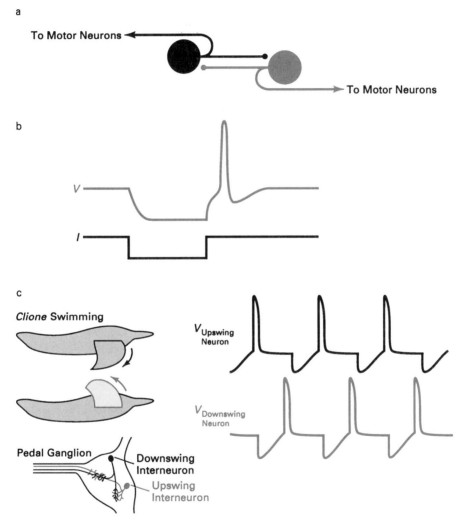

FIGURE 16-2. Reciprocal inhibition. *a*: Basic network that can generate alternating contraction and relaxation in two muscles. *b*: Anode break spike following a
hyperpolarizing current pulse. *c*: Interneuron activity during swimming in *Clione*
was studied by Satterlie (1985).

in a downswing neuron. Because of postinhibitory rebound, the downswing neuron generates an action potential at the end of the IPSP. This in turn triggers an IPSP in the upswing neuron. Thus a sustained ping-pong-like activity reverberates in this two neuron network, and is conveyed to the motor neurons that innervate the muscles in the parapodia, producing rhythmic swimming movements.

In theory, there is more than one way to design an oscillating network, even with only two neurons. For example, if individual neurons do not display postinhibitory rebound, but the firing of the individual neurons is subject to accommodation (see Fig. 2-11b), a circuit such as that in Figure 16-2a can produce alternating bursts of action potentials provided that a source of maintained excitation is provided to the cells. The important point here is that the specific electrical properties of the individual neurons are the major factors that determine (1) *whether* the network oscillates, (2) the exact *form* of the oscillations (single action potentials, bursts, or some other pattern), and (3) the exact *timing* of the rhythm. The latter is particularly important as it translates directly into the behavior of the animal.

Multineuron networks allow flexibility. Central pattern generators in most nervous systems are substantially more elaborate, and hence more versatile, than simply two groups of mutually inhibitory neurons. The more complex the network, however, the more difficult it is for experimentalists to unravel the factors that make the system work. Table 16-1 lists some of the systems that have been used successfully to analyze the detailed neuronal interactions that control a variety of rhythmic behaviors. Many studies of this type have been carried out using invertebrates, primarily because of the ease with which the intrinsic properties of invertebrate neurons can be analyzed and related to animal behaviors. Here, using the example of the crustacean stomatogastric ganglion, we shall summarize briefly some of the lessons that have been learned.

TABLE 16-1 Examples of Oscillating Neural Networks That Have Been Analyzed

Species	Location of Oscillator	Behavior Controlled
Clione	Pedal ganglion	Swimming
Tritonia, sea slug	Cerebral and pleural ganglia	Swimming
Panuliris, spiny lobster		
Homarus, lobster	Stomatogastric ganglion	Rhythmic stomach movements
Cancer, crabs		
Lobsters and crabs	Cardiac ganglion	Rhythmic contraction of heart muscle
Hirudo, leech	Segmental ganglia	Timing of heartbeat, swimming
Ichthyomyzon, lamprey	Spinal cord	Swimming

Rhythmic Neuronal Activity in Crustaceans

Although despised by gourmets, the stomachs of spiny lobsters and of crabs have provided pleasure to many neurobiologists. The stomach of such crustaceans is divided into three parts, the cardiac sac, the gastric mill, and the pylorus (Fig. 16-3a). Food enters the stomach through the esophagus and is

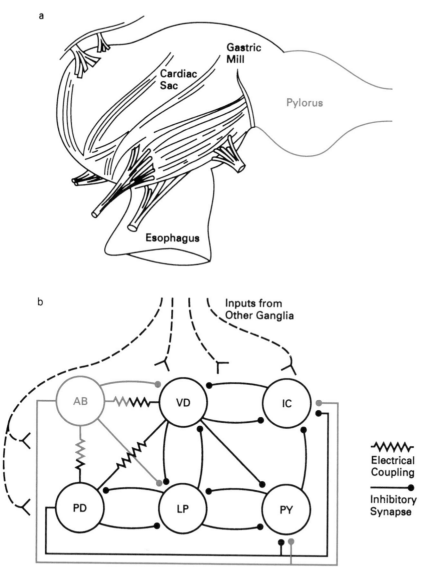

FIGURE 16-3. The lobster stomach. a: Diagram of the stomach (from Dickinson and Marder, 1989). b: Neuronal interactions in the pyloric network.

digested as it moves progressively through these three regions. Rhythmic contraction of muscles in all three regions contributes both to the physical disruption of the food and to movement through the stomach in a manner that resembles the chewing and swallowing of food by humans.

The stomatogastric ganglion. Muscles in the stomach are controlled by neurons in the *stomatogastric ganglion,* which contains 30 neurons. The three stomach regions are controlled by different sets of neurons. Initially, we shall consider only the central pattern generator for the pylorus, which consists of only 14 neurons. Synaptic connections between these are illustrated schematically in Figure 16-3b. The eight identical PY cells (lumped into a single neuron for simplicity in the figure), the two PD cells, and the VD, LP, and IC neurons are all motor neurons that directly innervate the pyloric muscles. Cell AB is an interneuron that makes connections only within the ganglion and sends information to the rest of the nervous system. Note that all the chemical synapses are inhibitory and that reciprocal inhibition between pairs of neurons is a dominant theme in the network. In addition some pairs of neurons are coupled by electrical synapses.

As shown in Figure 16-4, the rhythmic output of the pyloric neurons may be recorded both by intracellular microelectrodes in individual neurons and by extracellular electrodes placed on the nerves containing axons from motor neurons such as LP and IC to different sets of muscles. Three different phases of bursting are recorded in the different nerves.

The first important lesson about the rhythm generated by the pyloric circuit is that it depends on the presence of inputs from other parts of the nervous system. Traces such as those of Figure 16-4 are recorded only when the inputs from two other ganglia are intact. When the nerve from these ganglia to the stomatogastric ganglion is blocked, rhythmic bursting ceases. This is not because the incoming inputs generate any rhythmic activity themselves. Instead, the neurotransmitters used by inputs from these other ganglia appear to act as local hormones (see Chapter 8). When the inputs are activated and release their hormones, the pyloric neurons *acquire* the specific electrical properties needed to generate rhythmic bursts. The repetitive bursts of action potentials in the AB neuron resemble those of *Aplysia* neuron R15, which was discussed in Chapter 11. In contrast to neuron R15, however, this pyloric neuron requires the presence of modulatory substances from the other ganglia to generate these bursts. This neuron can therefore be termed a *conditional burster.*

The other neurons of the pyloric network are not endogenously bursting neurons. In response to a brief depolarization, however, they can generate a single long burst of action potentials. This burst occurs because of a sustained, regenerative depolarization that outlasts the brief stimulus. This regenerative depolarization is sometimes termed a *plateau potential* or *driver potential* (Fig. 16-5). Again, this plateau depolarization occurs only when inputs from other ganglia have been activated.

Despite the fact that reciprocal inhibition can generate rhythmic activity, this is not the major factor that shapes the pyloric rhythm. Rather, it is the intrinsic burstiness of individual neurons that drives the network. The syn-

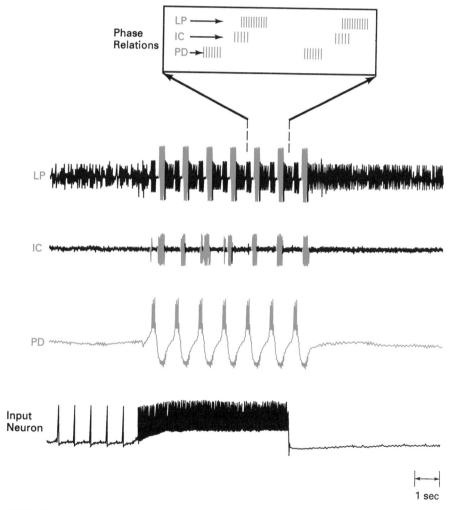

FIGURE 16-4. The pyloric rhythm. Recordings (made by Nusbaum and Marder, 1989) of rhythmic firing in three neurons (LP, IC, and PD) of the pyloric circuit during stimulation of a neuron with inputs to the ganglion. Because action potentials in the LP and IC neurons were detected with extracellular electrodes placed on nerves leading to the pyloric muscles, the activity of other neurons can also be seen in these recordings. The inset at the top shows phase relations among the three cells through two cycles.

aptic interactions and plateau potentials provide the appropriate phase relations and delays between the activities of different neurons. Only when the *intrinsic* electrical properties of the individual neurons in the network are considered is it possible to account for the timing and character of the bursts of action potentials that drive muscle contractions.

Modulatory neurotransmitters may "design" different networks. A period of dispassionate observation in a restaurant is sufficient to convince one that chewing and swallowing in humans are not simple processes, but can take many different dynamic patterns, depending on the nature of the food and the psychological state of the diner. So it is with lobsters. The rhythmic output of the pyloric circuit does not always follow the very stereotyped pattern described above. When recorded in intact animals the phase, timing, and amount of activity in different motor neurons can vary with time and with the pattern of behavior of the animal.

In addition to simply maintaining the rhythm, inputs from other ganglia serve to fashion different patterns of activity. As we shall see, this may occur because different transmitter substances act to effectively "rewire" the network into different configurations. A wealth of transmitters used by the inputs to the stomatogastric ganglion have been identified. These include acetylcholine, GABA, serotonin, dopamine, histamine, octopamine, and several neuropeptides including proctolin and a FMRFamide-like peptide. Because many of these act as local hormones, simple application of the transmitters to the external medium surrounding an isolated ganglion mimics the action of continuous firing in an input pathway. Here we shall compare the actions of two amines, serotonin and dopamine, when they are applied to the ganglion.

Figure 16-6a illustrates the effects of serotonin and dopamine on the

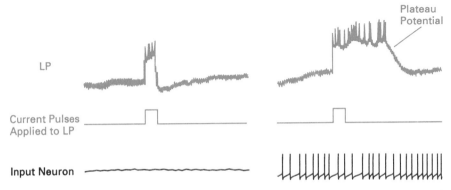

FIGURE 16-5. Induction of a plateau potential in the LP neuron by the activity of an input neuron (Moulins and Nagy, 1985).

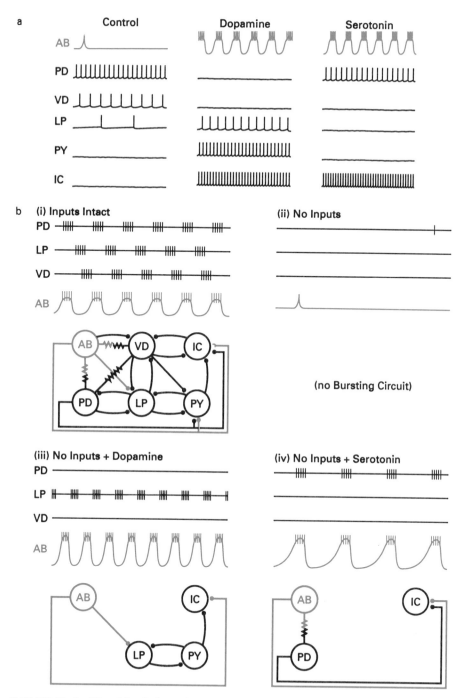

FIGURE 16-6. "Rewiring" the pyloric network. *a*: Actions of serotonin and dopamine on individual neurons in the pyloric circuit when synaptic connections with other cells have been eliminated. *b*: Actions of serotonin and dopamine on the intact pyloric circuit. Patterns of activity, together with the effective circuit diagram, are shown for four conditions: (i) "normal" rhythm with external inputs intact, (ii) no external inputs, (iii) no external inputs but with dopamine added to the ganglion, and (iv) no external inputs but with serotonin added (Harris-Warrick and Flamm, 1986).

properties of individual pyloric neurons when they are isolated from their synaptic inputs. Both of these agents induce endogenous bursting in inter-neuron AB. In contrast, PD motor neurons, which can fire at a low rate (but not burst) in the absence of synaptic input, are strongly inhibited by dopamine but not by serotonin. The VD neuron is strongly inhibited by both agents. LP and PY neurons are excited by dopamine but not by sero-tonin. Finally, neuron IC is excited by both agents. This inhibition or exci-tation of different neurons in the full network leads to a functional reor-ganization of the circuit. When either serotonin or dopamine is added to a stomatogastric ganglion that has been isolated from the neural inputs that would normally allow it to burst, they are able to reinstate the rhythm (Fig. 16-6b). The different amines, however, generate different rhythms. In fact, the two rhythms give the impression of being generated by very *different networks,* which, in a sense, they are.

This can be understood by a closer examination of the circuit. Neurons that are strongly inhibited, or simply not excited by serotonin, such as the VD, LP, and PY cells, are removed from the active circuit. Thus, in the presence of serotonin alone, the circuit is effectively driven by the endoge-nous activity of the AB–PD set of neurons (Fig. 16-6b). In the presence of dopamine, on the other hand, both the endogenous bursting of AB and reciprocal inhibition between the LP and PY cells shape the output of the circuit. The influence on the pattern and timing of impulses from motor neurons to muscles is therefore different for serotonin and for dopamine. Other neuroactive amines and peptides have also been found to induce characteristic configurations of the circuit that differ from those of sero-tonin and dopamine. Each of these, in turn, differs from those observed when the combined spectrum of modulatory inputs from neurons in the other ganglia is allowed to tinker with the active pyloric circuit. Thus, by altering the excitability of specific neurons, or the strength of individual connections, a wealth of different output patterns may be obtained.

Overlapping networks. The neurotransmitter-induced insertion and exci-sion of neurons into a circuit are not confined to a single network such as the pyloric. We have mentioned that the three different regions of the stom-ach are controlled by three different oscillating circuits in the stomatogas-tric ganglion. A neuron may participate in one or more of these circuits, depending on the state of inputs to the ganglion. For example, we have seen that neuron VD is an integral component of the pyloric network. It participates in the pyloric network, however, only when inputs from other ganglia allow this neuron to generate bursts driven by a plateau potential. These inputs are inhibited when certain sensory nerves are stimulated, and the VD neuron loses the ability to burst. When this happens, the change in its electrical properties allows it to become actively driven by neurons that generate the cardiac sac rhythm, a much slower oscillation than the pyloric.

These remarkable examples of the flexibility that can occur in networks of very small numbers of neurons emphasize the importance of the endogenous properties of a cell in determining the behavior of a network. Moreover, they illustrate that modulatory changes in the electrical properties of neurons can produce changes in the output of a circuit that, at first glance, might be thought to require a physical rearrangement of synaptic connections. The mechanisms of such modulation of neuronal electrical properties were covered in Chapter 11.

Command Systems of Neurons

Most animal behaviors do not persist day and night. Mechanisms must exist that allow activities such as walking, eating, swimming, and mating to be turned on and off. Furthermore, even relatively simple animal behaviors may require the coordinate activation or suppression of a number of apparently independent networks. These tasks are relegated to what are frequently termed *command systems* of neurons.

In some nervous systems, a single *command neuron* can exert control over relatively complex coordinated responses. The definition of a command neuron is that its activity should be both necessary and sufficient to trigger an entire coordinated behavior. For example, a flying cricket avoids high pitched ultrasound, similar to that emitted by a bat, by contracting a set of muscles that causes the animal to fly away from the direction of the sound. Stimulation of a single identified neuron is able to trigger this behavior in a flying cricket. Moreover, when the neuron is hyperpolarized, the animal fails to respond to the sounds. In most animals, however, important behavioral decisions are not likely to be entrusted to a single neuron. Rather, command systems of neurons weigh the pros and cons of a given course of action before committing the animal to a specific choice. As in the case of rhythmic networks, a thorough analysis of such systems would be out of place in this book. We shall, however, describe two systems of invertebrate neurons that preside over locomotor and reproductive behaviors, with an emphasis on the cellular properties of neurons in these command systems.

The swimming leech. The body of the medicinal leech *Hirudo medicinalis* is divided into segments (Fig. 16-7a). When it swims, it makes undulating motions with its body, in a manner generally similar to that of a fish or a snake. These movements result from the alternate contraction and relax-

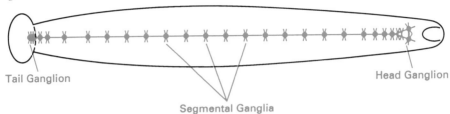

a

Tail Ganglion

Segmental Ganglia

Head Ganglion

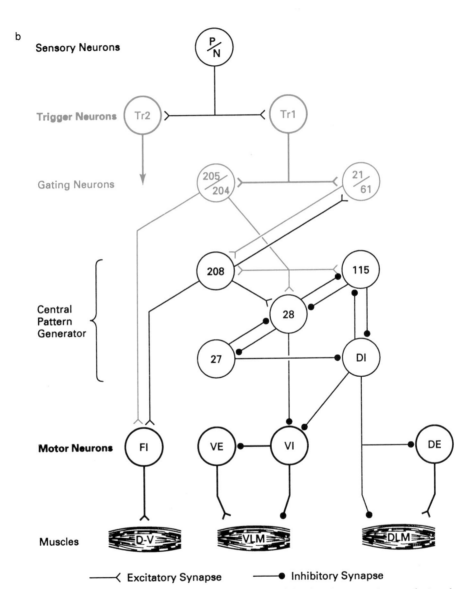

b

Sensory Neurons

P/N

Trigger Neurons   Tr2   Tr1

Gating Neurons   205/204   21/61

Central Pattern Generator

208   115

28

27   DI

Motor Neurons   FI   VE   VI   DE

Muscles   D-V   VLM   DLM

⊸< Excitatory Synapse   ⊸● Inhibitory Synapse

FIGURE 16-7. The leech. *a*: The nervous system of the leech comprises a chain of segmental ganglia. *b*: A simplified scheme of the network that controls swimming in the leech (modified from Friesen, 1989).

ation of muscles in the body wall of the animal. In addition to neurons that control the head and tail of the animal, the ventral nerve cord comprises 21 ganglia, each of which innervates the muscles in one segment of the body. During swimming, rhythmic bursts of action potentials in motor neurons in each ganglion are carefully timed so that during one undulation a wave of contraction travels from the front to the rear of the animal.

Figure 16-7b presents a simplified scheme of the network that controls swimming. Numbers and letters have been given to identified neurons in the circuit. As in the examples described earlier, swimming in the leech results from the rhythmic output of a central pattern generator. Neurons that comprise the central pattern generator are found in each segmental ganglion, and the activity in each segment must be coordinated with that in adjacent segments. In part because of this greater complexity, the role of the intrinsic properties of different neurons and the extent to which they may be modified is not yet understood nearly as well as in the stomatogastric ganglion. However, reciprocal inhibition certainly is involved in generating the rhythmic bursts. The output of central pattern generator neurons is conveyed directly to motor neurons that innervate three sets of muscles.

A bout of swimming can be evoked by a brief, strong mechanical stimulus administered to the body of the animal. There appear to be at least two levels of neurons that act on the information from sensory neurons in the body wall before the central pattern generator can be set into motion. These are the *trigger* neurons and the *gating* neurons. Stimulation of the sensory pathway excites the trigger neurons, and experimental stimulation of the trigger neurons alone is sufficient to cause all of the neuronal activity that produces swimming. The duration of the swim that is induced, however, substantially outlasts the brief period during which the trigger neurons are active. This is because the transient stimulation of the trigger neurons leads to a more prolonged and sustained period of firing in the gating neurons (Fig. 16-8). The role of the gating neurons is, in some ways, similar to that of the inputs to the stomatogastric ganglion described earlier. The rhythmic activity of the central pattern generator is sustained only as long as the gating neurons are active. The gating neurons do not, however, provide the rhythm itself.

There exist only a few trigger neurons, located in an anterior ganglion. They receive inputs from a large number of sensory neurons located along the length of the animal, and in turn project to a larger number of gating neurons, which are found in all of the segments. The trigger and gating neurons together may be considered to comprise the command system for swimming behavior.

A question that has yet to be answered is how the brief stimulation of trigger cells can lead to the sustained activity of the gating neurons. The answer is likely to lie in the actions of the neurotransmitters used in the synapses between the cells and in the membrane properties of the two types of cells. The cellular properties of these particular leech neurons have yet to yield their secrets, although it seems that long-term changes in membrane properties, of the kind discussed in Chapter 11, must occur. The beauty of leech swimming is that it provides a cellular explanation of how a transient behavioral stimulus can engage a more prolonged and relatively complex behavior, using a real network of neurons that can be identified at every level.

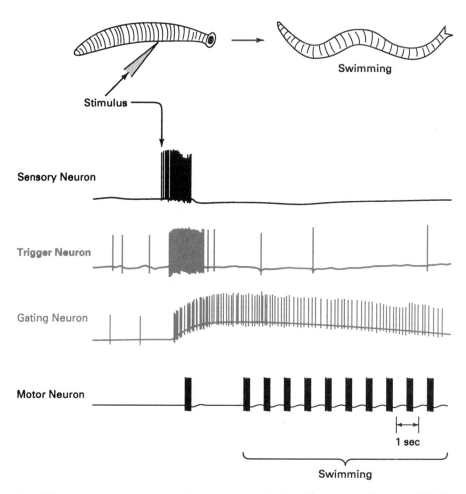

FIGURE 16-8. Trigger neurons. An experiment by Brodfuehrer and Friesen (1986) showing that transient activity in sensory neurons and in neuron Tr1 leads to a more prolonged and sustained period of firing in the gating neurons and motor output.

The bag cell neurons. We shall now turn to another command system of neurons, this time in the marine snail *Aplysia*. Two clusters of 200–400 cells each, located in the abdominal ganglion of this animal, control a sequence of very prolonged reproductive behaviors that lead to egg laying. Reproductive behaviors such as mating and egg laying are complex, even in *Aplysia*. For this reason the wiring diagram of the networks that control these behaviors is relatively poorly understood. In contrast, the cellular and molecular properties of the neurons in the command pathway for egg laying, termed the *bag cell neurons,* have been studied in substantial detail.

The bag cell neurons do not normally display any spontaneous electrical activity. In response to transient stimulation of an input from another ganglion, however, the cells depolarize and fire a long-lasting *discharge* of action potentials (Fig. 16-9a). Although the stimulus lasts only a few seconds, the evoked discharge usually persists for about 30 minutes. At the start of the discharge the neurons fire briskly for about 1 minute, after which they settle down to a slower period of firing, during which the action potentials become enhanced in height and width (we have already discussed the molecular mechanisms of these changes in action potential shape in Chapter 11). When a discharge occurs in an intact animal, it is followed by a stereotyped sequence of behaviors. If the animal is feeding, it abandons its food. It then seeks out a vertical substrate such as the side of a rock, and begins a characteristic sequence of head movements before depositing its eggs on the rock. These behaviors occur because of the action of neuropeptides released from the bag cell neurons during the discharge.

In Chapter 6, we discussed the structure of the precursor protein from which the neuroactive peptides are cleaved in the bag cell neurons (Fig. 6-6b). The major neuroactive peptide released by the bag cell neurons is *egg-laying hormone* (ELH), which, when injected into animals, induces egg laying and its associated behaviors. During a discharge, ELH is released locally onto other neurons in the abdominal ganglion. There it induces a change in the electrical properties of several identified neurons (Fig. 16-9a). ELH is also released directly into the blood from which it reaches peripheral targets and neurons in other ganglia, to influence the electrical properties of neurons in networks controlling activities such as feeding. In this way, ELH orchestrates changes in neural circuits that control different components of the evoked behaviors. In addition to ELH, several smaller peptides [*bag cell peptides* (BCPs)] are cleaved from the precursor protein (Fig. 6-6b). These also act as neurotransmitters, and alter the activity of other neurons in the abdominal ganglion. Moreover, the BCPs act at *autoreceptors* on the bag cell neurons themselves to further influence their excitability. Interestingly, during the production of neurotransmitter-filled secretory granules, the BCPs are not packaged into the same populations of granules as those containing ELH. It is possible, therefore, that the BCPs are released at different times or at different sites from those of ELH. The role of the BCPs in the behaviors is, however, not known.

At the end of the 30 minute discharge, it is not possible to stimulate another long-lasting discharge, although intense electrical stimulation can sometimes trigger short discharges. Recovery from this period of inhibition, sometimes termed the *refractory period* (not to be confused with the action potential refractory period described in Chapter 2), occurs gradually over about 18 hours. As the sequence of behaviors triggered by a discharge can last for several hours, the bag cell neurons are in the refractory state during these behaviors (Fig. 16-9b). Thus, the prolonged refractory period may pre-

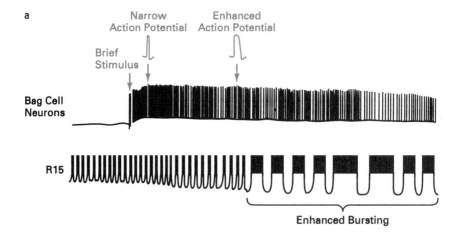

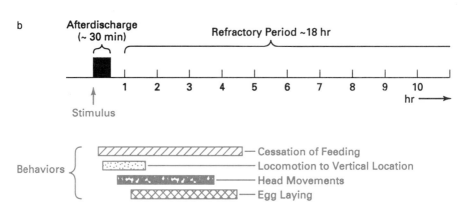

FIGURE 16-9. The bag cell neurons. *a*: Afterdischarge in bag cell neurons and the effects of ELH on neuron R15. *b*: Time scale of changes in excitability of the bag cell neurons (Conn and Kaczmarek, 1990).

vent the reinitiation of the behavioral sequence once it is underway, and also serve to limit the frequency with which the behaviors can be evoked. Thus, by undergoing a sequence of changes in their endogenous properties by the mechanisms discussed in Chapter 11, the bag cell neurons act as a sophisticated master switch for the sequence of behaviors leading to egg laying.

Hierarchies of behavior. An interesting concept, which has arisen from studies with another marine mollusc, *Pleurobranchaea*, is that behaviors are organized in a *hierarchical* manner. For example, it has been found that egg-laying behaviors will inhibit other behaviors including feeding, and that feeding takes precedence over mating (Fig. 16-10; many humans of course exhibit reversed priorities in their behavior). Because the wiring diagrams for many of these behaviors have been traced in *Pleurobranchaea*, the cellular correlates of these behavioral hierarchies can be examined. It is found that the command systems that are responsible for mating and egg laying can, when active, inhibit those networks responsible for feeding and locomotion. This example emphasizes the importance of considering the interactions among different neural networks in determining an animal's pattern of behavior.

Networks with Electrical Synapses

In the networks that we have described, many of the interactions among neurons occur through chemical synapses or local hormonal actions. Another important category that must not be overlooked is networks in which the cells interact via electrical synapses, mediated by gap junctions (see Chapter 6). In such networks, neurons are coupled electrically such that an action potential may propagate from one cell to the next with no chemical intermediary. In fact some of the cells discussed above, for example, cells AB and PD in the stomatogastric ganglion, are electrically coupled and therefore tend to fire together during bursts. Electrical synapses among individual bag cell neurons also act to preserve synchrony of firing in this population of neurons.

It is important to realize, however, that the role of electrical coupling may be more than simply to make two cells behave as one. Many groups of neurons that do not fire in synchrony have been found to be connected by electrical synapses, both in vertebrates and invertebrates. In some electrically coupled networks, action potentials can reverberate through the network, producing a sustained burst, which terminates when full synchrony is achieved. Such a burst has been found to underlie the activity of neurons in the command system for escape swimming in the sea slug *Tri-*

FIGURE 16-10. Behavioral hierarchies. This hierarchy in *Pleurobranchaea* was deduced by W. J. Davis (1979) and his colleagues.

tonia, for example. Other changes in the mode of operation of a network can occur when electrical synapses between neurons are subject to modulation by chemical synaptic inputs. For example, in the retina of vertebrates such as fishes and turtles, a class of neurons known as *horizontal cells* is coupled electrically. The neurotransmitter dopamine, acting through the second messenger cyclic AMP in these cells, produces a strong decrease in the strength of this coupling. This, in turn, causes a decrease in the area of the visual field to which an individual horizontal cell responds. Such modulatory effects of neurotransmitters on synaptic networks are likely to have profound effects on the way the network processes incoming information.

Summary

In previous chapters we have been discussing the properties of individual neurons, or pairs of neurons connected by a single synapse. However, complex interactions, involving both chemical and electrical synaptic connections among larger numbers of neurons, are required to generate most behaviors. Mathematical models, as well as studies in biological model systems, have provided insights concerning the organization of neurons into the neural networks that underlie particular behaviors. Some neurons can participate simultaneously in more than a single network, and the properties of a network may be modulated by the actions of neurotransmitters and hormones. In some cases the activation of a single command neuron or command system of neurons can trigger a complicated and long-lasting behavior. In the next chapter we shall consider some neuronal properties that may be involved in the phenomena of learning and memory.

Learning and memory

We saw in Chapter 16 that animals exhibit different kinds of behavior. There are, for example, *fixed-action patterns,* behaviors that always occur in a fixed and stereotyped manner once they are triggered. The trigger for a particular fixed-action pattern may arise from within the animal, as, for example, in the case of a chemical cue that appears at a certain time during development. Alternatively, fixed-action patterns may be triggered by a specific set of environmental conditions. It is thought that the neural circuitry underlying these invariant behaviors is more-or-less *hard wired,* that it is specified by the genome and is only to a limited extent subject to modulation. Many, but by no means all, behaviors in invertebrates and lower vertebrates tend to fall into this hard-wired category. We have seen, however, that modulation by neurotransmitters allows a considerable variety of outputs to be generated by a single hard-wired circuit.

This pattern changes as we move up the phylogenetic tree. Although many behaviors, particularly rhythmic ones such as breathing and locomotion, remain essentially hard wired in higher vertebrates including humans, many more examples of *adaptive behavior* begin to appear. In this chapter we will consider two closely related behavioral phenomena that are crucial for animals to survive, *learning* and *memory.* We may define learning in very broad terms as *a change in behavior as a result of experience,* and memory as *the ability to store and recall learned experiences.*

How the brain encodes, stores, and retrieves memories has fascinated not only scientists but the lay public for thousands of years. There is good reason for this wide interest in learning and memory. They are essential ingredients in defining an animal (or human being) as an individual. In addition, the complexity of the nervous system makes the understanding of such higher functions a challenging and exciting intellectual goal. Accordingly, many scientists from different disciplines have devoted their careers to the study of learning and memory, including the search for the *engram,* the physical memory trace in the brain; and technical and conceptual advances during the last few decades have produced substantial and exciting progress. An in-depth treatment of the various forms of learning and memory that have been defined and characterized by neuropsychologists is well beyond the scope of this book. We will focus on several simple kinds of behavioral phenomena, the mechanisms of which are beginning to be clarified with the techniques of cell and molecular biology.

Different Kinds of Learning and Memory

Before proceeding further we must define several different classes of behavioral modification exhibited by nervous systems. Historically, it has been useful to divide both learning and memory into two categories: *nonassociative* and *associative* learning and *short-term* and *long-term* memory (Fig. 17-1).

Nonassociative learning—habituation and sensitization. Habituation and sensitization are two simple forms of learning that involve a change in the intensity of response to a stimulus. Habituation can be defined simply as a *decrement* in the behavioral response during repeated presentations of the same stimulus. It is a form of learning that is observed in invertebrates, and in all vertebrate species including humans. An example might be the diligent student of cellular and molecular neurobiology who is trying to study when he/she is interrupted by some distracting noise, for example, a radio playing in the next room. Although initially the stimulus—the noise—interferes with the ability to concentrate, after many repetitions of the noise the nervous system stops responding, and the contemplation of ion channels can resume.

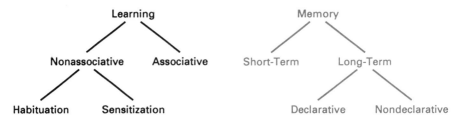

FIGURE 17-1. Different categories of learning and memory.

Sensitization is functionally the opposite of habituation, in that it can be defined as the *enhancement* of a reflex response by the introduction of a strong or noxious stimulus. It also differs from habituation in that the sensitizing stimulus is different from the stimulus that elicits the reflex. Furthermore there is no specificity—there is a general arousal of the nervous system, and all reflex pathways are strengthened. To carry our analogy further, if the diligent student is startled by the loud ringing of a telephone, he or she may subsequently be more distracted by the radio than when it was first turned on. Sensitization has important adaptive value, in that it allows a novel stimulus to alert an animal to possible predators and other potentially harmful stimuli in its environment.

Associative learning. Associative learning is more complex than habituation or sensitization, in that two stimuli must be closely associated in time for learning to occur. When a normally ineffective or neutral *conditioned* stimulus is paired temporally with a meaningful *unconditioned* stimulus, the animal learns to respond to the former as if it were the latter. The archetypal example of such *classical conditioning* is that of Pavlov's dogs, who learned to associate the ringing of a bell (the conditioned stimulus) with the presentation of food (the unconditioned stimulus), and would salivate in response to the bell alone. The unconditioned stimulus might be *reinforcing,* as in the example above, or *aversive,* for example, an electric shock, in which case the animal will take the conditioned stimulus as a cue and attempt to escape the aversive stimulus associated with it. In contrast with nonassociative learning, which does not involve any temporal relationship between stimuli, associative conditioning allows the animal to draw conclusions about causal relationships in its environment. There is another category of conditioning with different characteristics, called *operant conditioning,* which we will not consider here.

Short-term and long-term memories—different mechanisms. Another important concept, which has arisen from studies in invertebrates and a variety of vertebrates including humans, is that there are two temporally distinct forms of memory (Fig. 17-1). *Short-term memory* is the ability to acquire new information and retain it for periods of time ranging from a few seconds to some minutes. In contrast, *long-term memory* can involve retention for hours, days, years, or even a lifetime. Both associative and nonassociative learning can exhibit these two temporal components. An example of this distinction in real life involves looking up a new telephone number, and retaining it in short-term memory (often with rehearsal, for example, by repeating it continuously) long enough to dial it. A few minutes after dialing, the number will no longer be available in memory. It is only those few numbers that are dialed (and hence rehearsed) often that eventually find their way into long-term memory storage.

The finding that certain localized brain lesions disrupt long-term but not short-term memory suggests that different brain regions are involved in storage and retrieval in these two categories of memory. This will be discussed in more detail later in this chapter. Other experiments suggest that the molecular mechanisms that underlie short- and long-term memory are also different. For example, electroconvulsive shock selectively prevents the setting down of long-term memory traces. In addition, as we shall see below, inhibition of protein synthesis has no effect on the acquisition and short-term retrieval of a new behavior, but does inhibit the long-term memory.

Organization of Memory in the Brain: The Search for the Engram

To explore the biophysical and molecular mechanisms that contribute to learning and memory, it is necessary to identify the cells that participate in the memory trace. In fact the difficulty in locating the memory trace has been the greatest barrier to progress in understanding molecular mechanisms of learning and memory. The quest for the physical basis of the memory trace can be pursued at many levels of organization. The first step is localizing the engram to a particular organ. Although we take it for granted now that the brain is the right organ, there is evidence that the early Egyptians thought the heart and liver were the seat of human emotions and behavior. This conclusion was based on a simple experiment: if you remove the heart and liver from an animal, it stops behaving. By the time of Hippocrates a millennium later it was recognized by many that the heart and liver are necessary simply to keep the brain alive (bladder chauvinists may object to this simplification), and at the end of the nineteenth century Ramon y Cajal expressed the central role of the brain in lyrical and eloquent terms:

> To know the brain is the same thing as knowing the material course of thought and will, the same thing as discovering the intimate history of life in its perpetual duel with eternal forces, a history summarized and literally engraved in the defensive nervous coordination of the reflex, the instinct, and the association of ideas.

But localization of the memory trace to the brain was only the beginning. We shall see that narrowing things down further to a particular brain region, and to individual neurons within that region, has proven to be a much more formidable task for reasons that we are only now beginning to understand. However, there are several behavioral paradigms in vertebrate and invertebrate animals for which cellular correlates are now available, and we are learning much about the cellular, biophysical, and molecular mechanisms that may be responsible for some elementary forms of learning in these neural circuits.

Where in the brain is the memory trace? The great neuropsychologist Karl Lashley spent some three decades during the first half of this century trying to locate the engram in rodents. The basic approach was to train animals in a particular task, then to make lesions of the nervous system and ask whether the animals could still remember how to perform the task. The consistent result was that no single part of the brain was essential for long-term memory; impaired performance was proportional to the *extent* of the lesion but was not dependent on its *location.* In an influential 1950 paper entitled "In Search of the Engram," Lashley summarized his career-long search by concluding that there is no discrete memory trace, but that memories are distributed diffusely throughout the brain:

> This series of experiments has yielded much information about what and where the memory trace is *not.* It has discovered nothing directly of the real nature of the memory trace. I sometimes feel, in reviewing evidence on localization of the memory trace, that the necessary conclusion is that learning is just not possible. Nevertheless, in spite of such evidence against it, learning does sometimes occur.

We now have a good idea why Lashley's efforts were unsuccessful. The maze learning task that he used depends on many kinds of sensory information and cognitive functions that are processed and stored separately in different parts of the brain. Localized lesions may eliminate part of this information, but what remains is sufficient to allow the animals to perform the task, albeit less proficiently.

In thinking about the engram, it is important to distinguish among different kinds of long-term memory that have been elucidated (Fig. 17-1). Some of these distinctions have come from the study of human subjects who are amnesic as a result of accidental or surgical brain damage. *Nondeclarative* knowledge includes memory for skills and procedures—knowing *how*—for example, the rules of a game. Such knowledge depends on many different kinds of information that are processed and localized separately in different brain regions. It can be acquired even by severely amnesic human subjects. In contrast, *declarative* knowledge involves the memory of specific facts or events—knowing *that*—and new long-term memories of this class cannot be acquired by the amnesic patients. Some of the most striking evidence in support of these concepts comes from studies with amnesic patients who can remember new factual information for only a very short time, and cannot store new long-term declarative memories. However, when such patients are taught to read words in a mirror, they learn at a normal rate to carry out this complex task, and they retain the mirror-reading skill when retested months later. Interestingly the (nondeclarative) skill is retained although they do not remember the specific words themselves, or even the (declarative) fact that they have ever been trained to perform the task. When asked why they are able to perform so well they may reply that they are "just good at that sort of thing."

It is now evident from anatomical studies, either at autopsy or using *in vivo* imaging techniques, that many of these amnesic patients have suffered damage to the *limbic system*, a group of cortical structures that includes the *amygdala*, the *hippocampus*, and anatomically related structures (Fig. 17-2). When similar lesions are produced surgically in nonhuman primates, similar defects in declarative (but not in nondeclarative) memory tasks are observed. A wide variety of other animal studies have also implicated the hippocampus in certain kinds of learning and memory. It is believed that the long-term memory traces themselves are not stored there, but rather that the hippocampus participates in memory acquisition, and in establishing an enduring and retrievable memory elsewhere. Furthermore, the role of the hippocampus is limited to the acquisition and storage of declarative knowledge.

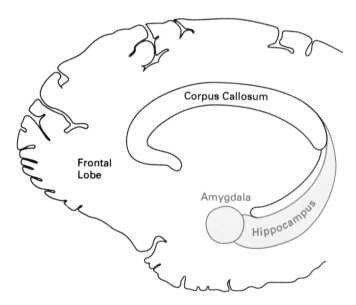

FIGURE 17-2. Structures in the brain important for long-term memory. Diagram after one drawn by Larry Squire depicting the medial aspect of the human brain to illustrate structures important in the setting down of long-term memories.

Other studies have suggested that the cerebellum is involved in certain kinds of learned behaviors. One of the most complete analyses is that carried out on a learned eyeblink response in the rabbit. Through lesioning studies it has been found that a portion of the cerebellum, ipsilateral (opposite) to the trained eye, is essential for the learning and retrieval of the trained response. This circuitry may also be important for the learning of other adaptive motor responses. The relevant pathways are becoming sufficiently well defined that cellular changes in the circuit, which may be associated with the learning and memory, can be explored. However, although this cerebellar circuitry is *essential* for the behavior, the precise location of the memory storage site itself has remained elusive.

Molecular Mechanisms of Memory

Even without detailed information about the localization of the memory trace, it is still possible to ask general questions about the molecular mechanisms that may be involved in learning and memory. Two distinct approaches to this question, the *interference* approach and the *correlation* approach, have been important historically. More recently, a powerful *genetic* approach using the fruit fly *Drosophila* has also been exploited. We will consider each of these approaches in turn.

The interference approach. The interference approach requires one to make a reasonable guess about the involvement of a particular molecular mechanism in learning and memory, and then use pharmacological agents that *interfere* with that mechanism to see whether acquisition and/or retrieval are affected. In the 1950s and 1960s a popular guess was that protein synthesis is required for long-term memory. As reasonably specific inhibitors of protein synthesis became available, investigators administered the inhibitors to animals subjected to several different training paradigms. Let us examine, as a typical example of this approach, learning in the goldfish.

Goldfish can be trained to avoid an electric shock by pairing the shock with light (Fig. 17-3a). In this classic example of a *conditioned avoidance response,* the fish learn to associate the light with the subsequent shock, and escape from the shock by swimming to the opposite side of the tank whenever they see the light. As shown in Figure 17-3b, the animals acquire the behavior rapidly, and retain it for many hours and even days after the training has ended. At various times before, during, or after training, the fish can be injected with *puromycin,* an inhibitor that blocks protein synthesis in their brains by more than 90%. If the puromycin is injected immediately prior to testing of the already trained animals, it is without effect, demonstrating that inhibition of protein synthesis does not alter *recall* or *performance* (Fig. 17-4a). On the other hand if the puromycin is injected immediately after training, the memory (as tested hours or days later) is inhibited completely (Fig. 17-4b). Injections at progressively later times within the first hour following training produce progressively less of an impairment. These results demonstrate that the laying down of the long-term memory trace requires protein synthesis during the first hour or so

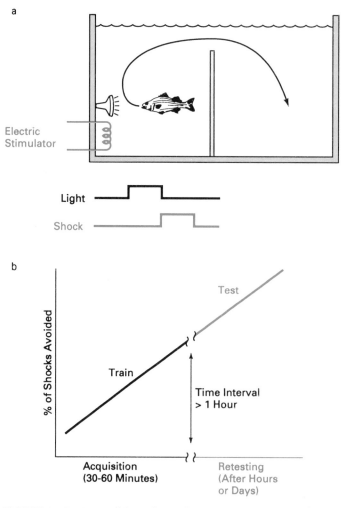

FIGURE 17-3. A conditioned avoidance response in goldfish. *a*: Bernard Agranoff found that when the turning on of a light is followed invariably by an electric shock, goldfish will swim over a barrier in response to the light, and hence avoid the shock. *b*: Acquisition and long-term memory of the light/shock association (see Agranoff et al., 1965).

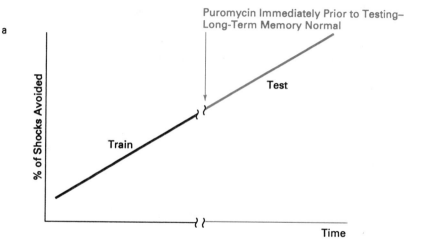

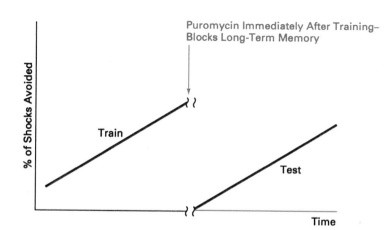

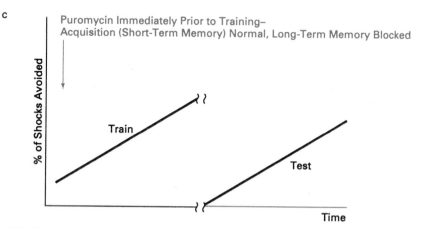

FIGURE 17-4. Protein synthesis is required for long-term memory. *a*: If puromycin is injected into the goldfish brain immediately prior to retesting, the performance of the animal is normal (compare with Fig. 17-3b). *b*: In contrast, if puromycin is injected immediately after the training period, there is no long-term memory. *c*: If puromycin is injected immediately prior to training, acquisition is normal but again there is no long-term memory (see Agranoff et al., 1965).

following training. Finally, if puromycin is injected immediately prior to training, the *acquisition* is completely normal, but again the long-term memory is blocked (Fig. 17-4c). Thus these experiments point to protein synthesis as being necessary for the establishment of long-term memory, but *not* for the short-term memory that is required during acquisition.

Although the precise timing of the short-term and long-term components of memory differs from one experimental system to another, these general conclusions hold for studies in mice, rats, birds, and other creatures in addition to fish. Indeed these experiments provide one of the most convincing pieces of evidence that short-term and long-term memory are distinct processes with different properties. In addition similar studies have been done with other protein synthesis inhibitors including *anisomycin* and *cycloheximide*. Although each of these pharmacological agents almost certainly will produce side effects in addition to inhibiting protein synthesis, it has been argued that these side effects will be different for the different inhibitors, and that it is most reasonable to attribute their actions on memory to the one effect they are known to have in common, namely disruption of protein synthesis.

The correlation approach. A complementary approach is to ask what molecular changes in the brain accompany learning and memory. Again this requires a guess about appropriate molecular mechanisms, and again protein synthesis has been a popular guess. This of course is very reasonable given the findings we have described above, and many investigators have measured patterns of protein (and RNA) synthesis in the brains of animals during, and at various times following, training. For the most part this approach has not been fruitful. One might expect the changes in protein synthesis that accompany the learning of a discrete behavioral task to be subtle, and accordingly difficult or perhaps impossible to measure against the large background of protein synthesis that occurs normally in the brain. Surprisingly enough, large changes in the incorporation of radioactive amino acids into proteins, as a function of training, are often observed in such experiments. It seems likely that these changes result from some epiphenomenon associated with the training, but do not provide any clues to the molecular basis of the memory trace itself.

On the other hand the central conclusion drawn from the interference approach, that protein synthesis is required for many kinds of long-term learning to occur, has not been seriously challenged. However we do not know the identities or roles of the proteins that are required. It is interesting that these 30-year-old approaches to the molecular underpinnings of learning and memory have enjoyed a revival in recent years. As we shall see below, there are several experimental systems in which the cellular correlates of certain behaviors have been identified, and both interference and correlation are being used to probe the details of the molecular mechanisms involved. Once again protein synthesis is one mechanism that has been implicated, but it is not the only one.

The genetic approach. The rationale for investigating the genetics of behavior is simple and straightforward: if one can generate mutants that exhibit aberrant patterns of behavior, an examination of the mutated gene might provide clues to the molecular mechanisms underlying the behavior. Again this approach is valid even if the engram has not been located. Such genetic studies of memory have focused on the fruit fly *Drosophila* because its genetics are better understood than those of any other multicellular organism. It is now well established that *Drosophila* can undergo both associative and nonassociative forms of learning, and a series of behavioral mutants of different types exist.

Fruit flies can learn to associate a particular odorant with an electric shock, and avoid the odorant in subsequent tests. The first behavioral mutant, *dunce,* was isolated when flies were treated with a chemical mutagen and their offspring screened in this behavioral assay. *Dunce* has normal sensory and motor capacities, but its learning is impaired. When the time course of memory decay is examined (Fig. 17-5), it is found that its short-term memory decays unusually rapidly, and there is little sign of long-term

FIGURE 17-5. Odorant-based shock avoidance in *Drosophila*. Decay of memory as a function of time after training, in normal flies and those carrying the *dunce* mutation. After a drawing by Tully (1987).

memory. However, when some low level of residual long-term memory is detected, it appears to decay at the same rate as in normal flies. This suggests that the primary effect of the *dunce* mutation is to severely attenuate short-term memory, and the loss of long-term memory may be a secondary effect.

What is the protein encoded by the *dunce* genetic locus? It is now accepted that the *dunce* gene codes for a form of the enzyme *cyclic AMP phosphodiesterase,* which contributes a large percentage of the cyclic AMP hydrolyzing activity in *Drosophila.* The levels of cyclic AMP in *dunce* are as much as sixfold higher than those in normal flies, suggesting that cyclic AMP may in some way be linked to memory formation. This suggestion is reinforced by the finding that another memory mutant, *rutabaga,* is also defective in cyclic AMP metabolism. Like *dunce, rutabaga* exhibits an abnormally rapid decay of short-term memory in various associative conditioning paradigms. However, in this mutant the phosphodiesterase activity is normal. Instead *rutabaga* exhibits a defect in a calcium/calmodulin-dependent subpopulation of the enzyme adenylate cyclase, which is responsible for the synthesis of cyclic AMP (see Chapter 10).

These findings contribute to the emerging picture that cyclic AMP plays an important role in learning and memory in the fruit fly. The picture, however, remains a little blurred. In *rutabaga* there is an apparent defect in cyclic AMP formation, whereas in *dunce* cyclic AMP levels are elevated, yet the learning impairments are similar in both mutants. Like the interference studies described above, the genetic approach has provided important clues about a molecular mechanism that contributes to learning and memory, but the final story is yet to be written.

Model Systems for the Cellular and Molecular Analysis of Learning and Memory

Why, one might ask, is the final story yet to be written? Why have these experimental approaches not yet provided a detailed account of the cellular and molecular mechanisms of learning and memory? The answer lies, at least in part, in the problem we have already discussed in some detail above: animals that exhibit interesting behaviors tend to have complicated nervous systems. One cannot study the memory trace without finding it first.

Since the time of Cajal and Sherrington it has been widely accepted that the synapse must be an important site of neuronal plasticity, and that plastic changes in synaptic efficacy might contribute to behavioral plasticity. It has been known for some 50 years that certain synapses in the peripheral nervous system can undergo changes in synaptic efficacy that can last for seconds or minutes (see Chapter 7). Such changes, in sympathetic ganglia and at the neuromuscular junction, have been thoroughly studied and are fairly well understood. But what does this have to do with learning and memory? One cannot claim that the neuromuscular junction or the sympathetic ganglion has the capacity to "behave." In this case the synapse that is modulated is accessible for study, but there is no behavioral correlate.

Thus it will be evident that workers interested in mechanisms of learning and memory are faced with a dilemma. The ideal model system would be a two neuron/one synapse organism with the behavioral repertoire of humans. Needless to say, this is not available. Accordingly many investigators have compromised on both sides of the issue, and have chosen model systems that exhibit reasonably sophisticated behavioral plasticity yet have reasonably accessible nervous systems. We shall now discuss one of these models, the gastropod mollusc, in some depth. Finally we shall consider an increasingly popular model for long-term learning in mammals, *long-term potentiation* in the hippocampus.

Model Systems I: The Gastropod Molluscs

The advantages of molluscan nervous systems for cellular neurobiology have already been emphasized in different contexts throughout this book. The gastropod molluscs not only provide accessible nervous systems with a relatively small number of large, identifiable neurons, they also exhibit a surprisingly varied repertoire of behaviors, including both nonassociative and associative learning. In Chapter 16 we saw how patterns of synaptic connections between identified neurons could explain the control of a variety of behaviors. Now we shall discuss the cellular and molecular analysis of short- and long-term memory in two gastropods, the marine snails *Aplysia* and *Hermissenda*.

Defensive withdrawal reflexes in Aplysia. *Aplysia* can withdraw from strong tactile stimuli, an effect that is analogous to reflex escape and withdrawal observed in vertebrates. The tail withdrawal reflex is a contraction of the tail musculature in response to a stimulus to the skin on the posterior portion of the animal. The gill and siphon withdrawal is a reflex evoked by touching the siphon or mantle shelf (these are external organs of the mantle cavity, a respiratory chamber that contains and protects the gill— Fig. 17-6). An analogous reflex response in humans would be the rapid withdrawal of the arm and hand from a hot stove. Both of the *Aplysia* withdrawal reflexes can be modified by experience, and they appear to be similar with respect to cellular and molecular mechanisms of the behavioral modifications. We focus here on the gill and siphon withdrawal.

The essential neuronal circuitry that underlies the gill and siphon withdrawal reflex has been more or less identified (Fig. 17-7). There are approximately 50 sensory neurons (SN) that have their sensory receptive fields in the skin of the siphon and mantle shelf. These sensory neurons make both monosynaptic and polysynaptic connections (the latter via interneurons, IN) with a group of motor neurons (MN). The latter in turn synapse directly onto the gill and siphon musculature, which contracts to produce the reflex withdrawal. The cell bodies of all of these neurons are within the *Aplysia* abdominal ganglion, but these identified central neurons do not tell

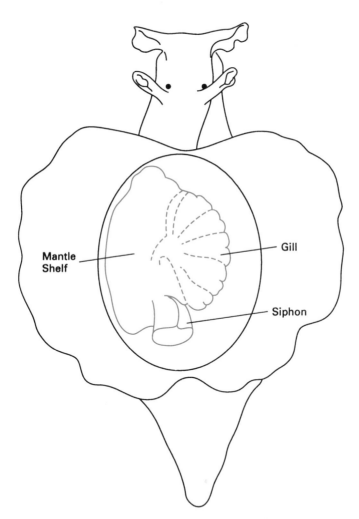

FIGURE 17-6. Drawing of an adult *Aplysia californica* to show the gill, siphon, and mantle shelf. A tactile stimulus to the mantle or siphon area results in contraction and withdrawal of the gill. After a drawing by Kandel (1979).

the entire story. There is also a peripheral circuit that participates in the gill and siphon withdrawal, but the extent of its contribution to the reflex and to the behavioral plasticity is not known.

Nonassociative plasticity in the gill withdrawal reflex—habituation and sensitization. This simple form of behavior can undergo both habituation and sensitization, as well as associative conditioning. When a weak tactile stimulus to the siphon is presented repeatedly, the reflex withdrawal is initially robust but becomes weaker with each subsequent stimulus (Fig. 17-8a). That is, the response habituates. An examination of the wiring diagram for the reflex (Fig. 17-7) reveals several potential loci for cellular plasticity that might account for this behavioral habituation. In principle the cellular change might lie in (1) the sensitivity of the sensory neurons to the stimulus, (2) the sensory or motor neuron firing patterns, spike amplitudes, or dura-

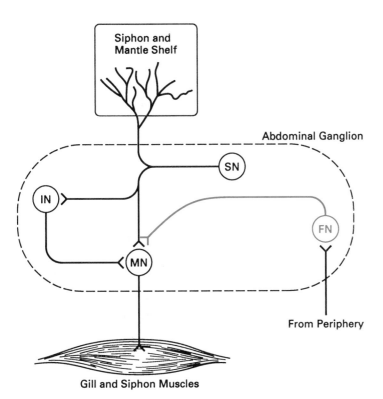

FIGURE 17-7. Wiring diagram for the gill withdrawal response. Sensory neurons (SN) synapse either directly, or indirectly via interneurons (IN), with the motor neurons (MN) that innervate the gill and siphon musculature. Facilitatory neurons (FN) receive synaptic input from the periphery and influence the SN to MN synapse.

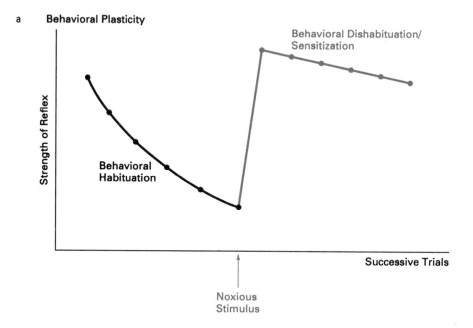

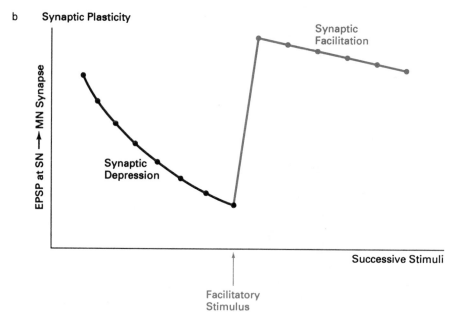

FIGURE 17-8. Modulation of the gill withdrawal reflex. *a*: The strength of the reflex can be measured by determining the amount of time the gill remains withdrawn following a stimulus. *b*: A change in the efficacy of the SN to MN synapse accompanies behavioral habituation and dishabituation/sensitization (see Kandel and Schwartz, 1982).

tions, or (3) the efficacy of the sensory-to-motor neuron or motor neuron-to-muscle synapses. By surveying the various components of the circuit in the behaving animal, Eric Kandel and his colleagues found that a decrement in synaptic transmission from the sensory to the motor neuron accompanies and might account for the habituation (Fig. 17-8b). This *synaptic depression,* the cellular correlate of behavioral habituation, appears to be *homosynaptic.* That is, it is a property of the sensory/motor pathway itself and does not require a contribution from any other neurons. In fact it can be evoked simply by repeated stimulation of the presynaptic sensory neuron. It is possible that there are also changes in the interneurons in the polysynaptic pathway, but this has not yet been investigated in detail.

When one delivers a strong noxious stimulus such as an electric shock to the animal's head or tail, a large increase in the gill and siphon withdrawal reflex occurs. Depending on the conditions of stimulation, this increase is termed either sensitization or *dishabituation.* As shown in Figure 17-8a, the reflex response is markedly enhanced, and this is accompanied by an increase in transmission at the sensory-to-motor neuron synapse (Fig. 17-8b). The *synaptic facilitation,* the cellular correlate of the enhanced reflex, is *heterosynaptic.* It results from the activation by the noxious stimulus of facilitatory neurons (FN) that synapse on the sensory neurons (Fig. 17-7), thereby altering the properties of the sensory-to-motor neuron synapse. Both depression and facilitation of the synapse can be observed in the isolated abdominal ganglion, the former by repeated stimulation of the sensory neuron with an intracellular electrode, and the latter by stimulation of a nerve trunk that contains the axons of the facilitatory neurons (Fig. 17-8b). With appropriately spaced stimuli, both the enhancement of the behavioral reflex and the heterosynaptic facilitation can be made to last for 24 hours or longer. In other words the animal exhibits long-term memory for this nonassociative behavioral modification, and the accompanying synaptic plasticity can also last for a very long time.

Associative plasticity in the gill withdrawal reflex—classical conditioning. The gill and siphon withdrawal reflex can be enhanced not only by nonassociative sensitization, but also by associative conditioning. The conditioning is carried out by pairing a mild tactile stimulus to the siphon, the conditioned stimulus, with a strong electric shock to the tail, the unconditioned stimulus. Prior to conditioning the conditioned stimulus elicits a weak withdrawal response, and the unconditioned stimulus elicits a powerful one. After the stimuli have been paired in time, the conditioned stimulus elicits a powerful response, and the conditioning can persist for days (Fig. 17-9a).

What are the cellular correlates of this associative behavioral plasticity? Again there is an increase in the efficacy of the sensory-to-motor neuron

a   **Behavioral Plasticity**

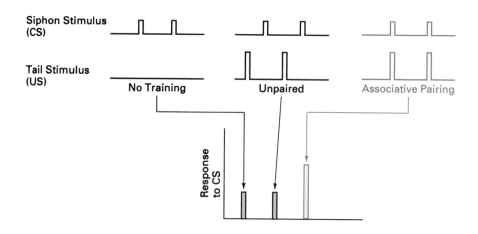

b   **Synaptic Plasticity**

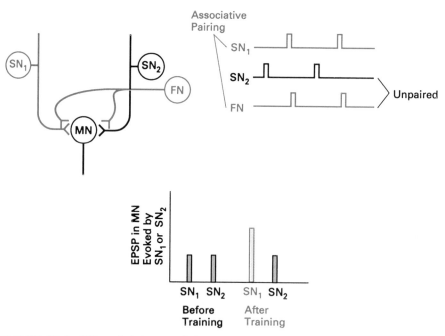

FIGURE 17-9. Gill and siphon withdrawal can be enhanced by associative conditioning. *a*: When a conditioned stimulus (CS) to the siphon is paired in time with an unconditioned stimulus (US) to the tail, the response to the conditioned stimulus is enhanced. *b*: An associative change in synaptic strength accompanies this behavioral response (see Kandel and Schwartz, 1982).

synapse, and again a similar associative change in synaptic efficacy can be elicited in the isolated nervous system (Fig. 17-9b). When action potentials in one sensory neuron (SN_1—the conditioned stimulus) are paired temporally with stimulation of nerves from the tail (FN—the unconditioned stimulus), after several trials there is a dramatic increase in the excitatory postsynaptic potential, in the motor neuron, that is evoked by the conditioned stimulus SN_1. However, the response to stimulation of *other* sensory neurons, which have *not* been paired with the unconditioned stimulus, remains unchanged (SN_2 in Fig. 17-9b). A very similar result is observed in neurons that participate in the tail withdrawal reflex.

A mechanism for synaptic plasticity—modulation of transmitter release. All of these plastic changes in synaptic strength involve a change in the amount of excitatory neurotransmitter released from the sensory neuron. Earlier classical work on mammalian and crustacean neuromuscular junctions, by Bernard Katz and Stephen Kuffler, respectively, had pointed to modulation of transmitter release as an important mechanism of synaptic plasticity. The technique of quantal analysis, pioneered by Katz (see Chapter 7), has been used to demonstrate that the synaptic depression that underlies short- and long-term behavioral habituation of the gill withdrawal reflex is accompanied by a *decrement* in transmitter release. Conversely, the synaptic facilitation that is responsible for short- and long-term sensitization results from an *enhancement* of transmitter release.

The increase in transmitter release from the sensory neurons during facilitation results in part from modulation of action potential duration by mechanisms discussed in Chapter 11. The transmitter released by the facilitatory neurons increases sensory neuron action potential duration, in part via a cyclic AMP-mediated decrease in the S potassium current (see Fig. 11-4). However, other ion currents and other second messenger systems may also be involved in modulating transmitter release from the sensory neurons.

Long-term facilitation requires protein synthesis. Short-term (lasting minutes) and long-term (lasting hours or days) facilitation of the sensory-to-motor neuron synapse share a common cellular mechanism—both involve an increase in the release of the sensory neuron neurotransmitter. Thus it might be expected that the details of the underlying molecular mechanisms would also be similar. However, the use of the interference approach, described above, reveals an important difference—the long-term facilitation of synaptic transmission, produced by application of a facilitatory transmitter, is blocked by inhibitors of RNA and protein synthesis, whereas the short-term facilitation is unaffected.

This finding is in agreement with the many earlier studies in vertebrates that suggested that there are different mechanisms underlying short-term and long-term memory, and that protein synthesis is essential for the latter but not the former. As in these previous studies, there is a critical period around the time of application of the facilitatory transmitter during which protein synthesis is required for long-term facilitation to occur. The proteins whose synthesis is required for the long-term facilitation have not yet been identified, but in this system they may be within reach.

Associative learning in Hermissenda. *Aplysia* and *Drosophila* are not the only invertebrates that exhibit learning and memory. Among the other creatures that learn, often associatively, and remember, often for a long time, are bees, locusts, and leeches, and several other gastropod molluscs including the garden slug *Limax* and the garden snail *Helix*. One other molluscan system in which the cellular and molecular correlates of long-term learning and memory have been explored in detail is the marine snail *Hermissenda*.

Hermissenda normally like to move toward light. They also normally decrease their velocity of movement when their vestibular organs are stimulated, for example, by rotation. When light (the conditioned stimulus) and rotation (the unconditioned stimulus) are paired temporally, the animals associate the two stimuli and, thereafter, move more slowly toward light. This learned behavior can be retained for days. The neuronal circuit that participates in the behavior exhibits changes that might be responsible for the long-term learning. An interesting and unusual feature of this system is that the behavior appears to be mediated by a change in the intrinsic *excitability* of one type of neuron in the circuit, the *B* photoreceptor, rather than by a change in synaptic efficacy alone (Fig. 17-10). The phosphorylation-mediated modulation of several different potassium channels has been implicated in this change in *B* cell excitability.

Model Systems II: Long-Term Potentiation

Can these kinds of cellular and molecular analyses be carried out in vertebrates? To recapitulate a point that we have already emphasized, the complexity of the vertebrate central nervous system and the lack of readily identifiable neurons present a formidable barrier to identifying the appropriate neurons to analyze. In only a very few experimental systems is the identification of the essential memory trace circuits at all within reach. Thus there has been an ongoing search for a system in vertebrate brain in which long-term changes in synaptic efficacy, evoked by experience, can be investigated. *Long-term potentiation* (LTP), a phenomenon that occurs at

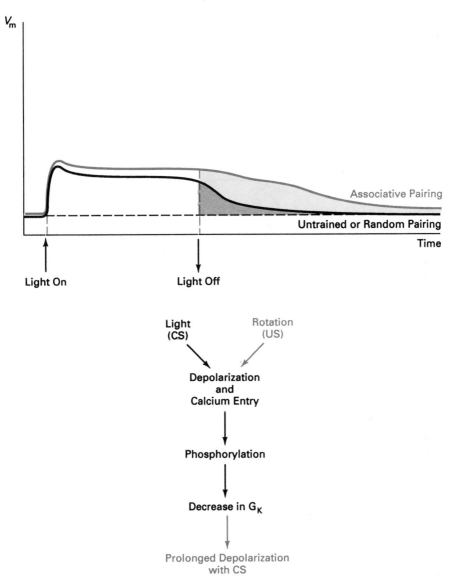

FIGURE 17-10. Associative learning in *Hermissenda*. The type B photoreceptor in the *Hermissenda* eye depolarizes in response to light. When light is paired in time with rotation of the animal, the depolarization in response to light is longer lasting. Some of the cellular and molecular changes that accompany this associative change have been elucidated *(bottom)*. Summarized by Crow (1988).

several different kinds of central and peripheral synapses but has been investigated most thoroughly in the hippocampus, is amenable to this kind of investigation. Although hippocampal LTP is not clearly associated with any known behavioral modification, it is a long-term increase in synaptic strength in a brain region known to be important for learning and memory.

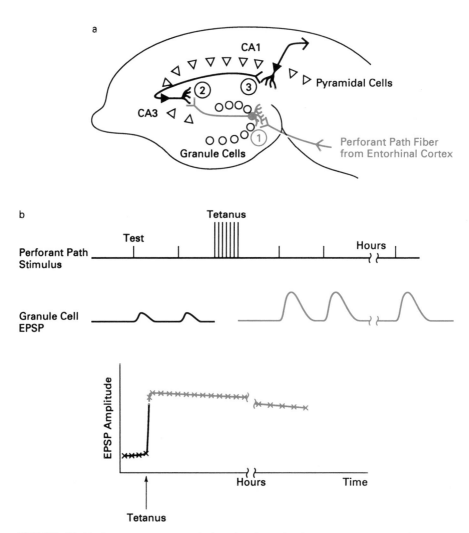

FIGURE 17-11. Long-term potentiation (LTP) in the hippocampus. *a:* Schematic drawing of a hippocampal slice. Fibers from the entorhinal cortex enter the hippocampus via the perforant path and synapse on dendrites of granule cell neurons (*1*). These in turn synapse on pyramidal cell neurons in the CA3 region of the hippocampus (*2*). The CA3 pyramidal cells synapse on other pyramidal cell neurons in the CA1 region (*3*). *b:* LTP of the perforant path to granule cell neuron synapse [excitatory postsynaptic potential (EPSP)] (modified from Nicoll et al., 1988). See also Bliss and Lomo, 1973).

In addition it has an associative component, and hence provides a model for the kinds of synaptic modulation that might be involved in long-term associative learning.

What is LTP? It is worth noting in passing that although the synaptic organization of the vertebrate brain may appear at first glance to be hopelessly complex, certain structures including the cerebellum and hippocampus actually exhibit an exquisitely precise organization. One cannot identify single cells as unique individuals as can be done in the gastropod molluscs. Nevertheless, particular *classes* of neurons and the synapses between them can indeed be recognized reliably. As shown in Figure 17-11a, there are three major synapses in the hippocampus at which LTP has been investigated. In 1973 it was shown in the rabbit that the strength of one of these synaptic connections, between the (presynaptic) *perforant fibers* and (postsynaptic) *granule cells*, can be markedly potentiated following a brief tetanic stimulus to the presynaptic axons (Fig. 17-11b). This potentiation, which can last as long as *several weeks* in intact animals, was subsequently demonstrated at the other two synapses as well.

The study of LTP has become much easier (and hence much more popular) in recent years because of the development of the *in vitro* hippocampal slice. The slice is a cross-section through the hippocampus, in which the pathways and synaptic organization depicted in Figure 17-11a remain intact. LTP can also be elicited in the slice and, although it cannot last for weeks because the slice dies within a matter of hours, it can persist as long as the slice does. In this preparation it is relatively easy to carry out intracellular recording (and even voltage clamp), and thus LTP can be examined in individual postsynaptic neurons, as well as in populations of postsynaptic neurons via an extracellular electrode. This and other technical advances have made it possible to investigate the mechanisms of temporally distinct components of LTP: *initiation, storage, and expression.*

Initiation of LTP. From the earliest experiments it was evident that the induction of LTP requires high-frequency stimulation of the presynaptic fibers at a stimulus intensity that is above a certain threshold. Both the high frequency and the strong stimulus (which activates a large number of presynaptic axons) are essential. Weak stimuli do not produce LTP even following tetanic stimulation, nor do strong stimuli if they are given at low frequency (Fig. 17-12a). It is known now that these conditions reflect the requirement for a large depolarization of the postsynaptic cell for LTP to be initiated. For example, LTP induction by a strong tetanic stimulus can be prevented by voltage clamping the postsynaptic cell to prevent the depolarization (Fig. 17-12b). However, although the postsynaptic depolarization is necessary, it is not in itself sufficient. For example, injection of depolarizing current via an intracellular electrode in the postsynaptic neuron can

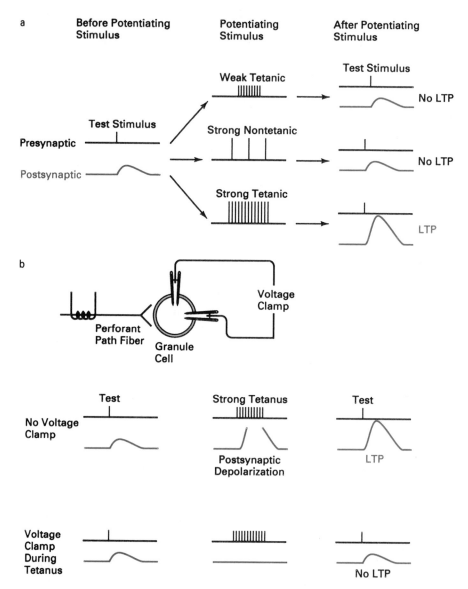

FIGURE 17-12. Presynaptic transmitter release together with postsynaptic depolarization are required for long-term potentiation. *a*: Neither a weak tetanic stimulus nor a strong nontetanic stimulus is capable of producing LTP. Only following a strong tetanic stimulus is enhancement of the EPSP amplitude seen. *b*: When the postsynaptic cell is voltage clamped to prevent depolarization during the tetanus, even a strong tetanic stimulus does not produce LTP. See Nicoll et al. (1988).

induce LTP, but only when it is paired with a weak synaptic input. Thus concurrent stimulation of the synaptic input, together with postsynaptic depolarization, is required for LTP induction.

The requirement for depolarization coupled with synaptic activation can be understood when we consider the neurotransmitter at the excitatory synapses that undergo LTP. There is substantial evidence that glutamate is the transmitter at these synapses, and this is not surprising when we recall (see Chapter 8) that glutamate is the major excitatory neurotransmitter in the mammalian brain. Remember also from Chapter 9 that glutamate binds to several different classes of postsynaptic receptor, and that the excitatory postsynaptic potential evoked by a weak stimulus is normally mediated by glutamate binding to the kainate/quisqualate class of receptor. Only the pronounced depolarization produced by a strong tetanic stimulus can relieve the magnesium block of the NMDA class of receptor, and allow calcium to flow through the NMDA receptor channel into the cell (see Fig. 9-7). This then explains why *both* depolarization and synaptic activation are required; neither alone can activate the NMDA receptor channel, and calcium entry though this channel is necessary for the initiation of LTP.

Associative LTP—Hebb's postulate. As we discussed in Chapter 15 in the context of synapse stabilization during development, Donald Hebb suggested that synaptic strength might be enhanced by concurrent activity in the pre- and postsynaptic neurons. He postulated further that this might provide a mechanism for associative learning. This idea has enjoyed a revival in recent years, in particular in the context of LTP. From the discussion above we can see that homosynaptic LTP fulfills the requirements of Hebb's postulate, in that simultaneous presynaptic activity (which results in glutamate release) and postsynaptic activity (in the form of membrane depolarization) are necessary for the long-lasting change in synaptic strength. There is also a heterosynaptic associative form of LTP, in which activity at one synapse can contribute to the generation of LTP at another. Consider the situation illustrated in Figure 17-13a, in which a weak and a strong input synapse on the same target neuron. We know from Figure 17-12a that stimulation of the weak input, even tetanic stimulation of this input, does not produce LTP, whereas tetanic stimulation of the strong input will generate homosynaptic LTP (Fig. 17-13b). Suppose now that the two inputs are tetanized simultaneously. The strong input can depolarize the postsynaptic cell, and this depolarization can spread to the site of the weak input. Since the latter is active and releasing glutamate at the time of the depolarization, both criteria necessary for producing LTP have been satisfied, and the result is LTP at the weak input (as well as at the strong—Fig. 17-13c). The requirements for temporal pairing of the two stimuli are identical to those required for associative learning paradigms.

Expression of LTP. These findings suggest that the induction of LTP is a postsynaptic phenomenon. What about storage and expression? Is there a long-lasting change in the release of neurotransmitter from the presynaptic cell, as has been observed in *Aplysia,* or a change in the sensitivity of the postsynaptic target, or perhaps both? Surprisingly very little is known about

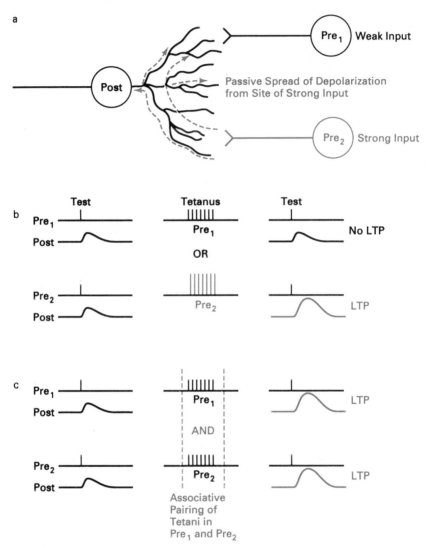

FIGURE 17-13. Associative long-term potentiation. *a:* Diagram depicting the spatial relationships between a strong and a weak presynaptic input. *b:* Stimulation of the strong input produces LTP, but stimulation of the weak input alone does not. *c:* When the strong and the weak inputs are paired, the depolarization produced by the strong input spreads to the site of the weak input, and contributes to the induction of LTP. See Nicoll et al. (1988).

this. There is some evidence for an increase in glutamate release, but it is not clear whether this is adequate to account for the potentiation of synaptic transmission, or whether the increase lasts as long as LTP does. In addition, such a presynaptic locus for maintenance would require that the postsynaptic cell in which LTP is initiated must communicate in some way with the presynaptic terminals, and it is not known how this might occur. Quantal analysis of transmitter release at synapses undergoing LTP might resolve this long-standing question, but unfortunately such an analysis is technically very difficult.

Molecular mechanisms of LTP—what does the calcium do? Simply knowing that calcium entry into the postsynaptic neuron is necessary for the induction of LTP does not provide us with a detailed molecular mechanism, since calcium does many things in cells. There is good evidence from pharmacological experiments that the actions of calcium during LTP initiation involve its interaction with calmodulin, but beyond that the picture is rather obscure. Protein kinase C translocation and activation accompany LTP, and inhibitor experiments suggest some role for protein kinase C, but the data are not yet definitive. It is intriguing that protein kinase C-mediated changes in the phosphorylation of the protein GAP-43 (see Chapter 14) have been reported to accompany LTP, suggesting that events similar to those occurring during neurite outgrowth contribute to this phenomenon. Structural changes in the postsynaptic cells have also been hypothesized to play a role in LTP, but here again no conclusive experiments are available.

One attractive hypothesis is that the multifunctional calcium/calmodulin-dependent protein kinase II is involved. This enzyme is highly concentrated in postsynaptic densities, so it is in the right location to contribute to modulation of synaptic efficacy. In addition, as shown in Figure 7–10, the enzyme can undergo autophosphorylation in the presence of calcium, after which time its activity becomes independent of calcium. Thus it provides a mechanism whereby a transient rise in postsynaptic calcium could trigger a long-lasting change in the properties of the postsynaptic cell (Fig. 17-14). As appealing as this "calcium switch" hypothesis is, it has only begun to be tested. In addition, kinase substrates whose phosphorylation might result in modulation of synaptic efficacy have yet to be identified.

There remain other mysteries about the LTP phenomenon. For example, it has not been possible to produce LTP by application of exogenous glutamate alone. It is not known if this is simply a technical problem, or whether it reflects the requirement for additional essential factors released during LTP. Moreover, the NMDA mechanism that we have described contributes to LTP only at two of the three best studied loci in the hippocampus (Fig. 17-11a). Although LTP can also be evoked at the granule cell to CA3 synapse (synapse 2 in Fig. 17-11a), the NMDA receptor does not play a role in this case. In addition, LTP occurs in several other brain regions, where NMDA receptors are not found, as well as at certain peripheral synapses. Thus, additional mechanisms of such long-lasting plasticity remain to be discovered.

Summary

The goal of neurobiologists, whether their experimental system is the single ion channel or the behaving human, is to understand how the brain works.

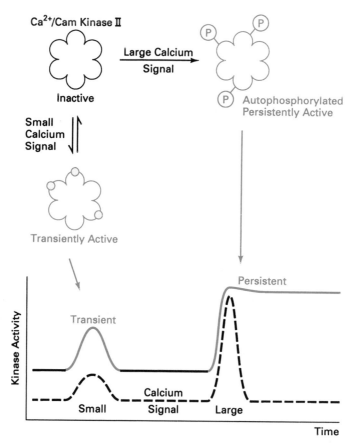

FIGURE 17-14. A hypothesis for the role of calcium in LTP. A small calcium signal can transiently activate Ca^{2+}/Cam kinase II in the postsynaptic cell. However, a large calcium signal, sufficient to stimulate autophosphorylation of the kinase, may allow the enzyme to be persistently active even after calcium levels have dropped. See Lisman and Goldring (1988) and Malenka et al. (1989).

One aspect of brain function that has most fascinated scientists and the lay public alike is learning and memory. Psychologists have described different kinds of learning and memory, and there is an ongoing search for the physical basis of these distinctions and for the cellular and molecular mechanisms responsible. Because of the complexity of most nervous systems, the search has focused to a large extent on animals with relatively simple nervous systems and on reduced preparations. In such systems it is sometimes possible to locate the neuronal pathways that participate in particular behaviors, and to identify changes in synaptic efficacy or neuronal membrane properties that are associated with learning and memory. A variety of approaches can also be used to obtain clues about the molecular changes that underlie the modulation of synaptic strength and membrane characteristics. Much remains to be understood. We can anticipate that the search for the cellular and molecular underpinnings of learning and memory will occupy neurobiologists for a long time to come.

Bibliography

Chapter 1

Recommended Reading

A. Peters, S. L. Palay, and H. De F. Webster. *The Fine Structure of the Nervous System: The Neurons and Supporting Cells.* Philadelphia: Saunders (1976).

S. Ramon y Cajal. Croonian Lecture. La fine structure des centres nerveux. *Proc. R. Soc. (London)* 55:444–468 (1894).

C. Sherrington. *The Integrative Action of the Nervous System.* Cambridge: Cambridge University Press (1948).

References

L. K. Kaczmarek, M. Finbow, J-P. Revel, and F. Strumwasser. The morphology and coupling of *Aplysia* bag cells within the abdominal ganglion and in cell culture. *J. Neurobiol.* 10:525–550 (1979).

M. D. Landis. Initial junctions between developing parallel fibers and Purkinje cells are different from mature synaptic junctions. *J. Comp. Neurol.* 260:513–525 (1987).

M. D. Landis and T. S. Reese. Differences in membrane structure between excitatory and inhibitory synapses in the cerebellar cortex. *J. Comp. Neurol.* 155:93–126 (1974).

O. Loewi. Uber humorale Ubertragbarkeit der Herznervenwirkung. *Pflugers Arch.* 189:239–242 (1921).

A. Matus. Microtubule-associated proteins: Their potential role in determining neuronal morphology. *Annu. Rev. Neurosci.* 11:29–44 (1988).

S. L. Palay and V. Chan-Palay. *Cerebellar Cortex: Cytology and Organization.* New York: Springer-Verlag (1974).

M. P. Sheetz, E. R. Steuer, and T. A. Schroer. The mechanism and regulation of fast axonal transport. *Trends Neurosci.* 12:474–478 (1989).

Chapter 2

Recommended Reading

B. Hille. *Ionic Channels of Excitable Membranes.* Sunderland, MA: Sinauer Associates (1984).

B. Katz. *Nerve, Muscle and Synapse.* New York: McGraw-Hill (1966).

C. F. Stevens. *Neurophysiology: A Primer.* New York: Wiley (1966).

Chapter 3

Recommended Reading

B. Hille. *Ionic Channels of Excitable Membranes.* Sunderland, MA: Sinauer Associates, (1984).

C. Miller. How ion channel proteins work. In *Neuromodulation: The Biochemical Control of Neuronal Excitability,* L. K. Kaczmarek and I. B. Levitan (Eds.). New York: Oxford University Press, pp. 39–63 (1987).

B. Sakmann and E. Neher (Eds.). *Single-Channel Recording.* New York: Plenum, (1983).

Reference

O. P. Hamill, A. Marty, E. Neher, B. Sakmann, and F. J. Sigworth. Improved patch-clamp techniques for high-resolution current recording from cells and cell-free membrane patches. *Pflugers Arch. 391*:85–100 (1981).

Chapter 4

Recommended Reading

D. J. Adams, S. J. Smith, and S. H. Thompson. Ionic currents in molluscan soma. *Annu. Rev. Neurosci. 3*:141–167 (1980).

C. M. Armstrong. Sodium channels and gating currents. *Physiol. Rev. 61*:644–683 (1981).

B. P. Bean. Classes of calcium channels in vertebrate cells. *Annu. Rev. Physiol. 51*:367–384 (1989).

B. Hille. *Ionic Channels of Excitable Membranes.* Sunderland, MA: Sinauer Associates (1984).

A. L. Hodgkin. Ionic movements and electrical activity in giant nerve fibres. *Proc. R. Soc. London Ser. B. 148*:1–37 (1958).

References

O. P. Hamill, A. Marty, E. Neher, B. Sakmann, and F. J. Sigworth. Improved patch-clamp techniques for high-resolution current recording from cells and cell-free membrane patches. *Pflugers Arch. 391*:85–100 (1981).

A. L. Hodgkin and A. F. Huxley. Currents carried by sodium and potassium ions through the membrane of the giant axon of *Loligo. J. Physiol. (London) 116*:449–472 (1952a).

A. L. Hodgkin and A. F. Huxley. The components of membrane conductance in the giant axon of *Loligo. J. Physiol. (London) 116*:473–496 (1952b).

A. L. Hodgkin and A. F. Huxley. The dual effect of membrane potential on sodium conductance in the giant axon of *Loligo. J. Physiol. (London): 116*:497–506 (1952c).

A. L. Hodgkin and A. F. Huxley. A quantitative description of membrane current and its application to conduction and excitation in nerve. *J. Physiol. (London) 117*:500–544 (1952d).

A. L. Hodgkin, A. F. Huxley, and B. Katz. Measurements of current-voltage relations in the membrane of the giant axon of *Loligo. J. Physiol. (London) 116*:424–448 (1952).

R. W. Meech and N. B. Standen. Potassium activation in *Helix aspersa* neurones under voltage clamp: A component mediated by calcium influx. *J. Physiol. (London) 249*:211–239 (1975).

P. H. Reinhart, S. Chung, and I. B. Levitan. A family of calcium-dependent potassium channels from rat brain. *Neuron 2*:1031–1041 (1989).

R. W. Tsien. Calcium currents in heart cells and neurons. In *Neuromodulation: The Biochemical Control of Neuronal Excitability,* L. K. Kaczmarek and I. B. Levitan (Eds.). New York: Oxford University Press, pp. 206–242 (1987).

Chapter 5

Recommended Reading

W. A. Catterall. Voltage-dependent gating of sodium channels: Correlating structure and function. *Trends Neurosci. 9*:7–10 (1986).

L. B. Salkoff, A, Butler, A. Wei, N. Scavarda, K. Baker, D. Pauron, and C. Smith. Molecular biology of the voltage-gated sodium channel. *Trends Neurosci. 10*:522–527 (1987).

References

C. M. Armstrong. Sodium channels and gating currents. *Physiol. Rev. 61*:644–683 (1981).

V. Auld, A. L. Goldin, D. S. Krafte, J. Marshall, J. M. Dunn, W. A. Catterall, H. A. Lester, N. Davidson, and R. J. Dunn. A rat brain Na$^+$ channel subunit with novel gating properties. *Neuron 1*:449–461 (1988).

W. A. Catterall. Molecular properties of voltage-sensitive sodium channels. *Annu. Rev. Biochem. 55*:953–985 (1986).

M. Noda, S. Shimizu, T. Tanabe, T. Takai, T. Kayano, T. Ikeda, H. Takahashi, H. Nakayama, Y. Kanaoka, N. Minamino, K. Kanagawa, H. Matsuo, M. A. Raftery, T. Hirose, S. Inayama, H. Hayashida, T. Miyata, and S. Numa. Primary structure of *Electrophorus electricus* sodium channel deduced from cDNA sequence. *Nature (London) 312*:121–127 (1984).

L. B. Salkoff and M. A. Tanouye. Genetics of ion channels. *Physiol. Rev. 66*:301–329 (1986).

T. L. Schwarz, B. L. Tempel, D. M. Papazian, Y. N. Jan, and L. Y. Jan. Multiple potassium-channel components are produced by alternative splicing at the *Shaker* locus in *Drosophila. Nature (London) 331*:137–142 (1988).

W. Stuhmer, F. Conti, H. Suzuki, X. Wang, M. Noda, N. Yahagi, H. Kubo, and S. Numa. Structural parts involved in activation and inactivation of the sodium channel. *Nature (London) 339*:597–603 (1989).

T. Tanabe, K. G. Beam, J. A. Powell, and S. Numa. Restoration of excitation-contraction coupling and slow calcium current in dysgenic muscle by dihydropyridine receptor complementary DNA. *Nature (London) 336*:134–139 (1988).

M. A. Tanouye and A. Ferrus. Action potentials in normal and *Shaker* mutant *Drosophila. J. Neurogenet. 2*:253–172 (1985).

Chapter 6

Recommended Reading

P. DeCamilli and R. Jahn. Pathways to regulated exocytosis in neurons. *Annu. Rev. Physiol. 52*:625–645 (1990).

R. B. Kelly. Pathways of protein secretion in eukaryotes. *Science 230*:25–32 (1985).

References

L. J. Breckenridge and W. Almers. Current through the fusion pore that forms during exocytosis of a secretory vesicle. *Nature (London) 328*:814–817 (1987).

J. M. Burt and D. C. Spray. Single channel events and gating behavior of the cardiac gap junction channel. *Proc. Natl. Acad. Sci. U.S.A. 85*:3431–3434 (1988).

L. K. Kaczmarek, M. Finbow, J-P. Revel, and F. Strumwasser. The morphology and coupling of *Aplysia* bag cells within the abdominal ganglion and in cell culture. *J. Neurobiol. 10*:525–550 (1979).

L. Makowski, D. L. D. Caspar, W. C. Phillips, and D. A. Goodenough. Gap junction structures. II. Analysis of the X-ray diffraction data. *J. Cell Biol. 74*:629–645 (1977).

E. Schwartz. Depolarization without calcium influx can release γ-aminobutyric acid from a retinal neuron. *Science 238*:350–355 (1987).

Chapter 7

Recommended Reading

R. Llinas. Calcium in synaptic transmission. *Sci. Am. 247*(4):56–65 (1982).

S. J. Smith and G. J. Augustine. Calcium ions, active zones and synaptic transmitter release. *Trends Neurosci. 11*:458–464 (1988).

R. S. Zucker. Neurotransmitter release and its modulation. In *Neuromodulation: The Biochemical Control of Neuronal Excitability*, L. K. Kaczmarek and I. B. Levitan (Eds.). New York: Oxford University Press, pp. 243–263 (1987).

References

G. J. Augustine, M. P. Charlton, and S. J. Smith. Calcium entry and transmitter release at voltage-clamped nerve terminals of squid. *J. Physiol. 369*:163–181 (1985).

I. A. Boyd and A. R. Martin. The end-plate potential in mammalian muscle, J. Physiol. 132:74–91 (1956).

G. Grynkiewicz, M. Poemie, and R. Y. Tsien. A new generation of Ca^{2+} indicators with greatly improved fluorescence properties. *J. Biol. Chem. 260*:3440–3450 (1985).

J. E. Heuser, T. S. Reese, M. J. Dennis, Y. Jan, L. Jan, and L. Evans. Synaptic vesicle exocytosis captured by quick freezing and correlated with quantal transmitter release. *J. Cell Biol. 81*:275–300 (1979).

R. Llinas, T. L. McGuinness, C. S. Leonard, M. Sugimori, and P. Greengard. Intraterminal injection of synapsin I or calcium/calmodulin-dependent protein kinase II alters neurotransmitter release at the squid giant synapse. *Proc. Natl. Acad. Sci. U.S.A. 82*:3035–3039 (1985).

Chapter 8

Recommended Reading

H. Akil, S. J. Watson, E. Young, M. E. Lewis, H. Khachaturian, and J. M. Walker. Endogenous opioids: Biology and function. *Annu. Rev. Neurosci. 7*:223–255 (1984).

F. E. Bloom. The functional significance of neurotransmitter diversity. *Am. J. Physiol. 246*:C184–C194 (1984).

J. R. Cooper, F. E. Bloom, and R. H. Roth. *The Biochemical Basis of Neuropharmacology,* 4th ed. New York: Oxford University Press (1982).

L.-M. Kow and D. W. Pfaff. Neuromodulatory actions of peptides. *Annu. Rev. Pharmacol. Toxicol. 28*:163–188 (1988).

References

B. Kanner and S. Shuldiner. Mechanism of transport and storage of neurotransmitters. *CRC Crit. Rev. Biochem. 22*:1–38 (1987).

A.-J. Silverman and E. A. Zimmerman. Magnocellular neurosecretory system. *Annu. Rev. Neurosci. 6*:357–380 (1983).

Chapter 9

Recommended Reading

A. C. Foster and G. E. Fagg. Acidic amino acid binding sites in mammalian neuronal membranes: Their characteristics and relationship to synaptic receptors. *Brain Res. Rev. 7*:103–164 (1984).

S. M. Goldin, E. G. Moczydlowski, and D. M. Papazian. Isolation and reconstitution of neuronal ion transport proteins. *Annu. Rev. Neurosci. 6*:419–446 (1983).

A. Karlin. The acetylcholine receptor. In *The Cell Surface and Neuronal Function,* C. W. Cotman, G. Poste, and G. L. Nicholson (Eds.). New York: Elsevier, pp. 191–260 (1980).

R. A. Nicoll, J. A. Kauer, and R. C. Malenka. The current excitement in long-term potentiation. *Neuron 1*:97–103 (1988).

P. R. Schofield, B. D. Shivers, and P. H. Seeburg. The role of receptor subtype diversity in the CNS. *Trends Neurosci. 13*:8–11 (1990).

References

E. A. Barnard, R. Miledi, and K. Sumikawa. Translation of exogenous messenger RNA coding for nicotinic acetylcholine receptors produces functional receptors in *Xenopus* oocytes. *Proc. R. Soc. London Ser. B 215*:241–246 (1982).

J. A. Dani. Site-directed mutagenesis and single-channel currents define the ionic channel of the nicotinic acetylcholine receptor. *Trends Neurosci. 12*:125–128 (1989).

E. R. Kandel. *Cellular Basis of Behavior: An Introduction to Behavioral Neurobiology.* San Francisco: Freeman (1976).

L. Nowak, P. Bregestovski, P. Ascher, A. Herbet, and A. Prochiantz. Magnesium gates glutamate-activated channels in mouse central neurones. *Nature (London) 307*:462–465 (1984).

B. Sakmann, C. Methfessel, M. Mishina, T. Takahashi, T. Takai, M. Kurasaki, K. Fukuda, and S. Numa. Role of acetylcholine receptor subunits in gating of the channel. *Nature (London) 318*:538–543 (1985).

Chapter 10

Recommended Reading

J. Axelrod, R. M. Burch, and C. L. Jelsema. Receptor-mediated activation of phospholipase A_2 via GTP-binding proteins: Arachidonic acid and its metabolites as second messengers. *Trends Neurosci. 11*:117–123 (1988).

M. J. Berridge. Inositol trisphosphate and diacylglycerol: Two interacting second messengers. *Annu. Rev. Biochem. 56*:159–193 (1987).

A. G. Gilman. G proteins: Transducers of receptor-generated signals. *Annu. Rev. Biochem. 56*:615–649 (1987).

R. L. Huganir. Biochemical mechanisms that regulate the properties of ion channels. In *Neuromodulation: The Biochemical Control of Neuronal Excitability,* L. K. Kaczmarek and I. B. Levitan (Eds.). New York: Oxford University Press, pp. 64–85 (1987).

M. B. Kennedy. Regulation of neuronal function by calcium. *Trends Neurosci. 12*:417–420 (1989).

Y. Nishizuka. Studies and perspective of protein kinase C. *Science 233*:305–312 (1986).

B. F. O'Dowd, R. J. Lefkowitz, and M. G. Caron. Structure of the adrenergic and related receptors. *Annu. Rev. Neurosci. 12*:67–83 (1989).

References

G. E. Breitwieser and G. Szabo. Uncoupling of cardiac muscarinic and β-adrenergic receptors from ion channels by a guanine nucleotide analogue. *Nature (London) 317*:538–540 (1985).

J. Codina, A. Yatani, D. Grenet, A. M. Brown, and L. Birnbaumer. The α subunit of the GTP binding protein G_K opens atrial potassium channels. *Science 236*:442–445 (1987).

D. E. Logothetis, Y. Kurachi, J. Galper, E. J. Neer, and D. E. Clapham. The $\beta\gamma$ subunits of GTP-binding proteins activate the muscarinic K^+ channel in heart. *Nature (London) 325*:321–326 (1987).

P. J. Pfaffinger, J. M. Martin, D. D. Hunter, N. M. Nathanson, and B. Hille. GTP-binding proteins couple cardiac muscarinic receptors to a K^+ channel. *Nature (London) 317:536–538* (1985).

Chapter 11

Recommended Reading

L. K. Kaczmarek and I. B. Levitan. *Neuromodulation: The Biochemical Control of Neuronal Excitability.* New York: Oxford University Press (1987).

M. B. Kennedy. Experimental approaches to understanding the role of protein phosphorylation in the regulation of neuronal function. *Annu. Rev. Neurosci.* 6:493–525 (1983).

I. B. Levitan. Phosphorylation of ion channels. *J. Membrane Biol. 87:177–190* (1985).

I. B. Levitan. Modulation of ion channels in neurons and other cells. *Annu. Rev. Neurosci. 11:119–136* (1988).

J. H. Schwartz and S. M. Greenberg. Molecular mechanisms for memory: Second-messenger induced modifications of protein kinases in nerve cells. *Annu. Rev. Neurosci. 10:459–476* (1987).

References

F. Belardetti and S. A. Siegelbaum. Up- and down-modulation of single K^+ channel function by distinct second messengers. *Trends Neurosci. 11:232–238* (1988).

S. A. DeRiemer, J. A. Strong, K. A. Albert, P. Greengard, and L. K. Kaczmarek. Enhancement of calcium current in *Aplysia* neurones by phorbol ester and protein kinase C. *Nature (London) 313:313–316* (1985).

K. Dunlap and G. D. Fischbach. Neurotransmitters decrease the Ca component of sensory neurone action potentials. *Nature (London) 276:837–838* (1978).

J. F. Hopfield, D. W. Tank, P. Greengard, and R. L. Huganir. Functional modulation of the nicotinic acetylcholine receptor by tyrosine phosphorylation. *Nature (London) 336:677–680* (1988).

R. Huganir and K. Miles. Protein phosphorylation of nicotinic acetylcholine receptors. *CRC Crit. Rev. Biochem. Mol. Biol. 24:183–215* (1989).

S. W. Jones and P. R. Adams. The M-current and other potassium currents of vertebrate neurons. In *Neuromodulation: The Biochemical Control of Neuronal Excitability,* L. K. Kaczmarek and I. B. Levitan (Eds.). New York: Oxford University Press, pp. 159–186 (1987).

E. R. Kandel. *Cellular Basis of Behavior: An Introduction to Behavioral Neurobiology.* San Francisco: Freeman (1976).

E. S. Levitan and I. B. Levitan. Serotonin acting via cyclic AMP enhances both the hyperpolarizing and depolarizing phases of bursting pacemaker activity in the *Aplysia* neuron R15. *J. Neurosci. 8:1152–1161* (1988).

J. A. Strong, A. P. Fox, R. W. Tsien, and L. K. Kaczmarek. Stimulation of protein kinase C recruits covert calcium channels in *Aplysia* bag cell neurons. *Nature (London) 325:714–717* (1987).

J. A. Strong and L. K. Kaczmarek. Potassium currents that regulate action potentials and repetitive firing. In *Neuromodulation: The Biochemical Control of Neuronal Excitability,* L. K. Kaczmarek and I. B. Levitan (Eds.). New York: Oxford University Press, pp. 119–137 (1987).

R. W. Tsien. Calcium currents in heart cells and neurons. In *Neuromodulation: The Biochemical Control of Neuronal Excitability,* L. K. Kaczmarek and I. B. Levitan (Eds.). New York: Oxford University Press, pp. 206–242 (1987).

Chapter 12

Recommended Reading

W. Baehr and M. L. Applebury. Exploring visual transduction with recombinant DNA techniques. *Trends Neurosci.* 9:198–203 (1986).

V. E. Dionne. How do you smell? The principle in question. *Trends Neurosci.* 11:188–189 (1988).

S. Kinnamon. Taste transduction: A diversity of mechanisms. *Trends Neurosci.* 11:491–496 (1988).

T. D. Lamb. Transduction in vertebrate photoreceptors: The roles of cyclic GMP and calcium. *Trends Neurosci.* 9:224–228 (1986).

References

J. J. Art and R. Fettiplace. Variation of membrane properties in hair cells isolated from the turtle cochlea. *J. Physiol.* 385:207–242 (1987).

A. C. Crawford and R. Fettiplace. An electrical tuning mechanism in turtle cochlear hair cells. *J. Physiol.* 312:377–412 (1981).

E. F. Fesenko, S. S. Kolesnikov, and A. L. Lyubarsky. Induction by cyclic GMP of cationic conductance in plasma membrane of retinal rod outer segments. *Nature (London)* 313:310–313 (1985).

L. W. Haynes, A. R. Kay, and K.-W. Yau. Single cyclic GMP-activated channel activity in excised patches of rod outer segment membrane. *Nature (London)* 321:66–70 (1986).

A. J. Hudspeth and R. S. Lewis. A model for electrical resonance and frequency tuning in saccular hair cells of the bull-frog, *Rana catesbeina. J. Physiol.* 400:275–297 (1988).

R. Kullberg. Stretch-activated ion channels in bacteria and animal cell membranes. *Trends Neurosci.* 10:387–388 (1987).

R. S. Lewis and A. J. Hudspeth. Frequency tuning and ionic conductances in hair cells of the bullfrog's sacculus. In *Hearing—Physiological Bases and Psychophysics,* R. Klinke and R. Hartmann (Eds.). Berlin: Springer-Verlag, pp. 17–24 (1983).

J. H. Martin. Somatic sensory system I: Receptor physiology and submodality coding. In *Principles of Neural Science,* E. R. Kandel and J. H. Schwartz (Eds.). New York: Elsevier North Holland, pp. 157–169 (1981).

Chapter 13

Recommended Reading

S. S. Easter, K. F. Barald, and B. M. Carlson. *From Message to Mind: Directions in Developmental Neurobiology.* Sunderland, MA: Sinauer Associates (1988).

D. Purves and J. W. Lichtman. *Principles of Neural Development.* Sunderland, MA: Sinauer Associates (1985).

References

S. Artavanis-Tsakonas. The molecular biology of the *Notch* locus and the fine tuning of differentiation in *Drosophila. Trends Genet. 4*:95–100 (1988).

D. Bar-Sagi and J. R. Feramisco. Microinjection of the *ras* oncogene protein into PC12 cells induces morphological differentiation. *Cell 42*:841–848 (1985).

K. Basler and E. Hafen. Control of photoreceptor cell fate by the *sevenless* protein requires a functional tyrosine kinase domain. *Cell 54*:299–311 (1988).

T. J. DeVoogt. Androgens can affect the morphology of mammalian CNS neurons in adulthood. *Trends Neurosci. 10*:341–342 (1987).

R. M. Evans. The steroid and thyroid hormone receptor superfamily. *Science 240*:889–895 (1988).

R. A. Gorski. Structural sex differences in the brain: Their origin and significance. In *Neural Control of Reproductive Function,* J. M. Lakoski, J. R. Perez-Polo, and D. K. Rassin (Eds.). New York: Liss, pp. 33–44 (1989).

M. E. Gurney. Hormonal control of cell form and number in the zebra finch song system. *J. Neurosci. 1*:658–673 (1981).

M. Jacobson. *Developmental Neurobiology.* New York: Plenum (1978).

F. Nottebohm. From bird song to neurogenesis. *Sci. Am. 260*:74–79 (1989).

F. Nottebohm. Hormonal regulation of synapses and cell number in the adult canary brain and its relevance to theories of long-term memory storage. In *Neural Control of Reproductive Function,* J. M. Lakoski, J. R. Perez-Polo, and D. K. Rassin (Eds.). New York: Liss, pp. 583–601 (1989).

J. Palka and M. Schubiger. Genes for neural differentiation. *Trends Neurosci. 11*:515–517 (1988).

P. Rakic. Neuron-glia relationship during ganglion cell migration in developing cerebellar cortex. A Golgi and electron microscopic study in *Macacus rhesus. J. Comp. Neurol. 141*:283–312 (1971).

J. R. Sanes. Roles of extracellular matrix in neural development. *Annu. Rev. Physiol. 45*:581–600 (1983).

S. L. Zipursky. Molecular and genetic analysis of *Drosophila* eye development: Sevenless, bride of sevenless and rough. *Trends Neurosci. 12*:183–189 (1989).

Chapter 14

Recommended Reading

J. Dodd and T. M. Jessell. Axon guidance and the patterning of neuronal projections in vertebrates. *Science 242*:692–699 (1988).

R. O. Hynes. Integrins: A family of cell surface receptors. *Cell 48*:549–554 (1987).

P. Patterson. On the importance of being inhibited, or saying no to growth cones. *Neuron 1*:263–267 (1988).

J. R. Sanes. Roles of extracellular matrix in neural development. *Annu. Rev. Physiol. 45*:581–600 (1983).

References

L. I. Benowitz and A. Routtenberg. A membrane phosphoprotein associated with neural development, axonal regeneration, phospholipid metabolism and synaptic plasticity. *Trends Neurosci. 10:*527–532 (1987).

P. Caroni and M. E. Schwab. Antibody against myelin-associated inhibitor of neurite growth neutralizes nonpermissive substrate properties of CNS white matter. *Neuron 1:*85–96 (1988).

G. M. Edelman: Cell adhesion molecules. *Science 219:*450–457 (1983).

P. Forscher and S. J. Smith. Actions of cytochalasins on the organization of actin filaments and microtubules in a neuronal growth cone. *J. Cell Biol. 107:*1505–1516 (1988).

A. L. Harrelson and C. S. Goodman. Growth cone guidance in insects: Fasciclin II is a member of the immunoglobulin superfamily. *Science 242:*700–708 (1988).

P. G. Haydon, D. P. McCobb, and S. B. Kater. Serotonin selectively inhibits growth cone dynamics and synaptogenesis of specific identified neurons. *Science 226:*561–564 (1984).

P. C. Letourneau. Fibronectin. In *Neuroscience Year,* G. Adelman (Ed.). Boston: Birkhauser, pp. 61–62 (1989).

L. Luckenbill-Edds. Laminin. In *Neuroscience Year,* G. Adelman (Ed.). Boston: Birkhauser, pp. 89–91 (1989).

M. Matsunaga, K. Hatta, A. Nagafuchi, and M. Takeichi. Guidance of optic nerve fibers by N-cadherin cell adhesion molecules. *Nature (London) 334:*62–64 (1988a).

M. Matsunaga, K. Hatta, and M. Takeichi. Role of N-cadherin cell adhesion molecules in the histogenesis of neural retina. *Neuron 1:*289–295 (1988b).

D. Purves and R. D. Hadley. Changes in the dendritic branching of adult mammalian neurones revealed by repeated imaging *in situ. Nature (London) 315:*404–406 (1985).

U. Rutishauser, A. Acheson, A. K. Hall, D. M. Mann, and J. Sunshine. The neural cell adhesion molecule (NCAM) as a regulator of cell-cell interactions. *Science 240:*53–57 (1988).

M. Westerfield and J. S. Eisen. Neuromuscular specificity: Pathfinding by identified motor growth cones in a vertebrate embryo. *Trends Neurosci. 11:*18–22, 1988.

M. X. Zuber, D. W. Goodman, L. R. Karns, and M. C. Fishman. The neuronal growth-associated protein GAP-43 induces filopodia in non-neuronal cells. *Science 244:*1193–1195 (1989).

Chapter 15

Recommended Reading

M. C. Brown. Sprouting of motor nerves in adult muscles: A recapitulation of ontogeny. *Trends Neurosci. 7:*10–14 (1984).

M. Constantine-Paton, H. T. Cline, and E. Debski. Patterned activity, synaptic convergence, and the NMDA receptor in developing visual pathways. *Annu. Rev. Neurosci. 13:*129–154 (1990).

J. T. Morgan and T. Curran. Stimulus-transcription coupling in neurons: Role of cellular immediate-early genes. *Trends Neurosci. 12*:459–462 (1989).

D. Purves and J. W. Lichtman. *Principles of Neural Development.* Sunderland, MA: Sinauer Associates (1985).

References

S. Bevan and J. H. Steinbach. The distribution of α-bungarotoxin binding sites on mammalian skeletal muscle developing *in vivo. J. Physiol. 267*:195–213 (1977).

A. J. Buller, J. C. Eccles, and R. M. Eccles. Interactions between motoneurons and muscles in respect of the characteristic speed of their responses. *J. Physiol. 150*:417–439 (1960).

M. Constantine-Paton and M. I. Law. Eye-specific termination bands in tecta of three-eyed frogs. *Science 202*:639–641 (1978).

P. Forscher, L. K. Kaczmarek, J. Buchanan, and S. J. Smith. Cyclic AMP induces changes in distribution of organelles within growth cones of *Aplysia* bag cell neurons. *J. Neurosci. 7*:3600–3611 (1987).

H. Fujisawa. Persistence of disorganized pathways on tortuous trajectories of regenerating retinal fibers in the adult newt *Cynops pyrrhogaster. Dev. Growth Differ. 23*:215–219 (1981).

D. O. Hebb. *The Organization of Behavior.* New York: Wiley (1949).

D. D. Hunt, V. Shah, J. P. Merlie, and J. R. Sanes. A laminin-like adhesive protein concentrated in the synaptic cleft of the neuromuscular junction. *Nature (London) 338*:229–234 (1989).

S. P. Hunt, A. Pini, and G. Evan. Induction of c-*fos*-like protein in spinal cord neurons following sensory stimulation. *Nature (London) 328*:632–634 (1987).

L. M. Marshall, J. R. Sanes, and U. J. McMahan. Reinnervation of original synaptic sites on muscle fiber basement membrane after disruption of the muscle cells. *Proc. Natl. Acad. Sci. U.S.A. 74*:3073–3077 (1977).

S. Salmons and F. A. Streter. Significance of impulse activity in the transformation of skeletal muscle type. *Nature (London) 263*:30–34 (1976).

J. R. Sanes. Cell lineage and the origin of muscle fiber types. *Trends Neurosci. 10*:219–221 (1987).

R. W. Sperry. Chemoaffinity in the orderly growth of nerve fiber patterns and connections. *Proc. Natl. Acad. Sci. U.S.A. 50*:703–710 (1963).

D. Trisler and F. Collins. Corresponding spatial gradients of TOP molecules in the developing retina and optic tectum. *Science 237*:1208–1209 (1987).

J. Walter, S. Henke-Fahle, and F. Bonhoeffer. Avoidance of posterior tectal membranes by temporal retinal axons. *Development 101*:909–913 (1987).

Chapter 16

Recommended Reading

P. A. Getting. Emerging principles governing the operation of neural networks. *Annu. Rev. Neurosci. 12*:185–204 (1989).

A. I. Selverston. Model Neural Networks and Behavior. New York: Plenum (1985).

References

P. D. Brodfuehrer and W. O. Friesen. From stimulation to undulation: A neuronal pathway for swimming in the leech. *Science* 234:1002–1004 (1986).

P. J. Conn and L. K. Kaczmarek. The bag cell neurons: A model system for the investigation of prolonged changes in animal behavior. *Mol. Neurobiol.* 3:237–273 (1990).

W. J. Davis. Behavioral hierarchies. *Trends Neurosci.* 2:5–7 (1979).

P. S. Dickinson and E. Marder. Peptidergic modulation of a multioscillator system in the lobster. I. Activation of the cardiac sac motor pattern by the neuropeptides proctolin and red pigment-concentrating hormone. *J. Neurophysiol.* 61:833–844 (1989).

W. O. Friesen. Neuronal control of leech swimming movements. In *Neuronal and Cellular Oscillators,* J. W. Jacklet (Ed.). New York: Dekker, pp. 269–316 (1989).

R. M. Harris-Warrick and R. E. Flamm. Chemical modulation of a small central pattern generator. *Trends Neurosci.* 9:432–437 (1986).

L. K. Kaczmarek. A model of cell firing patterns during epileptic seizures. *Biol. Cybernet.* 22:229–234 (1976).

M. Moulins and F. Nagy. Extrinsic inputs and flexibility in the motor output of the lobster pyloric neural network. In *Model Neural Networks and Behavior,* A. I. Selverston (Ed.). New York: Plenum, pp. 49–68 (1985).

M. P. Nusbaum and E. Marder. A modulatory proctolin-containing neuron (MPN). II. State-dependent modulation of rhythmic motor activity. *J. Neurosci.* 9:1600–1607 (1989).

R. A. Satterlie. Reciprocal inhibition and postinhibitory rebound produce reverberation in a locomotor pattern generator. *Science* 229:402–404 (1985).

R. A. Satterlie, M. LaBarbara, and A. N. Spencer. Swimming in the pteropod mollusc *Clione limacina.* I. Behavior and morphology. *J. Exp. Biol.* 116:189–204 (1985).

G. S. Stent, W. B. Kristan, W. O. Friesen, C. A. Ort, M. Poon, and R. L. Calabrese. Neuronal generation of the leech swimming movement. *Science* 200:1348–1357 (1978).

Chapter 17

Recommended Reading

Y. Dudai. *The Neurobiology of Memory: Concepts, Findings, Trends.* New York: Oxford University Press (1989).

E. R. Kandel and J. H. Schwartz. Molecular biology of learning: Modulation of transmitter release. *Science* 218:433–443 (1982).

L. R. Squire. *Memory and Brain.* New York: Oxford University Press (1987).

R. F. Thompson, T. W. Berger, and J. Madden IV. Cellular processes of learning and memory in the mammalian CNS. *Annu. Rev. Neurosci.* 6:447–491 (1983).

References

B. W. Agranoff, R. E. Davis, and J. J. Brink. Memory fixation in the goldfish. *Proc. Natl. Acad. Sci. U.S.A.* 54:788–793 (1965).

T. V. P. Bliss and T. Lomo. Long lasting potentiation of synaptic transmission in the dentate area of the anaesthetized rabbit following stimulation of the perforant path. *J. Physiol.* 232:331–356 (1973).

T. Crow. Cellular and molecular analysis of associative learning and memory in *Hermissenda. Trends Neurosci.* 11:136–142 (1988).

E. R. Kandel. *Behavioral Biology of Aplysia.* San Francisco: Freeman (1979).

K. S. Lashley. In search of the engram. *Soc. Exp. Biol. Symp.* 4:454–482 (1950).

J. E. Lisman and M. A. Goldring. Feasibility of long-term storage of graded information by the Ca^{2+}/calmodulin-dependent protein kinase molecules of the postsynaptic density. *Proc. Natl. Acad. Sci. U.S.A.* 85:5320–5324 (1988).

R. C. Malenka, J. A. Kauer, D. J. Perkel, and R. A. Nicoll. The impact of postsynaptic calcium on synaptic transmission—its role in long-term potentiation. *Trends Neurosci.* 12:444–450 (1989).

R. A. Nicoll, J. A. Kauer, and R. C. Malenka. The current excitement in long-term potentiation. *Neuron* 1:97–103 (1988).

T. Tully. *Drosophila* learning and memory revisited. *Trends Neurosci.* 10:330–335 (1987).

Index

AA. *See* Arachidonic acid
AB cell, 379–84, 391
Abdominal ganglion, 197, 255, 389, 411
Absolute refractory period, 39
Accommodation, 48, 378
Acetylcholine, 139, 142, 185, 233, 312, 370, 382
  at neuromuscular junction, 23, 26, 167, 357
Acetylcholine metabolism, 167–71, 183
Acetylcholine receptor, 195–97, 240, 267, 319
  junctional and extrajunctional, 362, 363
  muscarinic, 195–97, 208, 209, 219, 221, 225–27, 241, 264
  nicotinic, 63, 67, 146, 195–97, 205–15
  phosphorylation of, 269–72
  reconstitution of, 211
  supersensitivity of, 364
Acetylcholine release, 146–52
Acetylcholinesterase, 148, 170, 171, 175, 360, 364
Acetyl-CoA, 170
Aconitine, 108
Acquisition of behavior, 398, 400, 401, 404
ACTH, 189
Actin, 16, 143, 164, 278, 324, 344
Action potential
  amplitude and duration, 248–53, 256, 263, 272
  effects on postsynaptic properties, 365, 366
  in sensory cells, 274
  in Shaker flies, 120
  ionic mechanisms of, 73, 78, 79, 91, 96–98
  modulation of, 248–59, 272, 389, 413
  patterns of firing, 46–49, 51, 103–5, 261, 264, 265, 351
  propagation of, 41, 43, 92, 93
  properties of, 34, 38–44

role in rhythmic networks, 378–82, 387, 391
role in transmitter release, 155, 160–62, 355
tetrodotoxin block of, 86, 154, 354
Activational effect, 319
Activation gate, 89, 90, 112–16
Activation of receptors and ion channels, 72–74; *see also entries for individual receptors and ion channels*
Active circuit, 384
Active response, 38, 41
Active transport, 51, 175
Active zone, 146, 147, 155, 166
A-current, 101–4, 120–23
Adaptive behavior, 395
Adenohypophysis. *See* Anterior pituitary
Adenosine, 181
Adenylate cyclase, 224, 226, 229, 230, 232, 293, 406; *see also* Cyclic AMP
Adhesion, 299, 323, 327, 328, 330–35, 337, 340, 342, 349
Adhesion molecule, 299, 327, 331, 332, 335
Adrenaline, 172
Adrenal medulla, 139, 172, 310, 331
Adrenergic receptor, 172, 196, 197, 208, 223, 240–42, 271, 284, 289
Adrenergic receptor kinase, 241, 242
Aequorin, 155
Afterdischarge, 256, 260, 262
Afterhyperpolarization, 78, 103, 264, 265, 365
Aggregation, 364
Agonist, 201, 224, 236, 241, 244, 271
Agranoff, Bernard, 403, 404
AHP. *See* Afterhyperpolarization
Aldosterone, 316
All-or-none law, 38, 41, 43
Alternative splicing, 121